Freshwater Ecosystem of India

K. Vijaykumar

2022

Daya Publishing House®

A Division of

Astral International Pvt. Ltd.

New Delhi – 110 002

First Published, 1999

Reprinted, 2022

ISBN: 9788170352068 (H.B.)

Disclaimer:

Every possible effort has been made to ensure that the information contained in this book is accurate at the time of going to press, and the publisher and author cannot accept responsibility for any errors or omissions, however caused. No responsibility for loss or damage occasioned to any person acting, or refraining from action, as a result of the material in this publication can be accepted by the editor, the publisher or the author. The Publisher is not associated with any product or vendor mentioned in the book. The contents of this work are intended to further general scientific research, understanding and discussion only. Readers should consult with a specialist where appropriate.

Every effort has been made to trace the owners of copyright material used in this book, if any. The author and the publisher will be grateful for any omission brought to their notice for acknowledgment in the future editions of the book.

Published by : **Daya Publishing House®**
A Division of
Astral International Pvt. Ltd.
– ISO 9001:2015 Certified Company –
4736/23, Ansari Road, Darya Ganj
New Delhi-110 002
Ph. 011-43549197, 23278134
E-mail: info@astralint.com
Website: www.astralint.com

Dedicated
to
Prof. Ravindra Paul

Preface

An indisputable and inseparable bond exists between freshWaterbodies and man. Human civilization originated, developed and thrived in places within easy reach of freshwater sources. Although the primary concern of man was thought to be for meeting his basic requirements, it cannot be forgotten that water is basis for all primary life activities.

The role of water in nature is unique not only from the point of human consideration, but numerous of these organisms which make aquatic medium their abode. Understanding such aquatic requires sound knowledge not just of the organisms themselves but also these external influences of the medium which affects them, and limnology in its broader sense does just that the abundance of the water on the surface of earth has always been a source of developing interest of investigations for man. The total quantity of inland water of our biosphere about which estimates differ is a small fraction of the total hydrosphere which is of immense importance in governing the various life activities of all organisms.

Indian subcontinent with its strategic position presents unique and varied physical, geographical and meteorological features. The numerous freshwater habitats are represented by large rivers to small streams and great lakes to small ponds. The lentic habitat presents a variety of features because of there distribution ranging from high altitude to and plateau of our country. The organisms of these variety of lentic habitats have aroused immense interest of many limnologist and hydrobiologists.

This book depicts seventeen research/review paper on limnology, primary production, plankton, aquaculture and water management.

In the first chapter Kaushik and Saksena, discusses the physico-chemical parameters of Motijheel and Surajkund in Central India.

Vijaykumar *et al.* in second chapter made an attempt to assess limnological conditions and zooplanktonic life in a reservoir in Gulbarga and suggested the integrated steps for the management of Chandrampalli reservoir for essentially increasing the fishery activity for better exploitation.

A case study conducted by Muley and Gaikwad on the Shiroli reservoir reveals the physico-chemical parameters and emphasized the flora and fauna inhabiting the reservoir.

Angadi *et al.* made an attempt in the direction of understanding the influence of different physico-chemical parameters on the planktonic life in heavily polluted waterbody in Gulbarga.

Raghunathan and Revathi in their study on eutrophication of pond and reveals planktonic cladocera and its association in different Waterbodies of Tamil Nadu.

In the Seventh chapter, Arvind Kumar and Singh elaborates the role of rotifers as bioindicators in freshwater lentic ecosystem of Bihar.

Kaur *et al.* discussed the correlation of rotifers with physico-chemical factors at Harike reservoir, Punjab.

Khan and Alam did extensive work on cyclomorphosis in cladocera in relation to environmental factors and pollution status of a leechate reservoir receiving effluents from thermal power plant with reference to plankton population.

Kalyani and Charya in the Chapter of phytoplanktonic dynamics of Bhadrakali Lake Warangal reviews the planktonic abundance with monthly fluctuations to assess the nature of lake and to establish the algal role in bioindication.

Sarwar on the water quality of Ankar Lake, discusses the periphytic algal components of Kashmir Himalayan Lake; and the various physico - chemical and biological characteristics reveals the degraded and deteriorated nature of the water body.

Ismail in her chapter on the population biology of trematode species in freshwater fishes at Indore discusses the dynamics of infection relating to the immuno physiological status of the host fish.

Paul and Verma in their chapter on hydrobiological studies of the tropical standing Waterbodies with reference to primary production discuss the influence of nutrients and the gross primary productivity.

Khan *et al.* elaborate the study of water quality parameter in the river Ganga in relation to primary production and plankton.

In the sixteenth chapter Amita Saxena emphasized a detailed review on water quality management in aquaculture.

Pandey *et al.* delt with an ecological study with special reference to biological components on river ecosystem.

I am greatly indebted to Prof. T. R. C. Gupta and Prof. R. J. Katti, Aquatic Biology Department, Fisheries College, University of Agricultural Science, Mangalore, Prof. G. Sai Prasad, Fishery Science Department, Andhra Pradesh Agricultural University, Rajendra Nagar, Hyderabad who have nourished us with efficacious thoughts. Encouragements from Dr. Vasudev. H. Sedam, Dr. S. M. Gaddad, Mr. Darely S. Pattathil and Dr. Michael, Gulbarga University are gratefully acknowledged. The invaluable support by my wife Mrs. Anitha without who unstinted encouragement and help this venture could not have been completed successfully. Miss. Kuldeep Kaur for going through manuscript and preparation are greatly acknowledged. At last but not the least KV feels affection to his lovely daughter Nikita and Son K. V. Tarun who silently observed the work and do not disfigure the same.

K. Vijaykumar

List of Contributors

1. Alam, Aftab,
Fishery Science and Aquaculture Unit,
Department of Zoology,
Aligarh Muslim University,
Aligarh - 202 002,
U. P.

2. Alam, M. A,
Fishery Science and Aquaculture Unit,
Department of Zoology,
Aligarh Muslim University,
Aligarh - 202 002,
U. P.

3. Angadi, S. B,
Department of Botany,
Gulbarga University,
Gulbarga - 585 106,
Karnataka.

4. Arvind Kumar,
Environmental Biology Research Lab,
P. G. Department of Zoology,
S. K. University,
Dumka - 814 101,
Bihar.

5. Bath, K. S,
Department of Zoology,
Punjabi University,
Patiala - 147 002,
Punjab.

6. Das, P. K. L,
P. G. Deptt. of Zoology,
Purnea College,
Purnea - 854 301.

7. Dhillon, S. S,
Department of Zoology,
Punjabi University,
Patiala - 147 002,
Punjab.

8. Dubey, S. Vatsa,
P. G. Deptt. of Zoology,
Purnea College,
Purnea - 854 301.

9. Gaur, R.K,
Fishery Science and Aquaculture Unit,
Department of Zoology,
Aligarh Muslim University,
Aligarh - 202 002,
U. P.

10. Gaikwad, P. T,
Department of Zoology,
Shivaji University,
Kolhapur - 416 004,
Maharashtra.

11. Holkar Devendra,
Department of Zoology,
Gulbarga University,
Gulbarga - 585 106,
Karnataka.

12. Hussain, S.
P. G. Deptt. of Botany,
Purnea College,
Purnea - 854 301.

13. Ismail, Q,
Department of Zoology,
Govt. Girls P. G. College,
Moti Tabela,
Indore - 452 001,
M. P.

14. Kalyani, Y,
Department of Botany,
Kakatiya University,
Warangal - 506 009,
A. P.

15. Kaur, H,
Department of Zoology,
Punjabi University,
Patiala - 147 002,
Punjab.

16. Kaushik, S.
School of Studies in Zoology,
Jiwaji University,
Gwalior - 474 001,
M. P.

17. Khan Asif, A,
Fishery Science and Aquaculture Unit,
Department of Zoology,
Aligarh Muslim University,
Aligarh - 202 002,
U. P.

18. Kuldeep Kaur,
Department of Zoology,
Gulbarga University,
Gulbarga - 585 106,
Karnataka.

19. Lingannaiah, B,
Department of Botany,
Gulbarga University,
Gulbarga - 585 106,
Karnataka.

20. Mander, G,
Department of Zoology,
Punjabi University,
Patiala - 147 002,
Punjab.

21. Muley, D. V,
Department of Zoology,
Shivaji University,
Kolhapur - 416 004,
Maharashtra.

22. Pandey, B. N,
P. G. Department of Zoology,
Purnea College,
Purnea - 854 301.

23. Raghunathan, M. B,
Zoological Survey of India,
Southern Regional Station,
Chennai,
Tamil Nadu.

24. Ravindra Paul,
Department of Zoology,
Gulbarga University,
Gulbarga - 585 106,
Karnataka.

25. Revathi, K,
Department of Zoology,
Ethiraj College for Women,
Chennai - 600 105,
Tamil Nadu.

26. Saksena, D. N,
School of Studies in Zoology,
Jiwaji University,
Gwalior - 474 001,
M. P.

27. Sarwar, S. G,
Hydrobiology Research Lab,
S. P. College,
Sri Nagar - 190 001,
Kashmir.

28. Saxena Amita,
Department of Aquaculture,
College of Fisheries/FRTC,
GBPUAT,
Pantnagar - 263 145,
U. P.

29. Eshwarlal, Sedamkar,
Department of Botany,
Gulbarga University,
Gulbarga - 585 106,
Karnataka.

30. Singara Charya, M. A,
Department of Botany,
Kakatiya University,
Warangal - 506 009,
A. P.

31. Singh, A. K,
Environmental Biology Research Lab,
P. G. Department of Botany,
S. K. University,
Dumka - 814 101,
Bihar.

32. Verma, Satyanarayan,
Department of Zoology,
Gulbarga University,
Gulbarga-585 106,
Karnataka.

33. Vijaykumar. K,
Environmental Biology Research Unit,
Department of Zoology,
Gulbarga University,
Gulbarga - 585 106,
Karnataka.

Contents

Preface v

List of Contributors ix

1. Physico-Chemical Limnology of Certain Waterbodies of Central India 1
S. Kaushik and D.N. Saksena

2. Limnological Studies on Chandrampalli Reservoir, Gulbarga 59
K. Vijaykumar, Devendra Holkar and Kuldeep Kaur

3. Limnological Studies of Shiroli Reservoir - A Case Study 109
D.V. Muley and P.T. Gaikad

4. Limnological Studies of Jagat Tank: Gulbarga 133
S.B. Angadi, B. Lingannaiah and Eshwarlal Sedamkar

5. Limnological Studies of a Village Pond in Tamil Nadu 160
M.B. Raghunathan and K. Revathi

6. Planktonic Cladocera (Crustacea) and Species Associations 167
M.B. Raghunathan and K. Revathi

7. Role of Rotifers as Bioindicators in Freshwater Lentic Ecosystems 173
Arvind Kumar and A.K. Singh

8. Correlation of Rotifers with Physico-chemical Factors at Harike Reservoir (Punjab - India) 187
Kaur, H., Dhillon, S.S., Bath, K.S., and Mander, G.

9. Cyclomorphosis, the Morphological Responses in Cladocera (Waterfleas) to Certain Environmental Factors : A Review 193
Asif A. Khan and Aftab Alam

10. Phytoplanktonic Dynamics in Bhandrakali Lake, Warangal, Andhra Pradesh 208
Y. Kalyani and M.A. Singara Charya

11. Pollution Status of a Leachate Reservoir Receiving Effluents from a Thermal Power Plant with Reference to Plankton Population 226
Rajeev Kumar Gaur, Asif A. Khan, M. Afaque Alam and Aftab Alam

12. Water Quality and Periphytic Algal Component of Anchar Lake, Kashmir 237
S.G. Sarwar

13. Population Biology of Trematode Species in Freshwater Fishes at Indore (M.P.) 251
Q. Ismail

14. Hydrobiological Studies of a Tropical Standing Waterbody with Special Reference to Nutrient Status and Primary Production 255
Ravindra Paul and Satyanarayan Varma

15. A Comprehensive Study of Water Quality Parameters in the River Ganga Between Narora and 287

Kannauj : Primary Production and Plankton
Asif A. Khan, Aftab Alam and Rajeev K. Gaur

16. Water Quality Management in Aquaculture 295
Amita Saxena

17. Biomonitoring of Water Quality of River Ramjan (At Kishanganj) in Relation to its Impact on Biological Components 310
B.N. Pandey, P.K.L. Das Vatsa Dubey and S. Hussain

1

Physico-Chemical Limnology of Certain Waterbodies of Central India

S. Kaushik and D.N. Saksena

Introduction

The aquatic ecosystems are extremely important to mankind as they have various uses, including drinking water supply, irrigation, navigation, recreation etc. and are also source of organic productivity. They serve as cheap receptacles for all kinds of waste disposal. The freshwater ecosystems can be broadly divided into two habitats. Lentic habitats are standing waters including lakes, reservoirs, ponds, pools and swamps in which the water is without any flow in a definite direction. Nevertheless, a certain degree of waters movement may occur in these waters also due to wave action and internal currents. The lotic habitats are running waters systems like rivers and streams in which the water is continuously flowing in a particular direction.

The physical properties of water in any aquatic system are largely regulated by the existing meteorological conditions and chemical properties. The effects of physical forces such as light and heat are of great significance as they are solely responsible for certain phenomena, such as thermal stratification, chemical stratification, diurnal and seasonal qualitative and quantitative variations in the plankton, micro and macro organism, and also in the quality of water. The even increasing population and rapid industrial growth in the present era are contributing to a maximum extent in influencing the physico-chemical properties of most Waterbodies.

India has vast water spread area in the form of rivers, lakes, streams reservoirs etc. but this water area upto an extent of 70 per cent has gone polluted (Basu, 1986). The limnological studies of Indian waters in relation to the seasonal fluctuations of water characteristics and biological conditions of lakes, reservoir, tanks and ponds are scanty in comparison to the vast area of inland freshwater available. The study of seasonal condition of pond life in Punjab by Prasad (1916) was perhaps the first attempt towards the study of aquatic ecosystem in India. Latteron, Pruthi (1933) studied seasonal fluctuations in physical and chemical conditions of a tank located in Indian Museum compound, Calcutta while Gonzalves and Joshi (1946) tried to correlate abiotic and biotic factors of a tank at Bandra (Bombay).

The beginning of systematic limnological studies in India was made by Ganapati (1940, 1956) and his associates (Ganapati and Chacko, 1951; Ganapati *et al.*, 1953 and Ganapati and Sreenivasan, 1968). Since than lot of progress has been made toward the limnological studies in India (Tripathi and Srikandar, 1989, Sarkar, 1989 and Sreenivasan, 1991).

Eutrophication resulting due to excessive fertilization is manifested by excessive growths of planktonic and attached algae and aquatic macrophytes which may in turn have significance deleterious effects on the beneficial uses of lakes and aquatic systems. A recent report published by the Centre for Science and Environment, New Delhi (Anon, 1988) has already warned that all Indians lakes except two lakes, namely, the Kodaikanal in south and Mukerti in north-east are facing the danger of extinction.

The state of Madhya Pradesh is situated in the central part of India. The beginning of limnological studies in this state was made by Krishnamoorthy and Visweswara (1963) on Gandhi Sagar tank. Later on, several workers have conducted limnological work in various Waterbodies. The water quality parameters in relation to biotic factors in the lakes of Bhopal in particular and degradation of aquatic environment of Madhya Pradesh in general have been reviewed by Kulshrestha (1988, 1989). Limnological studies on the Waterbodies in and around Gwalior have also attracted the attention of various workers. The physico-chemical characteristics of Tekanpur tank (Verma, 1969); Chandanpura pond (Kaushik *et al.*, 1989); J. C. Mills pond (Kaushik and Saksena, 1989); Vivek Nagar pond (Kaushik *et al.*, 1990), certain tanks and ponds of Gwalior (Kaushik *et al.*, 1991a), Chambal tank (Kaushik *et al.*, 1991b) and Matsya Sarovar (Kaushik *et al.*, 1991c) have been studied.

In view of above observations, a detailed study of the physico-chemical characteristics of Motijheel, Surajkund and Ranital has been undertaken here on a comparative basis.

Study Area

Gwalior city is remarkable in having numerous reservoirs, tanks and baoris. Out of these Waterbodies namely Motijheel, situated at the msl of 205m and Surajkund and Ranital situated as the Gwalior Fort at the height of 300m msl were selected for the present study. The location of these Waterbodies has been shown in Fig.1. The description of Waterbodies has been given as under:

Motijheel

Motijheel is a masonry water tank having a water area of 88093.87 m^2 situated 7 km away from Gwalior at longitude 26° 12' N; latitude 76° 18' E on Agra - Bombay road. It was constructed about seventy four year back for pearl culture as the name suggests. The pearl culture experiment failed, hence it was handed over to Public Health Department in 1928 for its use as a storage tank to Gwalior city. Its water storage capacity is of approximately 32 million gallons and maximum depth is upto 2.03 m. It is somewhat rectangular in shape. All the four sides and bottom of tank are pucca type. Besides rainy water, it has a regular supply of water from Tighra reservior. The tank is heavily infested with macrovegetation such as *Hydrilla verticillata, Ceratophyllum demersum, Chara vulgaris, C. fragillis, Vallisneria spiralis, Potamogeton crispus* and *Zennichellia pelustrima*. The local people use tank water for nistar though the activities are limited.

Surajkund

Surajkund is supposed to be constructed by King Suraj Sen who laid the foundation of the Fort in 5 A.D. (perhaps even earlier) around a spring which according to a belief had some medicinal properties to cure skin diseases like leprosy. This tank was renovated by the Major General Rao Sardar Ganpat Rao Raghunath Rao Rajwade in 1914 A.D.

Octangular in shape (22x' X 232') Surajkund is situated on the road side of the Urwahi gate on the Western side of the Fort, near the Data Bandhi Chhor Gurudwara. It has a temple of Lord Shiva in the middle with a high stone pavement to it from the Eastern side. The worships to Lord Shiva in this temple are offered on almost religious occasion related to the Lord Shiva and Lord Surya. It receives water from the under ground springs and through rains. Its water storage capacity is of 10, 37, 116.4 gallons and maximum depth is upto 10.20m. This tank is also greatly infested with macrovegetation like *Hydrilla verticillata, Chara vulgaris, Lemna perpusilla* and *Typha angustata*. The tank water is used for bathing, washing the clothes and also for building construction. The side wall and bottom of the tank are pucca and rocky type respectively.

Ranital

Ranital is a rockcut perennial impoundment situated at the Southern part of the Fort hill near the Mahatma Gandhi statue. It was constructed during the regime of Raja Man Singh somewhere in 1500 A.D. and was used a swimming pool by the royal ladies. The water capacity of this tank is 1,13, 736.26 gallon with an area of 517.5 m^2. The maximum water depth at the corner of Western-Southern part is about 5.0 m. It received water from the underground springs and also through rains. The tank water is used for bathing, washing of cloths and building construction etc. but occasionally for drinking purposes. The macrovegetation is altogether absent in the tank water.

Materials and Methods

The water sample of Motijheel, Surajkund and Ranital were collected for its physico-chemical examination from August 1988 to July 1990. The monthly sampling of water was done in the last week of each month in the early hours of the day *i.e.*, between 6.00 to 9.00 A.M. Iodine treated PVC sampling bottle were used for the collection of water samples. Care was taken to avoid air bubbling during the sampling. Analytical technique as described standard

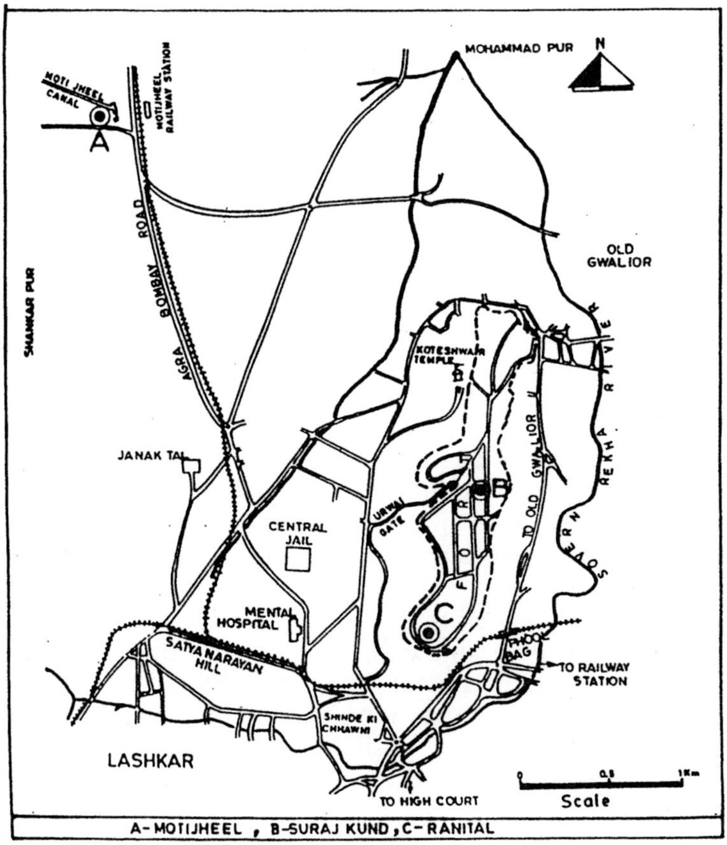

Fig. 1: Location map of Waterbodies selected for ecological study

methods for examination of water and waste water (APHA *et al.*, 1985) was adopted for the physico-chemical analysis. Some of the parameters, which may change due to time lapse were analysed at the sampling sites while turbidity, solids, conductivity, chemical oxygen demand, sodium, potassium and nutrients were determined in the laboratory. The coefficient of correlation between various physico-chemical characteristics was also calculated.

Results and Discussion

The mean and range alongwith standard error of various physico-chemical characteristics of Motijheel, Surajkund and Ranital during 1988-89 and 1989-90 have been given in Tables 1-3,Figs. 2-13. The perusal of data on physico-chemical characteristics suggests that Motijheel is mesotrophic, Surajkund is mesoeutrophic and Ranital is eutrophic as far as their trophic status is concerned.

The value of coefficient of correlation (γ) between various physico-chemical parameters was calculated and it was found that the ambient temperature with water temperature ; water temperature with conductivity, total alkalinity, chloride, chemical oxygen demand, sodium, potassium, silicate and phosphate; water depth with transparency; turbidity with total solids; total solids with total suspended solids, total dissolved solids; total suspended solids with total dissolved solids; total dissolved solids with conductivity, silicate, phosphate, calcium, magnesium, sodium, potassium; conductivity with calcium, magnesium, sodium, potassium, silicate, phosphate; pH with phosphate; total hardness with calcium and chloride with chemical oxygen demand were positively correlated, whereas, ambient and water temperature with dissolved oxygen; transparency with turbidity, total suspended solids; dissolved oxygen with chemical oxygen demand and free carbondioxide with pH are negatively correlated (Table-4).

Water is an important requirement of living beings. It has the property of dissolving large amount of substances in it. This property of water brings about changes in its physical as well as chemical properties of water and ultimately the pollution. Normally, the physical properties are governed by the fundamental geographics and chemical properties and are influenced by the cyclical pattern of nutrients involving both abiotic and biotic components of aquatic ecosystem. The stagnant Waterbodies have been variously classified (Forel, 1892). The ponds and tanks may be classified according to their origin, nature of water supply, location and water retention (Jhingran, 1982). The ponds with an area of 5000 m^2, 5000-10,000 m^2 and more than 10,000 m^2 respectively have been designated as small, medium and large (Ambasht and Shardendu, 1989). If this classification is applied to the tanks under study, Ranital fall under the category of small ponds, Surajkund under medium ponds and Motijheel under large ponds.

The physico-chemical characteristics are interrelated to each other and exert influence on the biological communities and biological production in Waterbodies and thus, it is essential to understand this interrelationship. The physical and chemical parameters observed in the Waterbodies under study have been discussed here in the light of available information in order to arrive at certain conclusions on their role in the structural and functional aspects of these Waterbodies.

Water Colour

Colour of water generally depends upon the production and degradation processes in the natural environment. It may be brought about by suspended matter, pH, climate, organic matter and substances like iron and mangnese, photosynthetic activity and co-precipitation of calcium carbonate and fluvic acid. Ganapati (1956) observed reddish colour of water in Errakuppam reservoir throughout the year indicating the presence of excessive inorganic suspended matter of iron compounds. In Devikund at Deoband, the colour of water was mostly greenish (Verma and Shukla, 1968) while in Tekanpur reservior water was transparent throughout the year except for monsoon months when it became turbid due to inflow of silt into the reservoir alongwith runoff water (Verma, 1969). In Ajwa reservoir in Western India, Ganapati (1970) recorded greenish to green colour of water. Roy (1977) also found that during rainy season water in a pond at Bhagalpur was grey and later on became deep green due to development of

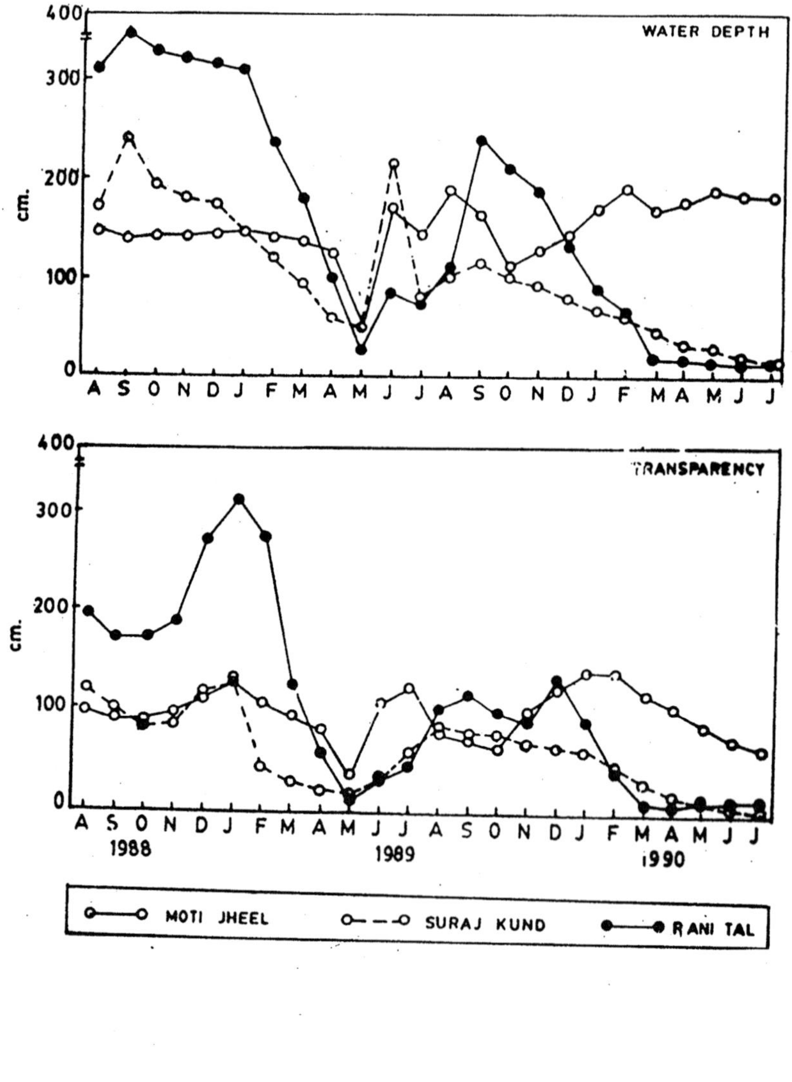

Fig. 2 : Seasonal variation in water depth and transparency in Motijheel, Surajkund and Ranital

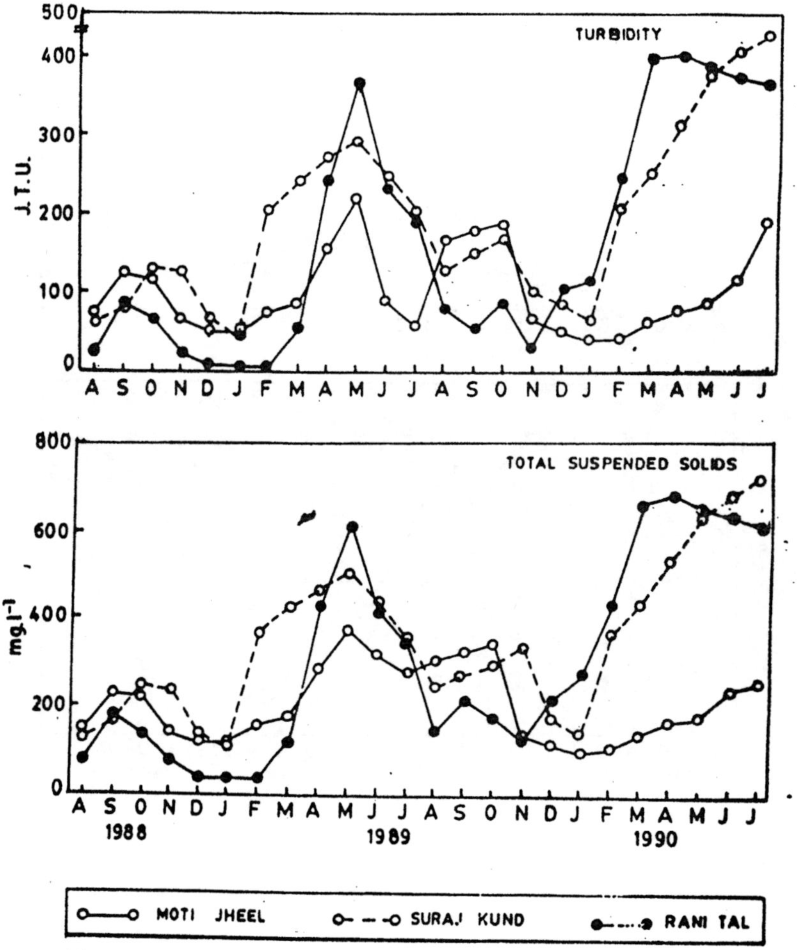

Fig. 3 : Seasonal variation in turbidity and total suspended solids in Motijheel,Surajkund and Ranital

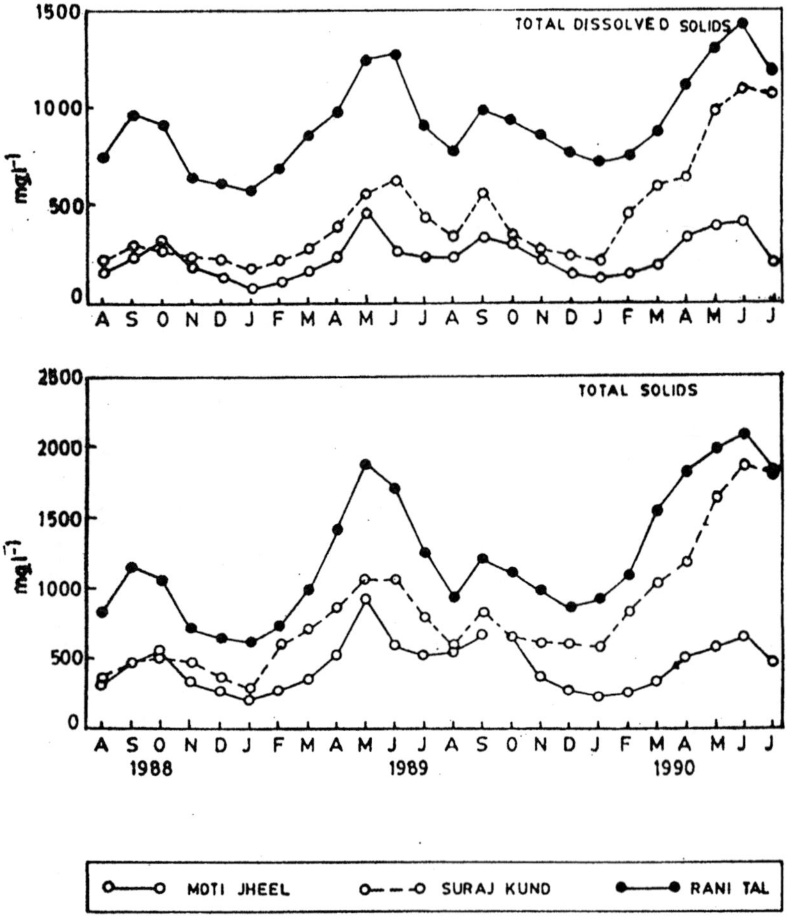

Fig. 4 : Seasonal variation in total dissolved solids and total solids in Motijheel, Surajkund and Ranital

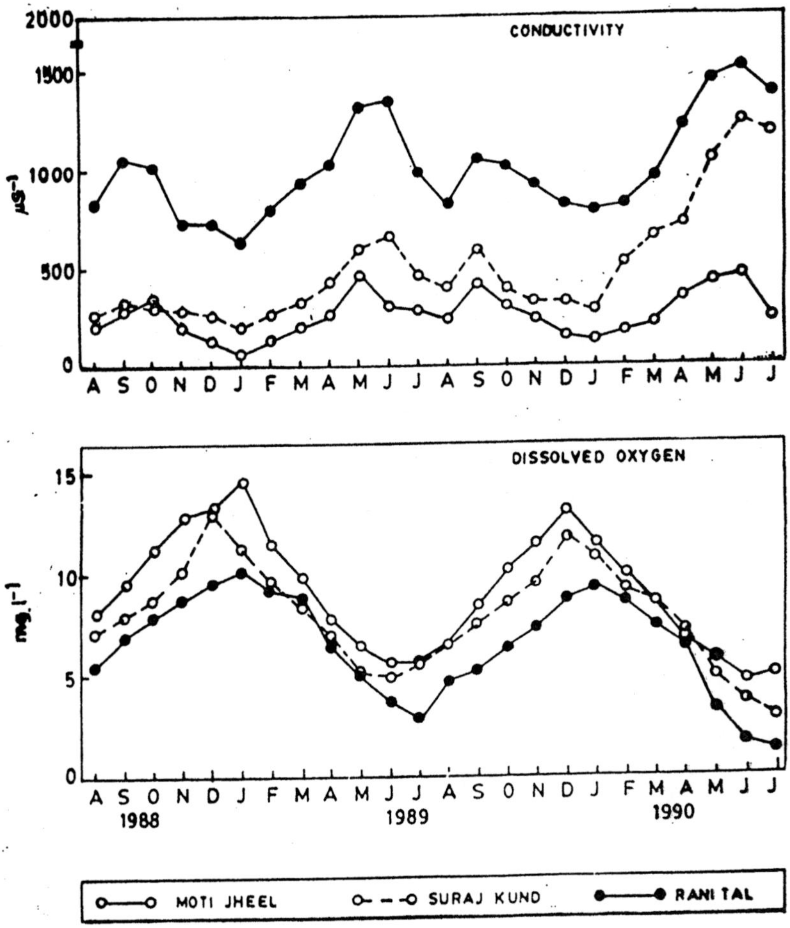

Fig. 5 : Seasonal variation in conductivity and dissolved oxygen in Motijheel, Surajkund and Ranital

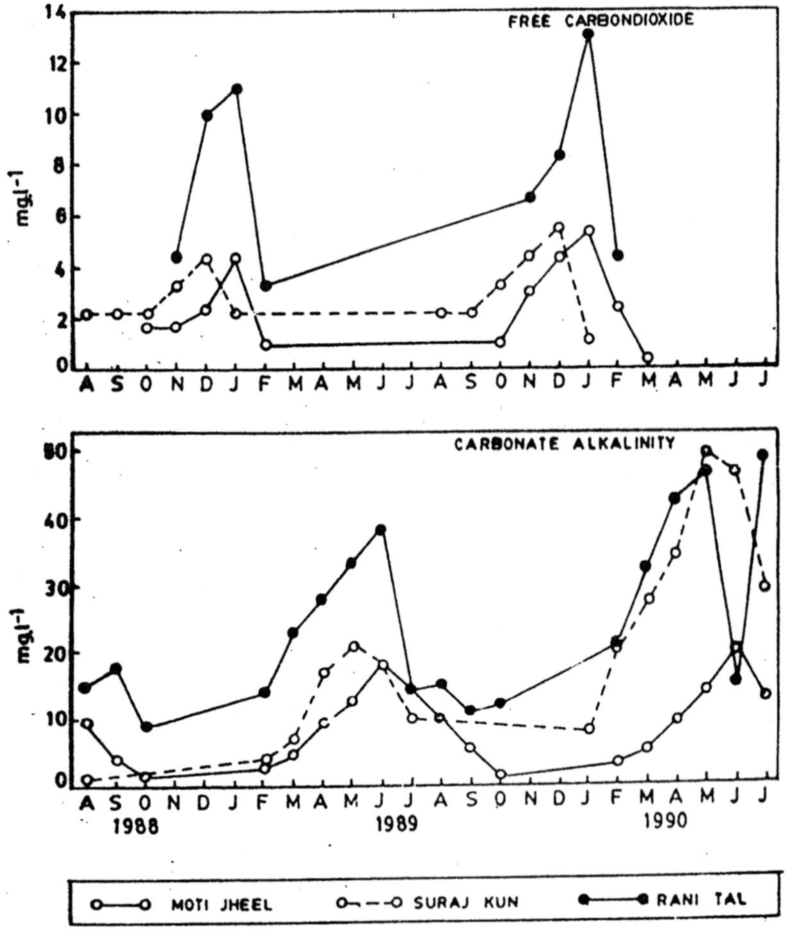

Fig. 6 : Seasonal variation in free carbondioxide and carbonate in Motijheel, Surajkund and Ranital

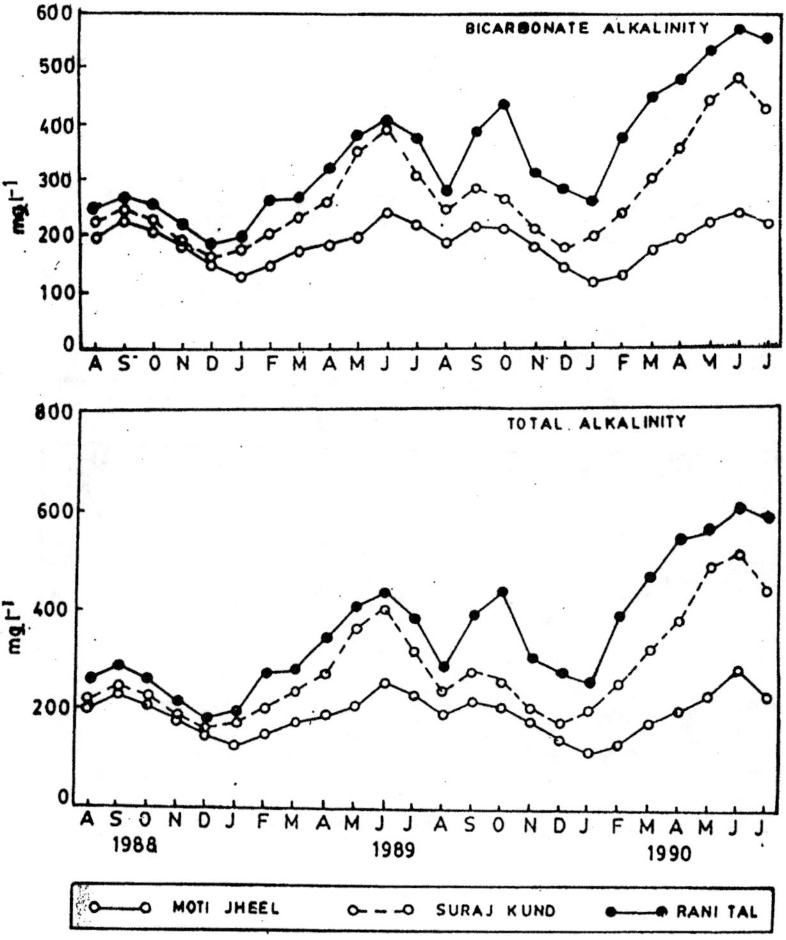

Fig. 7 : Seasonal variation in bicarbonate and total alkalinity in Motijheel, Surajkund and Ranital

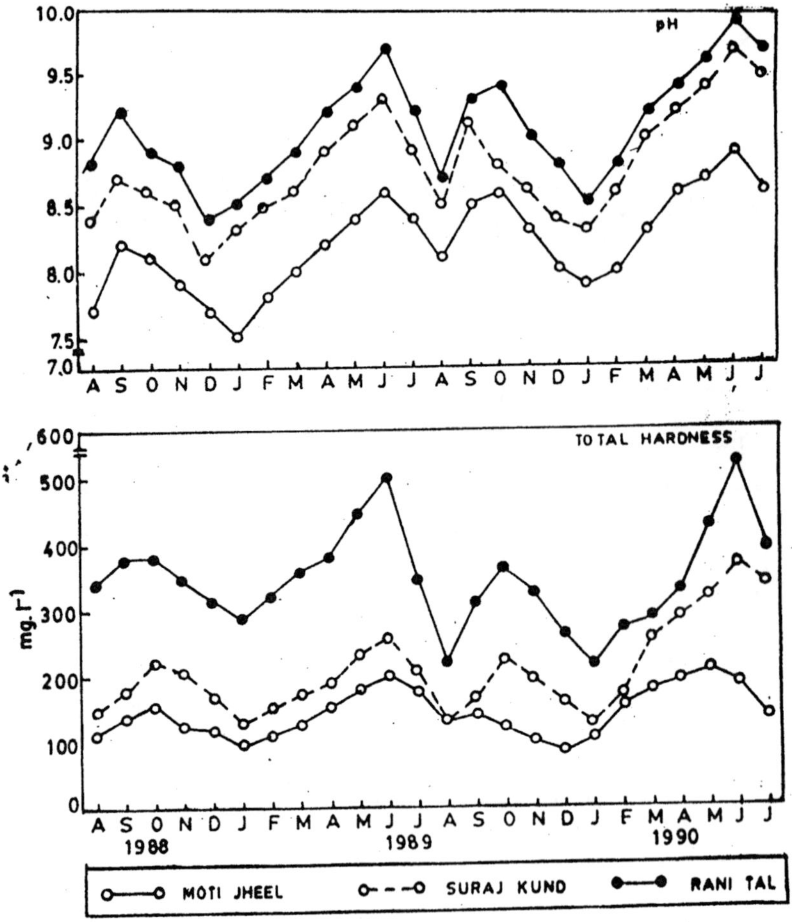

Fig. 8 : Seasonal variation in pH and total hardness in Motijheel, Surajkund and Ranital

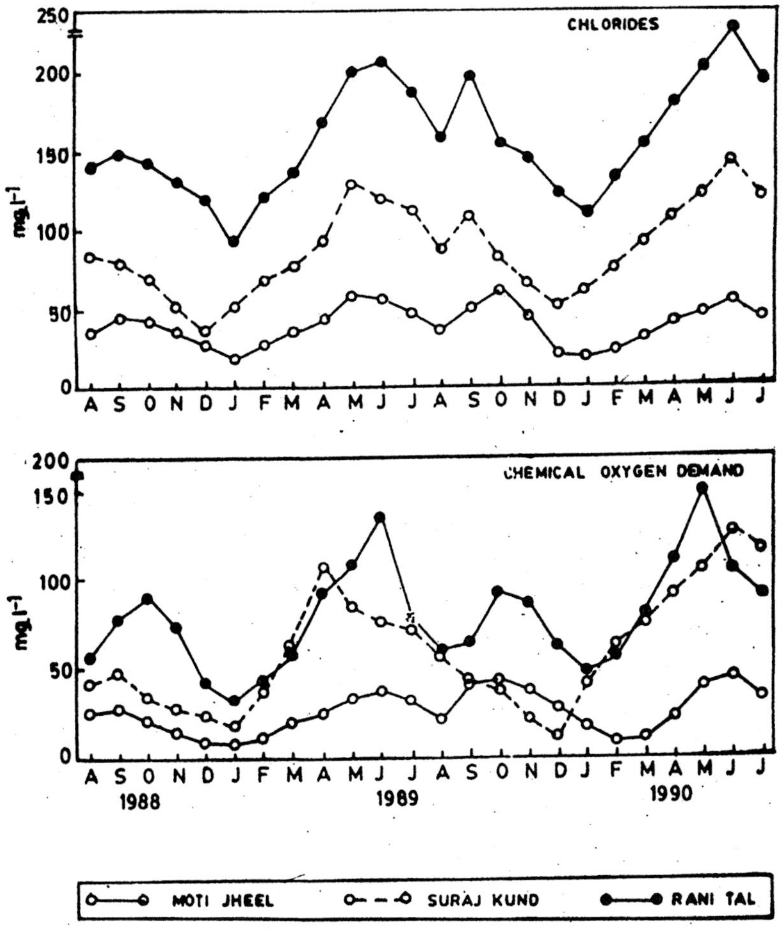

Fig. 9 : Seasonal variation in chloride and chemical oxygen demand in Motijheel, Surajkund and Ranital

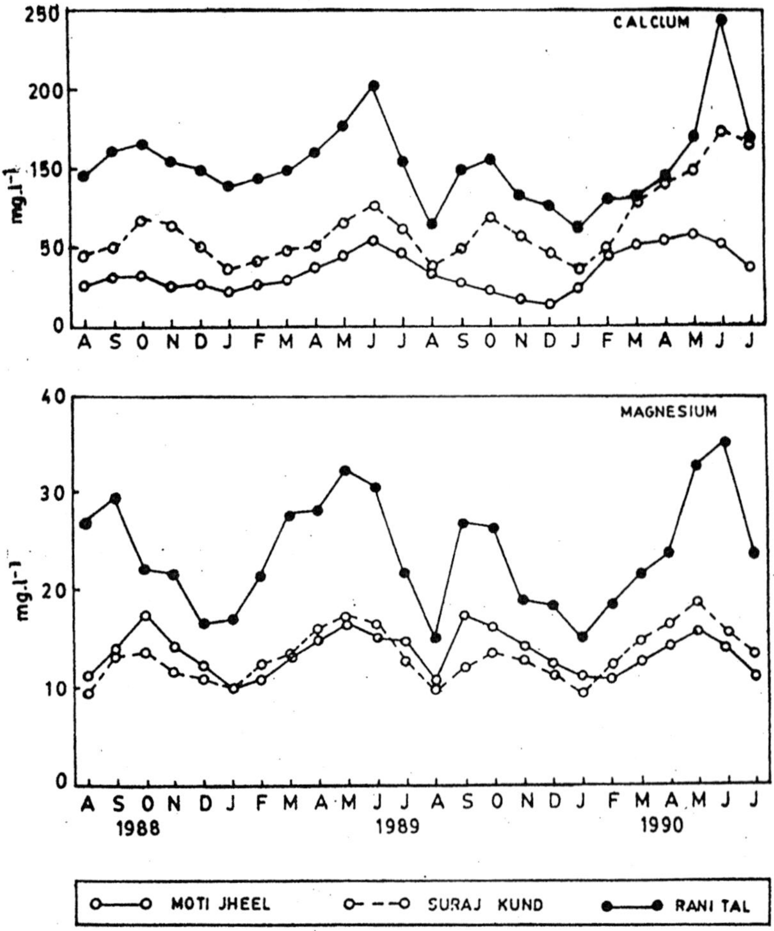

Fig. 10 : **Seasonal variation in calcium and magnesium in Moti jheel, Suraj kund and Ranital**

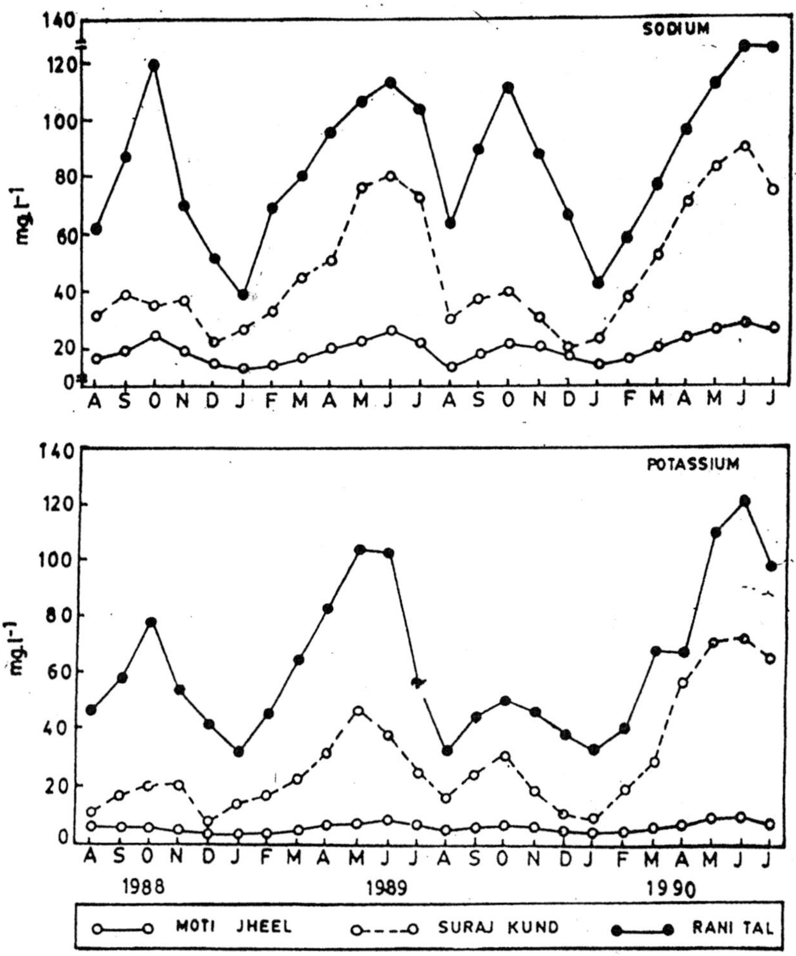

Fig. 11 : Seasonal variation in sodium and potassium in Motijheel, Surajkund and Ranital

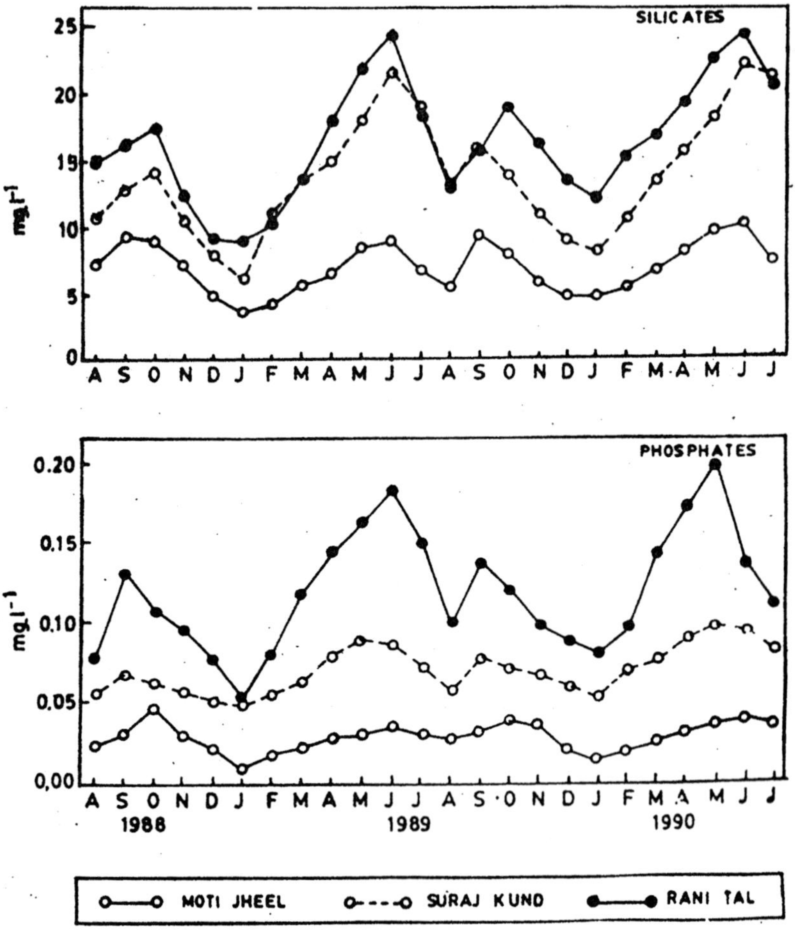

Fig. 12 : Seasonal variation in silicates and phosphates in Motijheel, Suraj kund and Ranital

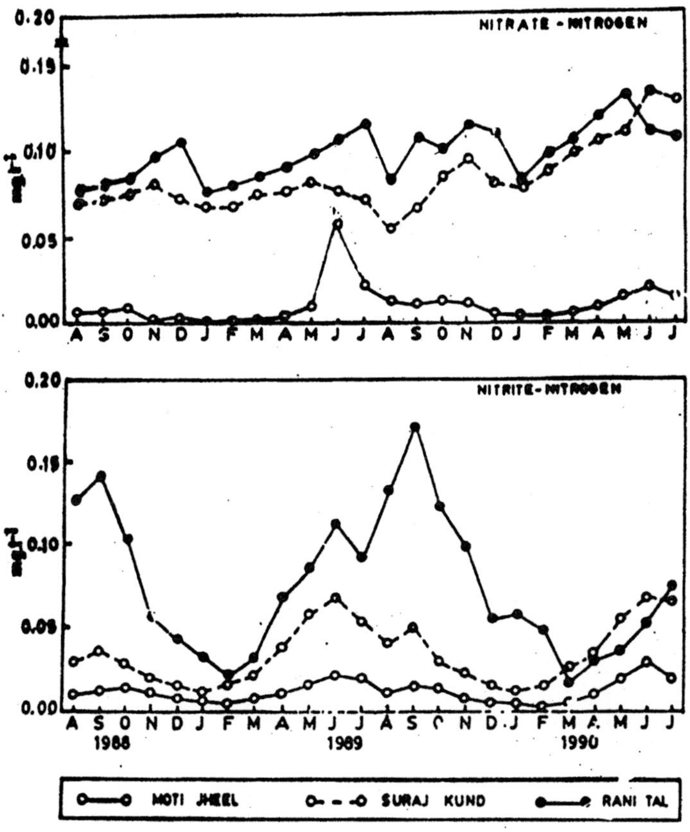

Fig. 13 : Seasonal variation in nitrate-N and nitrite-N in Motijheel, Surajkund and Ranital

Table 1 : Range, Mean Alongwith Standard Error of Physico-Chemical Characteristics of Water in Motijheel During 1988-89 and 1989-90

Physico-Chemical Characteristics		*1988-89*		*1989-90*	
		Range	*Mean and Standard Error*	*Range*	*Mean and Standard Error*
Water colour		Transparent green-Light bluish green		Transparent green-Light bluish green	
Ambient temperature	°C	12.67 - 40.00	28.05 ± 2.78	12.00 - 43.33	29.13 ± 2.99
Water temperature	°C	9.67 - 34.67	23.33 ± 2.46	9.00 - 33.33	22.80 ± 2.48
Water depth	cm	60.67 - 170.67	136.27 ± 7.48	111.66 - 189.66	165.49 ± 7.29
Transparency	cm	35.30 - 126.00	95.54 ± 6.81	62.00 - 139.30	94.97 ± 8.29
Turbidity	J.T.U.	48.30 - 155.60	96.03 ± 14.57	40.00 - 192.00	106.29 ± 17.18
Total solids	$mg.1^{-1}$	200.00 - 930.00	442.50 ± 58.12	230.00 - 670.00	464.16 ± 47.71
Total dissolved solids	$mg.1^{-1}$	80.00 - 460.00	219.16 ± 29.96	100.00 - 350.00	204.16 ± 26.03
Total suspen-ded solids	$mg.1^{-1}$	120.00 - 380.00	215.83 ± 24.69	130.73 - 420.00	260.14 ± 28.69
Conductivity	μ S	66.00 - 462.00	231.83 ± 31.40	132.00 - 462.00	281.33 ± 32.162
Dissolved oxygen	$mg.1^{-1}$	5.53 - 14.50	9.61 ± 0.87	4.59 - 12.97	8.39 ± 0.79
Free carbon-dioxide	$mg.1^{-1}$	Nil - 4.32	2.19 ± 0.36	Nil - 5.32	2.71 ± 0.55

Table 1 (Contd.) :

Carbonate alkalinity	$mg.1^{-1}$	Nil	-	18.67	8.59	±	1.70	Nil	-	20.00	9.03	±	1.70
Bicarbonate alkalinity	$mg.1^{-1}$	128.00	-	238.67	185.44	±	9.59	118.00	-	236.66	185.33	±	10.94
Total alkalinity	$mg.1^{-1}$	128.00	-	257.30	191.85	±	10.96	118.00	-	286.66	194.55	±	13.71
pH		7.70	-	8.41	8.06	±	0.09	7.98	-	8.94	8.42	±	0.09
Total hardness	$mg.1^{-1}$	98.00	-	201.30	141.74	±	9.24	85.33	-	207.33	145.16	±	11.289
Chlorides	$mg.1^{-1}$	17.90	-	57.60	38.65	±	3.42	17.98	-	61.65	38.86	±	4.00
Chemical Oxygen demand	$mg.1^{-1}$	8.00	-	37.30	22.20	±	2.80	9.38	-	45.33	29.21	±	3.70
Calcium	$mg.1^{-1}$	10.10	-	54.50	32.48	±	3.46	13.89	-	57.44	36.45	±	4.38
Magnesium	$mg.1^{-1}$	9.90	-	17.50	13.62	±	0.66	10.55	-	17.22	13.23	±	0.65
Sodium	$mg.1^{-1}$	12.30	-	25.50	18.51	±	1.21	12.33	-	27.08	19.28	±	1.41
Potassium	$mg.1^{-1}$	4.30	-	9.00	5.96	±	0.45	4.78	-	10.43	6.89	±	0.50
Silicates	$mg.1^{-1}$	3.52	-	9.22	6.69	±	0.56	4.56	-	9.99	6.97	±	0.55
Phosphates	$mg.1^{-1}$	0.010	-	0.046	0.027	±	0.002	0.014	-	0.038	0.029	±	0.002
Nitrate-nitrogen	$mg.1^{-1}$	0.001	-	0.060	0.010	±	0.012	0.003	-	0.020	0.110	±	0.001
Nitrite-nitrogen	$mg.1^{-1}$	0.005	-	0.022	0.011	±	0.001	0.003	-	0.028	0.012	±	0.002

Table 2 : Range, Mean Alongwith Standard Error of Physico-Chemical Characteristics of Water in Surajkund During 1988-89 and 1989-90

Physico-Chemical Characteristics		*1988-89*		*1989-90*	
		Range	*Mean and Standard Error*	*Range*	*Mean and Standard Error*
Water colour		Transparent green-green		Transparent green-green	
Ambient temperature	°C	15.50 - 41.50	30.66 ± 2.43	21.00 - 43.00	32.54 ± 2.08
Water temperature	°C	13.00 - 32.00	23.62 ± 2.08	15.50 - 35.00	25.95 ± 1.85
Water depth	cm	49.50 - 240.00	143.12 ± 17.98	15.00 - 115.05	63.95 ± 10.12
Transparency	cm	17.00 - 119.50	69.12 ± 11.76	4.00 - 84.00	45.91 ± 8.58
Turbidity	J.T.U.	44.00 - 293.50	164.20 ± 26.12	66.00 - 431.00	224.42 ± 37.42
Total solids	$mg.1^{-1}$	290.00 - 1070.00	632.50 ± 77.75	580.00 - 1860.00	1020.0 ± 141.41
Total dissolved solids	$mg.1^{-1}$	180.00 - 630.00	330.83 ± 41.58	220.00 - 1170.00	578.33 ± 96.29
Total suspended solids	$mg.1^{-1}$	110.00 - 510.00	301.66 ± 41.88	140.0 - 730.00	408.33 ± 57.62
Conductivity	μ S	198.00 - 660.00	364.75 ± 41.25	286.00 - 1254.00	645.33 ± 99.58
Dissolved oxygen	$mg.1^{-1}$	4.86 - 12.97	8.17 ± 0.71	2.86 - 11.75	7.54 ± 0.78
Free carbon-dioxide	$mg.1^{-1}$	Nil - 4.40	2.75 ± 0.26	Nil - 5.50	3.11 ± 0.46

Contd...

Table 2 (Contd.) :

Carbonate alkalinity	mg.1^{-1}	Nil - 21.00	11.14 ± 2.20	Nil - 49.00	30.42 ± 4.12
Bicarbonate alkalinity	mg.1^{-1}	162.00 - 391.00	245.25 ± 20.24	178.00 - 480.00	299.41 ± 29.14
Total alkalinity	mg.1^{-1}	162.00 - 409.00	252.25 ± 22.36	178.00 - 526.00	318.00 ± 34.39
pH		8.13 - 9.13	8.68 ± 0.08	8.35 - 9.70	8.96 ± 0.13
Total hardness	mg.1^{-1}	130.00 - 258.00	189.50 ± 10.87	127.00 - 369.00	229.16 ± 24.12
Chlorides	mg.1^{-1}	35.48 - 129.96	80.80 ± 8.32	51.48 - 144.95	93.62 ± 8.16
Chemical Oxygen demand	mg.1^{-1}	18.00 - 108.00	53.00 ± 7.98	12.00 - 128.00	66.33 ± 10.84
Calcium	mg.1^{-1}	35.67 - 66.92	54.36 ± 3.51	35.66 - 122.24	70.13 ± 8.61
Magnesium	mg.1^{-1}	9.49 - 17.29	13.10 ± 0.72	9.25 - 18.49	13.17 ± 0.78
Sodium	mg.1^{-1}	21.56 - 79.05	45.04 ± 5.71	19.21 - 88.77	48.01 ± 6.99
Potassium	mg.1^{-1}	7.60 - 46.85	22.68 ± 3.25	9.17 - 72.60	35.41 ± 6.88
Silicates	mg.1^{-1}	6.05 - 21.65	13.29 ± 1.32	8.02 - 22.50	14.28 ± 1.31
Phosphates	mg.1^{-1}	0.047 - 0.087	0.064 ± 0.003	0.052 - 0.096	0.073 ± 0.004
Nitrate-nitrogen	mg.1^{-1}	0.067 - 0.082	0.074 ± 0.001	0.055 - 0.134	0.093 ± 0.006
Nitrite-nitrogen	mg.1^{-1}	0.011 - 0.068	0.033 ± 0.005	0.012 - 0.067	0.084 ± 0.051

Table 3 : Range, Mean Alongwith Standard Error of Physico-Chemical Characteristics of Water in Ranital During 1988-89 and 1989-90

Physico-Chemical Characteristics		*1988-89*		*1989-90*	
		Range	*Mean and Standard Error*	*Range*	*Mean and Standard Error*
Water colour		Light green-Light blue		Light green-Light blue	
Ambient temperature	^0C	22.00 - 41.00	30.75 ± 1.79	21.00 - 42.50	33.66 ± 1.99
Water temperature	^0C	15.00 - 34.50	25.37 ± 1.77	18.50 - 36.00	27.50 ± 1.60
Water depth	cm	25.05 - 347.05	218.95 ± 34.46	13.00 - 240.00	92.60 ± 24.12
Transparency	cm	12.70 - 309.50	149.87 ± 29.03	7.50 - 132.50	58.12 ± 14.29
Turbidity	J.T.U.	4.00 - 369.00	109.00 ± 34.82	30.00 - 404.00	221.40 ± 44.90
Total solids	$mg.1^{-1}$	620.00 - 1870.00	1081.6 ± 119.21	920.00 - 2080.00	1362.50 ± 130.81
Total dissolved solids	$mg.1^{-1}$	580.00 - 1280.00	870.00 ± 66.86	720.00 - 1440.00	980.00 ± 68.49
Total suspended solids	$mg.1^{-1}$	40.00 - 620.00	211.66 ± 56.04	130.00 - 690.00	408.33 ± 67.10
Conductivity	µ S	627.00 - 1353.00	948.75 ± 65.63	825.00 - 1518.0	1067.00 ± 76.02
Dissolved oxygen	$mg.1^{-1}$	2.84 - 10.13	7.00 ± 0.69	1.22 - 9.30	5.80 ± 0.77
Free carbon-dioxide	$mg.1^{-1}$	Nil - 11.00	7.15 ± 1.11	Nil - 13.2	8.12 ± 1.07

Contd..

Table 3 (Contd.) :

Parameter	Unit	Range	Mean ± SE	Range	Mean ± SE
Carbonate alkalinity	$mg.l^{-1}$	Nil - 38.00	21.33 ± 2.83	Nil - 50.00	30.77 ± 4.69
Bicarbonate alkalinity	$mg.l^{-1}$	184.00 - 406.00	280.16 ± 21.16	259.00 - 566.00	407.50 ± 31.80
Total alkalinity	$mg.l^{-1}$	184.00 - 444.00	295.75 ± 24.39	259.00 - 616.00	433.25 ± 37.77
pH		8.42 - 9.74	9.00 ± 0.10	8.56 - 9.92	9.22 ± 0.12
Total hardness	$mg.l^{-1}$	287.00 - 503.00	366.75 ± 16.99	221.00 - 525.00	327.66 ± 25.67
Chlorides	$mg.l^{-1}$	92.46 - 206.45	149.33 ± 10.00	108.96 - 226.93	165.19 ± 10.35
Chemical Oxygen demand	$mg.l^{-1}$	32.00 - 136.00	73.91 ± 8.65	48.00 - 150.00	83.50 ± 8.29
Calcium	$mg.l^{-1}$	88.17 - 151.50	107.71 ± 4.96	61.72 - 193.10	97.15 ± 10.23
Magnesium	$mg.l^{-1}$	16.54 - 32.15	24.56 ± 1.49	14.84 - 34.82	22.76 ± 1.85
Sodium	$mg.l^{-1}$	38.38 - 119.06	82.33 ± 7.29	41.06 - 123.87	86.52 ± 7.80
Potassium	$mg.l^{-1}$	31.62 - 104.12	63.69 ± 6.81	32.55 - 121.25	62.31 ± 8.91
Silicates	$mg.l^{-1}$	8.85 - 24.40	15.37 ± 1.41	12.00 - 24.40	17.32 ± 1.12
Phosphates	$mg.l^{-1}$	0.053 - 0.182	0.104 ± 0.014	0.078 - 0.198	0.122 ± 0.010
Nitrate-nitrogen	$mg.l^{-1}$	0.076 - 0.115	0.091 ± 0.003	0.825 - 0.132	0.105 ± 0.004
Nitrite-nitrogen	$mg.l^{-1}$	0.021 - 0.141	0.075 ± 0.011	0.017 - 0.171	0.074 ± 0.013

Table - 4 : The Correlation Between Various Physico-Chemical Parameters as shown by the Value of Coefficient of Correlation (γ) in Motijheel, Surajkund and Ranital

Parameters	*Value of coefficient of of correlation (γ)*		
	Motijheel	*Surajkund*	*Ranital*
Ambient temperature Vs. Water temperature	0.981	0.945	0.926
Ambient temperature Vs. Dissolved oxygen	-0.929	-0.834	-0.926
Water temperature Vs. Turbidity	0.558	0.570	0.630
Water temperature Vs. Conductivity	0.758	0.615	0.835
Water temperature Vs. Dissolved oxygen	-0.909	-0.890	-0.914
Water temperature Vs. Total alkalinity	0.899	0.700	0.718
Water temperature Vs. Chloride	0.785	0.854	0.925
Water temperature Vs. Chemical oxygen demand	0.625	0.664	0.774
Water temperature Vs. Sodium	0.707	0.663	0.765
Water temperature Vs. Potassium	0.814	0.614	0.664
Water temperature Vs. Silicate	0.781	0.789	0.869
Water temperature Vs. Phosphate	0.652	0.682	0.755
Water depth Vs. Transparency	0.370	0.717	0.876
Transparency Vs. Turbidity	-0.919	-0.907	-0.846
Transparency Vs. Total suspended solids	-0.758	-0.912	-0.867
Transparency Vs. Conductivity	-0.692	-0.744	-0.728
Transparency Vs. Chemical oxygen demand	-0.603	-0.817	-0.710
Transparency Vs. Phosphate	-0.608	-0.782	-0.704
Turbidity Vs. Total solids	0.805	0.941	0.920

Contd.

Table-4 : Contd...

Parameters	*Value of coefficient of of correlation (γ)*		
	Motijheel	*Surajkund*	*Ranital*
Turbidity Vs. Total suspended solids	0.861	0.987	0.994
Turbidity Vs. Total dissolved solids	0.614	0.889	0.762
Total solids Vs. Total suspended solids	0.887	0.938	0.926
Total solids Vs. Total dissolved solids	0.921	0.979	0.949
Total suspended solids Vs. Total dissolved solids	0.652	0.880	0.771
Total dissolved solids Vs. Conductivity	0.983	0.998	0.992
Total dissolved solids Vs. Calcium	0.566	0.903	0.759
Total dissolved solids Vs. Magnesium	0.769	0.631	0.798
Total dissolved solids Vs. Sodium	0.749	0.847	0.854
Total dissolved solids Vs. Potassium	0.787	0.932	0.886
Total dissolved solids Vs. Silicate	0.843	0.819	0.948
Total dissolved solids Vs. Phosphate	0.753	0.856	0.816
Conductivity Vs. Calcium	0.585	0.904	0.753
Conductivity Vs. Magnesium	0.777	0.610	0.778
Conductivity Vs. Sodium	0.762	0.828	0.866
Conductivity Vs. Potassium	0.817	0.929	0.892
Conductivity Vs. Silicate	0.892	0.799	0.937
Conductivity Vs. Phosphate	0.765	0.843	0.802
Dissolved oxygen Vs. pH	-0.676	-0.863	-0.825
Dissolved oxygen Vs. Chemical oxygen demand	-0.586	-0.842	-0.661

Contd.

Table-4 : Contd...

Parameters	*Value of coefficient of of correlation (γ)*		
	Motijheel	*Surajkund*	*Ranital*
Dissolved oxygen Vs. Phosphate	-0.505	-0.788	-0.631
Free Carbondioxide Vs. Total alkalinity	-0.759	-0.640	-0.558
Free Carbondioxide Vs. pH	-0.487	-0.599	-0.684
Total alkalinity Vs. pH	0.691	0.954	886
Total alkalinity Vs. Phosphate	0.800	0.921	0.713
pH Vs. Chloride	0.706	0.924	0.913
pH Vs. Chemical oxygen demand	0.732	0.857	0.846
pH Vs. Magnesium	0.518	0.767	0.786
pH Vs. Phosphate	0.685	0.944	0.813
Total hardness Vs. Calcium	0.965	0.988	0.949
Chloride Vs. Chemical oxygen demand	0.831	0.854	0.789
Chloride Vs. Sodium	0.764	0.904	0.823
Chloride Vs. Potassium	0.766	0.834	0.771
Chloride Vs. Phosphate	0.822	0.892	0.822
Chemical oxygen demand Vs. Sodium	0.680	0.869	0.809
Chemical oxygen demand Vs. Potassium	0.793	0.856	0.819
Chemical oxygen demand Vs. Phosphate	0.675	0.834	0.871
Calcium Vs. Sodium	0.615	0.818	0.726
Calcium Vs. Potassium	0.707	0.913	0.803
Magnesium Vs. Sodium	0.624	0.809	0.708

Contd.

Table-4 : Contd...

Parameters	*Value of coefficient of of correlation (γ)*		
	Motijheel	*Surajkund*	*Ranital*
Magnesium Vs. Silicate	0.734	0.703	0.750
Magnesium Vs. Phosphate	0.740	0.0	0.696
Sodium Vs. Potassium	0.884	0.896	0.814
Sodium Vs. Silicate	0.811	0.909	0.876
Sodium Vs. Phosphate	0.848	0.898	0.710

phytoplankton. Rai and Dutta Munshi (1979) reported transparent, grey and brownish water in Makhana, Bhathwa Pokar and Barela chaur of Dharbhanga respectively. The brown colour of water is also imparted due to decaying leaves. The colour of Ranital was found to be darker throughout in comparison to Motijheel and Surajkund. Motijheel had transparent water during winter, greyish during early monsoon and green during post-monsoon and summer seasons while Surajkund and Ranital had transparent green colour in winter, green in rainy and deep green during summer seasons. The growth of phytoplankton was found to be greater in all the Waterbodies but suspended solids were maximum in Ranital during summer season.

Ambient Temperature

The seasonal cycle of temperature is superimposed by the air masses over the land mass. Climatic conditions (Welch, 1952); Hutchinson, 1957 and Kant and Raina, 1990), time of collection, climate, solar radiation and topography of water (Ganapati, 1956), solar radiation (Munawar, 1970) have an impact on the ambient temperature. The variation in air temperature at Idukii reservoir, Kerala fluctuated between 23.5^0 C and 30.6^0 C (Khatri, 1985) and at a fish pond in Dharbhanga between 28.0^{0}C and 31.0^{0}C (Dudani *et al.*, 1987). In Kashmir valley also very small variation (22.0^{0}C to

26.4^{0}C) in ambient temperature were noted at different Waterbodies (Yousuf and Shah, 1988). In the present study, the ambient temperature at Motijheel varied from 12.0^{0}C to 43.8^{0}C, at Surajkund from 15.5^{0}C to 43.0^{0}C and at Ranital from 21.5^{0}C to 42.5^{0}C. Maximum atmospheric temperature during summer and minimum during winter were observed in all the Waterbodies. The similar phenomenon has been reported by several workers (Chourasia and Adoni, 1985; Ramanibai and Ravichandran, 1987; Ambasht and Shardendu, 1989; Ghosh and George, 1989 and Fasihuddin and Kumari, 1990).

Water Temperature

Temperature of water depends upon water depth besides solar radiation, climate and topography. It may be mentioned that no other single factor has so much profound direct or indirect/ much influence on physico-chemical, biological, matabolic and physiological behaviour of aquatic ecosystem than temperature (Welch, 1982). It also reflects on the dynamics of living organisms (Chandler, 1942). In Indian subcontinent, most of the Waterbodies had their temperature lying between 7.8^{0}C and 38.5^{0}C (Qadri and Yousuf, 1980; Singhal *et al.*, 1986; Ramanibai and Ravichandran, 1987; Valecha and Bhatnagar, 1988; Ghosh and George 1989 and Kant and Raina, 1990).

In Motijheel the water temperature fluctuated between 9.0^{0}C and 34.76^{0}C; in Surajkund between 15.0^{0}C and 35.0^{0}C and in Ranital between 17.0^{0}C and 36.0^{0}C . It was observed that waterbody such as Surajkund and Ranital located at higher altitude had warmer water. When the classification of Venkateswarlu (1983) is applied to these Waterbodies, the Surajkund and Ranital fall under the category of meso-euthermal and Motijheel under meso-thermal Waterbodies. The water temperature has been found to increase correspondingly with the increasing atmospheric temperature. This phenomenon was, however, also observed in the Waterbodies under study. In warmer months water temperature increased due to atmospheric temperature and long day length. During rainy season a considerable fall in the water temperature was observed in all the tanks due to reduced solar radiation, presence of clounds, comparatively low atmospheric temperature and fast winds. Low water temperature during winter season may be attributed to the shorter photoperiod and decreased atmospheric temperature. Kant and Raina (1990) have shown that temperature of water during winter are under the influence of high specific heat of the water and over turn condition of the pond. A similar seasonal trend of water temperature was exhibited in the Waterbodies under question as has been already reported for most of the Indian waters (Ganapati, 1956; Sreenivasan, 1974; Singhal *et al.*. 1986; Kaushik *et al.*,1989; Kant and Raina, 1990 and Fasihauddin and Kumari, 1990). Moderately high temperature of water has been found to favour active multiplication of plankton resulting in low transparency (Kant and Anand, 1978).

Water Depth

Water level plays an important role in governing the water quality. Shallow waters get heated up and thus increase their temperature and consequently affect the biological and chemical reactions taking place. The fluctuation in water level is mainly due to climatic factors *e.g.*, evaporation of water due to increased atmospheric temperature during the day time, wind velocity, rainfall and humidity (Welch, 1952). Nasar and Dutta Munshi (1974) expressed that the water level is inversely related to the temperature. Water use is also an important factor in bringing out fluctuations in the water level. Motijheel had maximal depth in the month of June during 1988-89 and in the month of February during 1989-90. This variation was brought about due to its water use and influx of water from Tighra reservoir through a canal other than rains. In Surajkund and Ranital which are exclusively rain fed had greater water depth during monsoon months. Water depth decreased slowly through winter but quite fastly during summer due to evaporation, water use for building construction and also because of percolation losses of water as suggested by Ganapati (1956).

Transparency

Transparent waters allow more light penetration which has far reaching effects on all aquatic organisms, including their development, distribution and behaviour etc. Colour of water, concentration of suspended organic and inorganic particles, turbulence and concentration of phytoplankton and zooplankton affect the transparency of water and thus also influence the primary productivity and fish food organisms. Light penetration is dependent upon meteorological condition and low transparency can influence other physical factors. Sreenivasan (1968) have also emphasized that waters with low transparency are not suitable for fish growth. Dense populations of algae of algae and bacteria in the metalimmion and upper part of hypolimmion decreased the transparency of waters.

Seasonally, the fluctuatuation in the transparency of Indian waters had been quite variable. In the present study, secchi disc transparency varied from 35.3 cm to 139.3 cm in Motijheel, from 4.0 cm to 126.0 cm in Surajkund and from 7.5 cm to 309.0 cm in Ranital. The transparency was found to be low twice in a year; once during rainy season when silt and clay was brought into the tanks along with the run off water and later in summer due to evaporation of water, rich concentration of nutrients, increasing temperature and development of plankton. Chourasia and Adoni (1985), Kaushik *et al.* (1989, 1990) and Kant and Raina (1990) have also reported similar trend with regards to transparency of water in different Waterbodies in the country.

Turbidity

The turbidity is an expression of light scattering and light absorbing properties of the water and is caused by the presence of suspended matter, such as clay, silt, colloidal organic particles, plankton etc. This may also be caused by bottom sediments. This has been recognised as a evaluable limiting factor in the biological productivity of the Waterbodies. Turbidity usually hampered the light penetration in the water and is a most critical factor regulating the growth of chlorophyll bearing organisms. In the present study, out of three Waterbodies studied, the water in Surajkund was most turbid, Ranital and Motijheel follow a descending order with biomodal pattern of seasonal variation in turbidity values. During monsoon season silt, clay and other suspended particles contribute to the turbidity while during summer season low water level due to evaporation, decaying vegetation, organic matter and planktonic growth are responsible for the turbidity. High turbidity during rainy season is evitable. Zafar (1966) has shown that water temperature and turbidity have a positive correlation. It is because of the fact that particles in the water absorb heat and radiate the heat in surrounding water. In the Waterbodies studied, the turbidity was observed upto 224.42 JTU in Surajkund, 221.20 JTU in Ranital and 113.03 JTU in Motijheel. Sukumaran *et al.* (1984) have reported turbidity upto 100 $mg.1^{-1}$ in Milghatta tank and in Hutcharayankera tank in spite of the fact that water was generally clear. In managed tanks in Haryana Singhal *et al.* (1986) have reported turbidity between 15.0 $mg.1^{-1}$ to 67.5 $mg.1^{-1}$.

Total Suspended Solids

The total suspended solids in Waterbodies are contributed by particles of different sizes ranging from coarse to fine colloidal particles of various organic complexes and plankton. Total suspended solids ranging from 40.00 $mg.1^{-1}$ to 560.00 $mg.1^{-1}$ in Chandanpura pond (Kaushik *et al.*, 1989); 160.00 $mg.1^{-1}$ to 780.00 $mg.1^{-1}$ in J.C. Mill pond (Kaushik and Saksena, 1989); 100.00 $mg.1^{-1}$ to 420.00 $mg.1^{-1}$ in Chambal tal (Kaushik *et al.*, 1991 b); 60.00 $mg.1^{-1}$ to 66.75 $mg.1^{-1}$ in Ravindra sarovar (Shastree *et al.*, 1991); and 80.00 $mg.1^{-1}$ to 260.00 $mg.1^{-1}$ in Ganga sagar tank and 40.00 $mg.1^{-1}$ to 1400.00 $mg.1^{-1}$ in Viveknagar pond (Kaushik *et al.*, 1991a). Comparatively lower values of total suspended solids in lake Rankala and Kotithirth were reported by Kadam (1990). The total suspended solids were highest in Surajkund and lowest in Motijheel. They were found to be high during monsoon season; may be because of silt, clay and other particle entering into the Waterbodies along with the run off water. During winter, the total suspended solids were decreased due to sedimentation but again in summer season the total suspended solids were increased due to lower water level and abundance of phytoplankton density. The similar phenomenon have been observed by several authors.

Total Dissolved Solids

The total dissolved solids in water comprise mainly of inorganic salts and small amount of organic matter. Generally, carbonate, bicarbonate, chloride, sulphate, nitrate, sodium, potassium, calcium and magnesium contribute total dissolved solids. The total dissolved solids in water originate from natural sources and depend upon location, geological nature of the pond basin, drainage, rainfall, bottom deposit and inflowing water.

The total dissolved solids in Indian water vary very greatly upto 2416.0 $mg.1^{-1}$. In the present study, the total dissolved solids also showed high value during summer and post monsoon season with highest value of 460.00 $mg.1^{-1}$ in Motijheel; 1170.00 $mg.1^{-1}$ in Surajkund and 1440.00 $mg.1^{-1}$ in Ranital. Similar findings with regards to the seasonal variation of total dissolved solids have been reported by Zafar (1964); Seenayya (1971); Kaushik *et al.* (1989, 1991 b) and Shastree *et al.* (1991). It may be mentioned that the excess amount of total dissolved solids in water disturbed the ecological balance due to osmotic regulation and suffocation in aquatic fauna even in the presence of fair amount of dissolved oxygen.

Total Solids

Total solids in water are due to inorganic substance, organic matter, suspended particles silt, clay and plankton etc. High total solids reduce the light penetration and effect water quality indirectly and imbalance in aquatic life. Polluted waters have high total solids. A maximum of total solids *i.e.*, 91.0 $mg.1^{-1}$ in Errakuppam reservoir (Ganapati, 1956); 512.00 $mg.1^{-1}$ in Devi kund and 294.00 $mg.1^{-1}$ in Kamla Nehru tank (Verma and Shukla, 1968); 1552.00 $mg.1^{-1}$ in a Haryana pond (Singhal *et al.*, 1986); 500.9 $mg.1^{-1}$ in Osman sagar and 772.5 $mg.1^{-1}$ in Mir Alam lake (Mohan, 1987); 1060.00 $mg.1^{-1}$ in Chandanpura pond (Kaushik *et al.*, 1989); 2320 $mg.1^{-1}$ in J.C. Mill pond (Kaushik and Saksena, 1989); 1665.00 $mg.1^{-1}$ in Kotitirth and 905.00 $mg.1^{-1}$ in Rankala lake (Kadam, 1990); 1040.00 $mg.1^{-1}$ in Chambal tal (Kaushik *et al.*, 1991b); 285.00 $mg.1^{-1}$ in Ravindra sarovar (Shastree *et al.*, 1991) and 980.00 $mg.1^{-1}$ in Ganga sagar tank and 1900.00 $mg.1^{-1}$ in Vivek nagar pond (Kaushik *et al.*, 1991a), have been observed. In the present study total solids were found upto 930.00 $mg.1^{-1}$ in Motijheel; 1860.00 $mg.1^{-1}$ in Surajkund and 2080.00 $mg.1^{-1}$ in Ranital. The greater quantity of total solids is related to a high evaporation rate due to warm weather.

Specific Conductivity

A well marked seasonal pattern of specific conductivity in the Waterbodies under study was noted. The specific conductivity was found to increase during summer season due to surface evaporation of water thereby increasing concentration of salts. During early rainy season, the specific conductivity values showed a gradual decline, may be because of dilution by the rain water but in the post monsoon season, the specific conductivity was increased while during colder months it was found to decrease because of sedimentation and utilization of minerals by growing phytoplankton and macrophytes. Similar trend of variation in specific conductivity was also observed by Sreenivasan *et al.* (1964) and Singhal *et al.* (1986) while Zutshi and Vass (1978) have shown somewhat reversed pattern.

The specific conductivity was found to fluctuate between 66 µS to 462 µS in Motijheel; 198 µS to 1234.0 µS in Surajkund and 627 µS to 1518 µS in Ranital which fall within the range observed for Indian waters. Olsen (1950), Vollenweider and Frei (1953) and Dunn (1962) have used specific conductivity for assessing trophic status of lakes. The Waterbodies under study can be classified as meso-eutrophic according to the criteria of Olsen (1950) and Vollenweider and Frei (1953) but are eutrophic according to the criterion of Dunn (1962).

Dissolved Oxygen

A great variety of gases are found dissolved in natural waters. Out of these gases, oxygen is the most significant one because it is a regulators of metabolic processes of the organisms and also the community as a whole. It acts as an indicator of trophic status and magnitude of eutrophication. The saturation level of dissolved oxygen in the aquatic habitats is much less than in the atmosphere. The rate at which atmospheric oxygen passes across the air-water interphase and becomes dissolved in water depends upon temperature, partial pressure of gases in the atmosphere, concentration of dissolved salts, wave action, relative solubility, pollution, inflowing under groundwater, photosynthetic activity of plants and respiration by bacteria, plants and animals in the water (Zutshi and Vass, 1978).

George (1961) has maintained that a concentration of 1.4 $mg.1^{-1}$ oxygen is sufficient to maintain life in water. The range of dissolved oxygen is quite great in small Waterbodies. In Indian waters (Singhal *et al.*, 1986; Valecha and Bhatnagar, 1988; Yousuf and Shah, 1988; Kant and Raina, 1990; Shastree *et al.*, 1991 and Kaushik *et al.*, 1991a,b) has been found to be quite variable and it varies greatly from one waterbody to the other in the same area. In the present investigation, however, the dissolved oxygen was found to vary from 4.59 $mg.1^{-1}$ to 14.5 $mg.1^{-1}$ in Motijheel; 2.83 $mg.1^{-1}$ to 12.97 $mg.1^{-1}$ in Surajkund and 1.22 $mg.1^{-1}$ to 10.13 $mg.1^{-1}$ in Ranital with maximum concentration during winter season in all the Waterbodies. High value of dissolved oxygen in colder months have also been reported by several workers (Kaushik *et al.*, 1989 and Prasad, 1990), while low dissolved oxygen during winter season has been reported by Verma and Shukla (1968); and Bisht and Das (1985). In summer season the dissolved oxygen was decreased due to increasing water temperature (Zutshi and Vass, 1978 and Kaushik *et al.*, 1989, 1991b); while Qadri and Yousuf (1978) and Bisht and Das (1985) have reported higher value of dissolved oxygen in summer season. In monsoon months, the dissolved oxygen was again increased in all the Waterbodies probably due to circulation and mixing by inflowing water during monsoon rains. The similar phenomenon have been observed by several authors. Kant and Raina (1990) have, however, reported decreased oxygen values during monsoon season. The dissolved oxygen was quite enough to support life in all the Waterbodies studied. The dissolved oxygen had negative correlation with parameters like temperature, pH and hardness etc.

Free Carbondioxide

The free carbondioxide contributes to the fitness of natural water as it serves to buffer the environment against rapid shifts in the acidity or alkalinity and also regulates biological processes in aquatic communities and can form many compounds. Free carbondioxide in water is derived from many sources such as atmosphere, respiration by the organisms, bacterial decomposition of organic matter etc. The rains also absorb small amounts of the gas and deliver it to the waters on which it falls. Unni (1972) emphasised that the rate of changes in free carbondioxide concentration is considerable due to decomposition of organic matter at the bottom. The presence or absence of the free carbondioxide in the surface water is mostly governed by its utilization by algae during photosynthesis and also through its diffusion from air (Sreenivasan, 1974). Free carbondioxide may be present throughout the year (Ganapati, 1956; Verma, 1969; Sreenivasan, 1972; Chourasia and Adoni, 1985 and Singhal *et al.*, 1986); in few successive months of a year (Verma and Shukla, 1968; Qadri and Yousuf, 1978; Kaushik *et al.*, 1989 and Prasad, 1990) or present sporadically (Hussainy, 1967; Bisht and Das, 1985; Shardendu and Ambasht, 1988; Ghosh and George, 1989; Kant and Raina, 1990 and Fasihuddin and Kumari, 1990). Ganapati (1960) has reported its complete absence in certain tropical waters. The absence of carbondioxide is due to complete utilization of carbondioxide during photosynthesis by algae or carbonates present did not allow the carbondioxide to be produced in the bottom and column to reach to the surface. Sometimes, continued acidification releases carbondioxide and carbonic acid from the bicarbonates accompanied by loss of carbondioxide from the system. In the present investigation, the free carbondioxide was totally absent from the month of March to September during 1989-90 and from April to September during 1989-90 in Motijheel, February to July and from March to October during from both the years in Surajkund and Ranital respectively. Generally, the free carbondioxide is observed in winter months but in Surajkund it was also present during rainy season when the tank had a covering of *Lemma paucicostata* hindering light penetration and thus reducing the rate of photosynthetic activity. Similar observations were also made by Kaushik *et al.*, (1989) in respect of Chandanpura pond. Greater free carbondioxide in Ranital is attributed to the presence of decomposable organic matter in the bottom as suggested by Unni (1972). Reid and Wood (1976) have classified soft, medium-hard and hard waters on the basis of the amount of free carbondioxide present. In the above classification Motijheel and Surajkund are placed under soft while Ranital under medium-hard Waterbodies.

Carbonate Alkalinity

The chemical buffering of the system is brought about by carbondioxide - bicarbonate - carbonate complex. This system actually forms the carbon sources of phytoplankton during carbon assimilation. The total alkalinity of water is mainly due to the cations of calcium, magnesium, sodium, potassium and from combined either as carbonate and bicarbonate or occasionally as hydroxide. The carbonate alkalinity in many waters is found to be absent (Ganapati, 1956; Verma, 1969; Sreenivasan, 1972 and Singhal *et al.*, 1986). This absence is caused by low photosynthetic rate of phytoplankton (Jana and Sarkar, 1971; Qadri and Yousuf, 1980 and Yousuf and Shah, 1988). Khatavkar *et al.*, (1990) have reported it maximum load upto 330.0 $mg.l^{-1}$ in Ambedkar tank in Southwestern Maharashtra. In the present study the carbonate alkalinity in all the Waterbodies was recorded at occasions when the free carbondioxide was absent, but not in sufficiently large amounts. It varied from 1.33 $mg.l^{-1}$ to 20.00 $mg.l^{-1}$ in Motijheel, 1.0 $mg.l^{-1}$ to 49.0 $mg.l^{-1}$ in Surajkund and 9.0 $mg.l^{-1}$ to 50.0 $mg.l^{-1}$ in Ranital. The observations of Verma and Shukla (1968), Sreenivasan (1972), Shardendu and Ambasht (1988), Kaushik *et al.*, (1989), Siddiqui *et al.* (1990), Fasihuddin and Kumari (1990), Kant and Raina (1990) and Prasad (1990) were repeated once again in this study also with regards to the seasonal variation of carbonate alkalinity.

Bicarbonate Alkalinity

The bicarbonate alkalinity is mainly constituted to the total alkalinity and is invariably present in waters in which photosynthesis is actively taking place. Decreased bicarbonate alkalinity during summer season and increased alkalinity during rainy season has been reported by Zutshi and Vass (1978) and Chourasia and Adoni (1985). But increased bicarbonate alkalinity during winter season and decreased level in summer and rainy have also been observed by several workers (Qadri and Yousuf, 1980). The bicarbonate alkalinity in Motijheel, Surajkund and Ranital exhibited a biomodal pattern, first peak of low magnitude appeared in the post monsoon seasons and other during late summer months with higher magnitude. Similar trend was also observed by Ganapati (1956), Verma (1969), Bisht and Das (1985), Kaushik *et al.*, (1989, 1990, 1991a, b).

Total Alkalinity

Alkalinity of water is its capacity to neutralize acid and is characterized by the presence of hydroxyl (OH^-) ions capable of combining with hydrogen (H^+) ions in solution. Alkalinity, therefore, is a measure of buffering capacity of the water. Natural bodies of water in the tropics usually show a wide range of fluctuations in total alkalinity values depending upon the location, season, plankton population, rainfall, washerman's activity and nature of bottom deposits etc. Alkalinity is important for aquatic life in freshwater system because it equilibriated pH changes that occur naturally as a result of photosynthetic activity of the cholorophyll bearing vegetation. Both the carbonates and bicarbonates form the components of alkalinity. The range of total alkalinity in Indian waters may be found from 40 $mg.1^{-1}$ to over 1000 $mg.1^{-1}$ (Jhingran, 1982).

In the present study, the total alkalinity of Motijheel was comparatively low *i.e.*, ranging from 118.0 $mg.1^{-1}$ to 286.66 $mg.1^{-1}$ and in Surajkund and Ranital, total alkalinity was higher *i.e.*, from 162.00 $mg.1^{-1}$ to 526.00 $mg.1^{-1}$ and 184 $mg.1^{-1}$ to 616 $mg.1^{-1}$ respectively. The total alkalinity of water is dependent upon the concentration of anions (Kaushik *et al.*, 1991a). The alkalinity in Ranital along with higher total dissolved solids and conductivity value explained above observations. In the present study the total alkalinity had gone down with rainfall and gone up with evaporation of water. This phenomena has also been recorded by Verma (1969), Hussainy (1967), Singhal *et al.* (1986) and Kaushik *et al.* (1989).

Alkalinity as a measure of productivity which is believed to be so by many workers. Moyle (1946) noticed that most of the Waterbodies having a total alkalinity greater than 200 $mg.1^{-1}$ were highly productive. Speence (1964) divided south Scottish lakes into three major categories based on alkalinity *viz.*, (i) nutrient poor with alkalinity ranging from 1.0 to 15.0 $mg.1^{-1}$ (ii) moderately rich with alkalinity ranging from 16.0 to 60.0 $mg.1^{-1}$ and (iii) nutrient rich with alkalinity greater than 60.0 mg.1^{-1}. Alikunhi (1957) also stated that in highly productive waters, the alkalinity is ought to be over 100 mg.1^{-1}. The total alkalinity in Motijheel, Surajkund and Ranital was always greater than 100 mg.1^{-1} and thus can be said as nutrient rich and highly productive. But Motijheel though had alkalinity over 100 mg.1^{-1} upto 286.66 mg.1^{-1} exhibited a lower production in comparison to other Waterbodies studied.

pH

The pH of any aqueous system is suggestive of the acid-base equilibrium achieved by various dissolved compounds. Hydrogen ion concentration of natural waters is an important environmental factor, the variation of which, among other causes are linked with the chemical changes, species composition and life processes of animal and plant communities inhabiting them. pH is generally considered as an index for suitability of the environment. Webber and Stumm (1963) have concluded that the pH of the raw water sources mostly lies within the range 6.5 to 8.5. A inland waters in India lie in the alkaline range without much variation (Sreenivasan, 1972; Singhal *et al.*, 1986; Ghosh and George, 1989; Shastree *et al.*, 1991). pH in Motijheel was recorded in the range of 7.53 to 8.94, in Surajkund it ranged from 8.13 to 8.70 and in Ranital from 8.4 to 9.92 confirming the above observations. Venkateswarlu (1983) classified the waters into five categories based on the pH. (i) Acidobiontic (pH below 5.5), (ii) Acidophilous (pH from 5.5 to 6.5), (iii) In different (pH from 6.5 to 7.5), (iv) Alkaliphilous (pH from 7.5 to 9.0), and (v) Alkalibiontic (pH above 9.0). If Venkateswarlu's (1983) criteria is applied to the Waterbodies in question, Motijheel is alkaliphilous while Surajkund and Ranital are alkaliphilous to alkalibiontic in nature.

The pH was low during winter season while in summer it was high. With rains the pH got reduced and then again increased in post-monsoon, period. Lowered pH in winter season with short day period was controlled by the decreased photosynthetic activity. Dead vegetation upon decay and absence of sunlight increased the concentration of carbondioxide and decreased pH. In summer increased photosynthetic regulated the pH towards alkaline side. Similar observations have been made by Singhal *et al.*, (1986).

A fall in the pH value in monsoon season was recorded by Shardendu and Ambashit (1988), Kaushik *et al.*, (1989, 1991b). The cloudy atmosphere also reduce the pH considerably.

Alikunhi (1957) and Sreenivasan (1976) have demonstrated that the pH between 6.5 to 8.5 with large variation plays a role in the productivity of the waters. Spence (1967) has stated that alkalinity and pH are closely connected with an accurate measure of the tropic status of lake water. On the basis of Spence (1967) assumption, Motijheel, Surajkund and Ranital are in increasing order of eutrophy.

Ellis (1937) suggested that waters with hydrogen ions ranging between 6.5 to 9.0 are most suitable for fish production. Verma and Shukla (1968) believed that alkaline waters support large amount of biota. Bell (1971) has stated that pH ranges of 6.5 to 9.0 appears to provide adequate protection to the life of freshwater fish and bottom dwelling invertebrate fish food organisms.

Total Hardness

The total hardness is not a specific constituent of water but is a variable and complex mixture of cations and anions and is predominantly contributed by calcium and magnesium along with minor contribution by strontium, barium, zinc, manganese, aluminium and iron. It may be of temporary (carbonate) or permanent (non-carbonate) type. It has also been suggested that whenever hardness is lower than alkalinity, it is due to carbonates. Mairs (1966) has found that the hardness and alkalinity are closely related to each other. Like alkalinity, the hardness is an important parameter in decreasing the toxic effect of poisonous elements.

Sawyer (1960) classified waters on the basis of hardness into three categories (i) soft with hardness from 0.0 to 75.0 $mg.1^{-1}$, (ii) moderately hard with hardness form 75.0 to 150.0 $mg.1^{-1}$ and (iii) hard with hardness from 150.0 to 300.0 $mg.1^{-1}$. Thus Vihar lake at Bombay (Hussainy, 1967); Amarvanthy, Manimuthar, Aliyar, Perinchani, Tirurmoorthy, Sandynulla and Pykare reservoirs at Madras (Sreenivasan, 1970) and Swetganga in West Bengal (Jana and Sarkar, 1971) are soft; stanley, Bhavani sagar,

Vidur, Gomukhi and Poondi reservioirs at madras (Sreenivasan, 1970) and Bhimtal at Nainital (Bisht and Das, 1985) are soft to moderately hard; Krishnagiri reservoir at Madras (Sreenivasan, 1970), lower lake at Bhopal (Valecha *et al.*, 1987) are moderately hard; Santhaur reservoir at Madras (Sreenivasan, 1970) is moderately hard to hard; Odathurai tank in Coimbatore (Sreenivasan, 1972) is hard, and Nainital lake at Nainital (Bisht and Das, 1985), certain Waterbodies of Kashmir (Yousuf and Shah, 1988) and Chandanpura pond and Vivek Nagar pond at Gwalior (Kaushik *et al.*, 1989, 1990, 1991a) are hard to very hard, according to the classification of Sawyer (1960). In the present study, the water in Motijheel and Surajkund it is moderately hard to very hard but in Ranital water is hard to very hard. since, the hardness is greater than total alkalinity, it is presumed that the hardness is contributed mainly by bicarbonate. Unni (1983) has suggested that total hardness can be used as an indicator for classifying domestic pollution in water.

Chlorides

Chlorides in waters are generally due to the salts of sodium, potassium and calcium. They constitute approximately 0.05 per cent of the lithosphere and are attributed to the dissolution of salt deposits. It may also be contributed by sewage discharges, contamination from reflux leachates, discharges of effluents from contamination from reflux leachates, discharges of effluents from chemical industries and irrigation drainage to the natural waters.

In the present study, Motijheel had chloride concentration from 17.98 $mg.l^{-1}$ to 57.65 $mg.l^{-1}$, Surajkund from 35.48 $mg.l^{-1}$ to 144.95 $mg.l^{-1}$ while Ranital had highest concentration of chloride from 92.46 $mg.l^{-1}$ to 226.93 $mg.l^{-1}$ with higher concentration in summer season in all the tanks. Lower concentration of chloride in winter season can be attributed to the sedimentation, and human disturbance. Kaushik *et al.* (1989, 1990, 1991b) and Kant and Raina (1990) had similar view regarding seasonal variation of chloride in water. In summer season, increased chlorides due to high evaporation rate because of high ambient temperature and maximum human activities have been reported by several workers (Chourasia and Adoni, 1985; Kaushik *et al.*, (1989, 1991a, b). In monsoon months it again increased in all the Waterbodies is probably due to run off water which might have brought animal and human wastes and salts from the catchment area. Kaushik (1989, 1991a, b) also observed increased chloride level in ponds during rainy season.

Verma and Shukla (1970) have pointed out that higher values of chlorides are observed when a maximum growth of phytoplankton, zooplankton and bottom biota occurred but Sarkar and Rai (1964) have revealed that chlorides are not used in plant growth and thus accumulated in large amounts due to input of pollution load. Increasing concentration of chloride is an indication of eutrophy.

Chemical Oxygen Demand

The chemical oxygen demand is a measure of the oxygen equivalent of the organic matter content of water that is susceptible to oxidation by a strong chemical oxidant (APHA *et al.*, 1985) and thus is a reliable parameter for judging the extent of pollution in water. But it does not suggest about the degradability of the waste nor it indicates the rate at which biological oxidation would proceed and hence the rate at which oxygen would be required. Therefore, chemical oxygen demand of water increases with increasing organic matter concentration. The oxygen utilization by living planktonic communities increases proportionally with increased chemical oxygen demand (Boyd, 1973).

Most open Indian ponds have higher chemical oxygen demand with maximum values in summer season (Shardendu and Ambasht, 1988; Kaushik *et al.*, 1989, 1991b) but masonry ponds and lakes have comparatively lower chemical oxygen demand (Kaushik *et al.*, 1991c). In the present study, chemical oxygen demand ranged from 8.00 $mg.1^{-1}$ to 45.3 $mg.1^{-1}$ in Motijheel; from 12.0 $mg.1^{-1}$ to 128.0 $mg.1^{-1}$ in Surajkund and from 32.0 $mg.1^{-1}$ to 150.0 $mg.1^{-1}$ in Ranital with maximum values during summer months. With increased temperature, the oxygen consumption by the living planktonic communities is also increased (Boyd, 1973), and increased organic matter may require more oxygen to oxidise under increased thermal condition as suggested by Shardendu and Ambasht (1988) and Kaushik *et al.* (1991b).

Calcium

The content of calcium is one of the variables in freshwater on which faunistic differences can be based as calcium serves as micronutrient to most of the organisms and is required in small quantity for phytoplanktonic growth. The lack of this element may play a role in promoting dystophy, decreasing the rate at which organic substances are precipitated, mineralised and recycled for use by the primary producers. Concentration upto 1800 $mg.1^{-1}$ has no adverse physiological effects in human beings (Lehr *et al.*, 1980). In Indian Waterbodies calcium content fluctuated from 11.22 $mg.1^{-1}$ to 390.67 $mg.1^{-1}$. The calcium content varied from 13.89 $mg.1^{-1}$ to 57.44 $mg.1^{-1}$ in Motijheel, 35.66 $mg.1^{-1 \, to}$ 122.24 $mg.1^{-1}$ in Surajkund and 61.72 $mg.1^{-1}$ to 153.10 $mg.1^{-1}$ in Ranital with a biomodal annual pattern; first peak of low magnitude appreared after the rains while second peak of high magnitude was evident in late summer in all the Waterbodies studied. Increased calcium level after rains is attributed to the surface run off to the Waterbodies during rainy season. In winter season due to sedimentation and utilization of calcium by plankton and other organisms its content was low but during summer, increased calcium level in water is probably due to evaporation of water and decomposition of dead plants and animals. High concentration of calcium in summer (Ramanibai and Ravichandran, 1987 and Kaushik *et al.*, 1991b) and low calcium level in winter season (Zutshi and Vass, 1978 and Mohan and Reddy, 1987) have been reported. Ramanibai and Ravichandran (1987) and Kant and Raina (1990) have, however, noted lowest value during rainy season, Khan and Siddiqui (1974) has described pond Moat in Marhata forest, Aligarh as calcium poor (6.0 to 10.0 $mg.1^{-1}$) during May to August, good (17.0 to 27.2 $mg.1^{-1}$) during April and December and very good (41.0 to 86.0 $mg.1^{-1}$) during October to March calcium concentration.

The wide variation in calcium content in natural waters lead to a noted German limnologist, Ohle (1938) to give the following classification of Waterbodies.

(i) Waters with Ca > 10 mg.1^{-1} - Poor

(ii) Waters with Ca 10 mg.1^{-1} to 25 mg.1^{-1}- Medium

(iii) Waters with Ca < 25 mg.1^{-1} - Rich

According to above classification, Motijheel water falls under the category of waters with medium to rich calcium level while Surajkund and Ranital under the category of calcium rich waters. As suggested by Ramanibai and Ravichandran (1987) relatively higher proportion of calcium in the surrounding rocks and soils might have also contributed to the rich calcium level in Surajkund and Ranital.

Magnesium

According to Wetzel (1975), magnesium is required by the flora to build its chlorophyll and in enzymatic transformation, especially transphosphorylations of algae, fungi and bacteria. The depletion of magnesium acts as limiting factor for the growth of phytoplankton and reduces the number of phytoplankton. Magnesium is generally present in water as bicarbonate and in this form it resembles calcium bicarbonate in reactions with water.

The range of magnesium is Indian waters may be found from 2.45 mg.1^{-1} to 107.0 mg.1^{-1} (Goel *et al.*, 1986; Khatavkar *et al.*, 1990). In the present investigation the magnesium content of water varied from 9.91mg.1^{-1} to 17.54 mg.1^{-1} in Motijheel, 9.25 mg.1^{-1} to 18.49 mg.1^{-1} in Surajkund and 14.84 mg.1^{-1} to 34.82 mg.1^{-1} in Ranital. In Surajkund and Ranital, magnesium was found to be maximum during summer but in Motijheel an irregular pattern was observed. Khan and Siddiqui (1974) have observed magnesium content ranging from 9.9 mg.1^{-1} to 24.6 mg.1^{-1} with minimum in November and maximum in February in a fish pond at Aligarh. A high concentration of magnesium during summer months which decreased in monsoon as well as in winter has been reported by Shastree *et al.* (1991).

Sodium

Sodium is widely present, sometimes in substantial quantities, in water, plants rocks and many food. Considerable amount of sodium is excreted by human beings and thus is a common constituent of domestic sewage. Sodium is metabolised only by blue green algae (Jhingran, 1982).

The sodium level in Waterbodies is quite variable. Wide range of seasonal fluctuation in sodium have been reported by Khan Siddqui (1974) in a Aligarh pond, Goel *et al.* (1986) and Khatavkar *et al.* (1990) in certain lentic water in Southwestern Maharashtra but a narrow range of variation was reported in Dal lake (Zutshi and Vass, 1978), Osman sagar and Mir Alam lakes (Mohan and Zafar, 1986). In the present investigation, sodium content fluctuated from 12.33 $mg.1^{-1}$ to 27.08 $mg.1^{-1}$ in Motijheel; 19.21 $mg.1^{-1}$ to 88.77 $mg.1^{-1}$ in Surajkund and 38.38 $mg.1^{-1}$ to 123.87 $mg.1^{-1}$ in Ranital with a wide range of seasonal fluctuation. A higher value of sodium was exhibited by all the Waterbodies under the study during summer season and a lower value was obtained in winter season through an increase in sodium level was observed in post-monsoon season. Mohan and Zafar (1986) have pointed out that the rains bring large quantities of sodium from the catchment area which also sometimes seems true in the present study. Evaporation of water is a significant factor in increasing sodium level during summer season.

Potassium

Potassium is also a naturally occurring element which remains far lesser in concentration than calcium, magnesium and sodium in freshwaters. Geological, meteorological and agricultural factors are responsible for its occurrence in water (mohan and Reddy, 1987). Under low potassium condition, the growth rate and photosynthesis of algae especially blue green algae becomes poor and respiration is increased (Wetzel, 1975).

The potassium level in aquatic habitats is quite variable from 0.2 $mg.1^{-1}$ to 35.0 $mg.1^{-1}$. In the present study, the content of potassium varied from 3.88 $mg.1^{-1}$ to 10.43 $mg.1^{-1}$ in Motijheel; 7.60 $mg.1^{-1}$ to 72.60 $mg.1^{-1}$ in surajkund and 31.62 $mg.1^{-1}$ to 121.25 $mg.1^{-1}$ in Ranital with still higher concentration in second year. Mohan and Zafar (1986) and Mohan and Reddy (1987) also demonstrated that the potassium is fast accumulated with the passage of time and doubling up every fifteen years.

Khan and Siddqui (1974) and Zutshi and Vass (1978) could not observe any seasonal variation in the potassium content, in a fish pond of Aligarh; in Dal lake of Kashmir and in Sawarkar sarovar and Laxman Talliya of Gwalior respectively, whereas, in the present study a clear cut seasonal pattern of potassium content in all the three Waterbodies was noted with higher concentration in late summer and in monsoon season due to evaporation and input through run off water respectively. Similar seasonal variation in potassium content of water has also been observed by Mohan and Reddy (1987). Sedimentation utilization of potassium by biota caused decrease in its content during winter season.

Silicates

Silica in the form of orthosilicic acid ($SiOH_4$) is an essential macronutrient for the growth of diatom. The major source of silica in natural waters is in decomposition of alumino-silicate minerals in drainage basins. Silica occurs in freshwater as undefined dissolved silicic acid, as colloidal suspended soils and as a silica oxide. The silicic acid is weaker than carbonic acid and can be removed from its hydrolysed compounds (Silicates) in the presence of carbondioxide or bicarbonates. It is then held in water as silicic acid in dissolved or colloidal forms. Thus, the silica varied directly with carbondioxide and bicarbonates in water and its concentration is linked with the density of diatom population because most of silicate taken up by the diatom cells is polymerised intracellularly and deposited as hydrated amorphous silica frustules (Vyas and Kumar, 1968). Synthesis of DNA polymerase and chlorophyll synthesis are some of the physiological processes which have been shown to be dependent on silicate. In tropical region where decomposition of organic matter is complete silica is likely to be more mobile (Hutchinson, 1957). The abundance of silica is regulated by the redox potential and is possibly caused by the reduction of ferric silicate formed in the superficial layers of bottom mud when the latter comes in contact with oxygenated water (Jhingran, 1982). Generally silicates are higher in concentration in surface waters and may reach upto 79.5 $mg.1^{-1}$ in open lakes (Jhingran, 1982). Silica content of natural waters depends upon the nature of the host rocks, physical and chemical characteristics of water including temperature, carbondioxide, bicarbonates, rate of evaporation of water, rains and decomposition of organic matter(Hutchinson, 1957; Seenayya, 1971). Jana and Sarkar (1971) commented that dissolved silica content enters into the water in huge amounts from the fissures of thermal springs. Trivedy and Goel (1986) have also reported that silicates in water vary form 1.0 $mg.1^{-1}$ to 30.0 $mg.1^{-1}$ but thermal springs may have them upto 100 $mg.1^{-1}$.

In most Indian Waterbodies, silicate varied from its absence to 42.0 $mg.1^{-1}$ (Ganapati, 1956; Vyas and Kumar, 1968; Sreenivasan, 1970, 1972; Jana and Sarkar, 1971; Seenayya, 1971; Khan and Siddiqui , 1974; Zutshi and Vass, 1978; Mohan and Zafar, 1986; Mohan, 1987; Ramanibai and Ravichandran, 1987 and Kant and Raina, 1990). In the present study silicate content was found to fluctuate from 3.52 $mg.1^{-1}$ to 9.99 $mg.1^{-1}$ in Motijheel; from 6.05 $mg.1^{-1}$ to 22.20 $mg.1^{-1}$ in Surajkund and from 8.85 $mg.1^{-1}$ to 24.40 $mg.1^{-1}$ in Ranital. Ranital had a higher concentration of silicates than Surajkund and Motijheel which also had rich amounts of chlorides and organic matter as observed by Mohan (1987). Silicate contents have exhibited a marked seasonal variation with highest values in late summer and monsoon months and with lower concentration in winter season in the tanks under study. Higher silicate concentration during summer season may be due to increased water temperature, higher evaporation rate, rich chloride content and chemical oxygen demand, and the release of nutrients due to decomposition as suggested by Khan and Siddiqui (1974), Ramanibai and Ravichandran (1987) and Mohan (1987). However, the dependence of silica content in water on chloride and rich organic matter has been questioned as they become involved in the formations of stable chemical complexes such as polychlorinated biphenyls, catechol and querectol which are esters of silicic acid. Occurrence of low silica in winter season was found to be related with their continuous utilization by photoplankton specially diatoms and macrophytes of waterbody and less decomposition activity, due to low water temperature and sedimentation. Similar pattern has also noted by Ganapati (1956). Addition of silicate particles from surrounding catchment area through rains increased silicates level in water during rainy season as shown by Vyas and Kumar (1968), Jana and Sarkar (1971), Khan and siddiqui (1974) and Ramanibai and Ravichandran (1971).

Phosphate

Phosphorus, necessary for the fertility, is generally recognised as a key nutrient in the productivity of the water. Hutchinson (1957) demonstrated that phosphorus is one of the major nutrients responsible for biological productivity. The main supply of phosphorus in natural waters comes from the weathering of phosphorus bearing rocks, leaching of soils of the catchment area by rain, cattle dung and night soil (Jhingran, 1982). Other factors regulating phosphorus in Waterbodies are the rate at which inorganic phosphate is lost to the mud, bacterial activity, sewage contamination, agriculture fertilizers, industrial effluents, environmental temperature, pH of water and soil, depth of waterbody, bottom fauna, aquatic vegetation, phytoplankton, aquatic fauna, dead eggs of aquatic animals, migrating fauna and excreta of bird and other animals. With the growth and age of the pond, accumulation of organic matter in the sediment also influences phosphorus fertility (Sarkar, 1989). It is estimated that upto 7.8 per cent of the total phosphorus was present in different Waterbodies while more than 10 per cent of the total phosphorus load in the waterbody was locked in fish and zooplankton and about 4 per cent of total phosphorus was found to be absorbed by phytoplankton (Golterman *et al.*, 1969). Cowan and Lee (1976) found that higher (< 30 per cent) percentage of phosphorus may be absorbed by the phytoplankton.

In aquatic ecosystem, phosphorus is lost due to its retrogradation in acid and alkaline soils forming insoluble ferric, calcium and aluminium phosphates and absorption on mud sediments or a ferric hydroxide gel in the soil. The percentage and rate of phosphorus release is regulated by the interaction of nutrients, soil texture, abiotic and biotic factors, pollution potential and biological activity (Sarkar, 1989). The phosphorus in mud is released from iron (Ferric) - phosphorus complex which gets reduced to soluble ferrous form in the absence of oxygen and thus liberates the phosphorus. The phosphorus concentration was inversely proportionate to the oxygen content in the tanks under study.

Lee *et al.*, (1981) have classified the Waterbodies on the basis of phosphorus contents into five categories. In the present study, Motijheel is mesotrophic with phosphorus level ranging from 0.010 $mg.1^{-1}$ to 0.055 $mg.1^{-1}$ while Surajkund and Ranital are eutrophic with phosphorus level lying between 0.047 $mg.1^{-1}$ to 0.096 $mg.1^{-1}$ and 0.053 $mg.1^{-1}$ to 0.198 $mg.1^{-1}$ respectively.

Zafar (1996) has observed an increase in the phosphate value with increased depth. Qadri and Yousuf (1980) reported that in Malpur lake phosphates were found in minute quantities throughout the year with highest concentration in June and complete absence in March. In the present study, phosphate content exhibited distinct seasonal variation with highest level in late summer and monsoon months as compared to other seasons. During summer season phosphate content was highest and was related with high wind speed, decrease in water level effecting concentration due to higher evaporation rate and the release of nutrients due to decomposition which increased with the raised temperature. The release of phosphorus in water under higher thermal conditions. Higher concentration of phosphorus in summer may also be attributed to high turnover rates and decomposition of large algal population in addition to wind recycling. Chourasia and Adoni (1985), Singhal *et al.* (1986), Kaushik *et al.* (1991b), Prasad (1990) and Kant and Raina (1990) also confirmed this view. Low phosphate level during winter season may be due to its utilization in macrophytic growth and its sedimentation in form of ferric complexes in soil due to low calcium level in the water and low water temperature (Seenayya, 1971; Khan and siddiqui, 1974; Sarkar, 1989; Kaushik *et al.*, 1989, 1990). Zutshi and Vass (1978), however, reported high phosphorus content in winter season.

Nitrate-Nitrogen

Because of its role in the synthesis, maintenance of proteins and productivity of water, the nitrogen is important in aquatic ecosystem. The nitrogen in water occurs as bound forms like nitrate, nitrite, ammonia and organic nitrogen *viz.*, urea, amino acids etc. In aquatic ecosystem nitrogen level is regulated through precipitation, atmospheric solution and volatilization under meteorological process; sedimentation influence, effluent and groundwater movement under geologic process, nitrogen fixation, denitrification uptake growth, decay, hydrophytes pumping and fish and weed removal under biologic process (Toetz, 1976). Nitrates are products of oxidation of organic nitrogen by the bacteria present in soil and water where sufficient oxygen is present (WHO, 1984).

High concentration of nitrates are useful in irrigation but their entry into the water resources increases the growth of nuisance algae and trigger eutrophication and pollution (Trivedy and Goel, 1986). 150 $mg.l^{-1}$ nitrate concentration has been considered as critical beyond which blooms can normally be expected. In the present study nitrogen fixing algae (cyanophycean) were higher in summer when maximum phosphate concentration was also observed. The high rates of nitrogen fixation are being predicted in lakes containing high concentration of phosphates and organic nitrogen.

A wide range of variation in the concentration of nitrate nitrogen from its absence to presence in traces to 251.2 $mg.l^{-1}$ in Indian waters has been reported but in the present study, the level of nitrates was found to fluctuate from 0.0015 $mg.l^{-1}$ to 0.0609 $mg.l^{-1}$ in Motijheel; 0.0550 $mg.l^{-1}$ to 0.1340 $mg.l^{-1}$ in Surajkund and 0.0762 $mg.l^{-1}$ to 0.1352 $mg.l^{-1}$ in Ranital. In Motijheel, Surajkund and Ranital, the nitrates were present in higher concentration during summer season and during monsoon season and lowest in winter season. Similar seasonal pattern was also observed by several other workers (Vyas and Kumar, 1960, Verma and Shukla, 1968; Chourasia and Adoni, 1985). Nitrate has been said to be independent of oxygen (Shastree *et al.*, 1991) which is true in the present study also.

Nitrite-Nitrogen

Nitrites occur fairly widely but generally at much lower levels than nitrates and are also formed due to incomplete bacterial oxidation of organic nitrogen. Nitrites are not stable end products and their absence or presence in small quantity may not owe much significance but the concentration of nitrite in the water depends on the relative abundance of nitrifying and denitrifying organisms and their activities. The presence of appreciable amount of nitrites in the water has been due to sewage contamination, fertilizer use, decaying vegetable and animal matter, municipal wastes, industrial discharges, discharge from automobile exhausts, leachates from refuse dumps and atmospheric wash out.

The level of nitrites in Indian Waterbodies is less variable from absence to 0.14 $mg.1^{-1}$. In the present study the level of nitrites ranged from 0.003 $mg.1^{-1}$ to 0.028 $mg.1^{-1}$ in Motijheel; from 0.011 $mg.1^{-1}$ to 0.068 $mg.1^{-1}$ in Surajkund and 0.017 $mg.1^{-1}$ to 0.171 $mg.1^{-1}$ in Ranital.

A high value of nitrite exhibited by all the Waterbodies under study during summer season except during monsoon in Ranital in second year of study when the nitrites were maximum and lowest values of nitrites were reported in winter season with an increase in post-monsoon season. During warmer months nitrite level was higher due to increased water temperature, low oxygen content, rich chloride, chemical oxygen demand and decomposition of organic matter and during monsoon season (Ranital) may be due to natural run off water. Qadri and Yousuf (1980), Goel *et al.* .(1986) and Kaushik *et al.* . (1991b) have had similar trend in nitrite content in water and confirmed observations of this study. During winter season lower concentration of nitrite may be due to low water temperature and chemical oxygen demand, higher dissolved oxygen, less decomposition of organic matter and greater sedimentation rate.

Considering the over all physico-chemical nature of Waterbodies, it is inferred that Motijheel categorised is mesotrophic, Surajkund as meso-eutrophic and Ranital as eutrophic.

Motijheel is currently being used for storage of water which is supplied after treatment to the Gwalior city while other two tanks are lying fallow with limited water use including that of use bathing, washing the clothes and occasionally for drinking as well as building construction. Since, Motijheel has plenty of phytoplanktonic organisms (Kaushik, 1992; unpublished) which may lead to the formation of carcinogenic trihalomethanes by the interaction of free chlorine and organic matter (phytoplankton) during chlorination and which should be kept under check. Like Motijheel, Surajkund and Ranital are also having good amount of available nutrients, well growth of plankton, sufficient primary production and abundant insect population. In view of these observations, it is suggested that all these Waterbodies can be profitably stocked with culturable fish species especially carps in Motijheel and a mixture of carps, cat fishes and murrels in

Surajkund and Ranital as carps are planktivorous and also utilize other material available while cat fishes and murrels are carnivorous feeding mainly on insects and other benthic organisms.

Acknowledgments

We express our grateful thanks to late Dr. J. Bahadur, and the then Professor and Head, School of Studies in Zoology and Rector, Jiwaji University, Gwalior and to Dr. R. Mathur, Professor and Head, School of Studies in Zoology for providing all laboratory facilities and encouragement.

References

Alikunhi, K.H. (1957), Fish culture in India. Farm Bulletin of Indian Council Agricultural Research, No. 20, pp-1 -150.

Ambasht, R.S. and Shardendu (1989), Morphometry and physico-biotic characters of Varanasi ponds. Proc. Nat. Acad. Sci. India, *59(B)* : 421 - 426.

Anon (1988) : Hamara Paryavaran, Gandhi Santi Pratishthan, New Delhi.

APHA, AWWA and WPCF (1985) : Standard methods for the examination of water and waste water. 16th edition. American Public Health Association, Washington, D.C.

Basel, B. (1986) : Environment protection a many faceted problem. Yojana, *30 (10)* : 4-6.

Bell, H.L (1971) : Effect of low pH on the survival and imergence of aquatic insects. Water Res., *5* : 313.

Bisht, R.S and Das, S.M (1985) : Abiotic parameters and aquatic insects of two cold water Kumaon lakes. Ind. J. Ecol., *12* : 365-369.

Boyd, C.E. (1973) : The chemical oxygen demand of waters and biological materials from ponds, Trans. Amer. Fish. Soc., *3* : 606 - 611.

Chandler, D.C. (1942) : Limnological studies of western lake Erie. III. Phytoplankton and physical chemical data from November 1939 to November 1940. Ohio J. Sci., *42* : 24 - 44.

Chourasia, S.K. and Adoni, A.D. (1985) : Zooplankton dynamics in a shallow eutrophic lake. In proc. Nat. Symp. Pure and Appl. Limnilogy (Ed. A.D. Adoni). Bull. Bot. Soc. Sagar, *32* : 30 - 39.

Cowen, W.F. and Lee, G.F. (1976) : Algal nutrient availability and limitation in lake Ontario during IFYGL. Part I. Available phosphorus in urban runoff and lake Ontario Tributary waters, U.S. EPA-600/3-76-094.

Dudani, V.K. Singh, R. Kumar, S. and Pandey, S. (1987) : Diurnal Variation in physico-chemical parameters of fish pond of Darbhanga. J. Hydrobiol., *3* : 5-7.

Dunn, B. (1962) : Hydrology of the Iron dequott Greek basis Rochester, New York. Open File Research Paper No.40, U.S. Geological Survey and New York State Department of Health.

Ellis, M.M. (1937) : Detection and measurement of stream pollution. U. S. Bur. Fish. Bull., *22* : 367-437.

Fasihuddin, M.D. and Kumari, J. (1990), Seasonal variations in physico-chemical properties and plankton periodicity in a fresh water fish pond at Bhagalpur, India. Environment and Ecology, *8* : 929-932.

Forel, F.A. (1892) : Le Leman monographic limnologique. F. Rouge Lausanne. Tome 1 Geographic, hydrographie-geologie, Climatologie hydrologie.

Ganapati, S.V. (1940) : The ecology of a temple tank containing a permanent bloom of *microcystis aeruginosa* (Kutz.) Henfr. J. Bombay Nat. Hist. Soc, *42* : 65-67.

Ganapati, S.V. (1956) : Hydrobiological investigations of the Hope reservoirs and of Thambaraparni River at Papanasam Tirunalveli Dt. Madras State Indian Geogr. J., *31* : 1-2.

Ganapati, S.V. (1960) : Ecology of tropical waters. Proc. Symp. Alology, ICAR, New Delhi, pp 204-218.

Ganapati, S.V. (1970) : Energy availing relationship in natural aquatic bio-systems in India. Trop. Ecol., *11* : 49-68.

Ganapati, S.V. Alikunhi, K.H. and Zobeeri, A.R.K. (1953) : Hydrobiology of carps spawning in
the river Cauvery at Hoggainkal during July-August, 1946. Ind. Corn. J., *8* : 376-379.

Ganapati, S.V. and Chacko, P.I. (1951) : An investigation of the river Godavari and the effects of paper mill pollution at Rajamundry. Proc. 1 P.F.C. Sec. II.5.

Ganapati, S.V. and Sreenivasan, A. (1968) : Aspects of limnology, primary production and fisheries in the Stanley reservior, Madras State Hydrobiologia, *32* : 551-569.

George, M.G. (1961) : Observation on the rotifers from shallow ponds in Delhi. Curr. Sci., *30* : 265-269.

Ghosh, A. and George, J.P. (1989) : Studies on the abiotic factors and zooplankton in a polluted urban reservoir, Hussain sagar hyderabad impact on water quality and embryonic development of fishes. Indian J. Environ. Hlth., *37* : 49-59.

Goel, P.K., Khatavkar, S.P., Kulkarni, A.Y. and Trivedy, R.K. (1986) : Limnological studies of a few fresh Waterbodies in Southwestern Maharashtra with special reference to their chemistry and phytoplankton. Poll. Res., *5* : 79-84.

Golterman, H.C., Bakels, L.C. and Mogelin, J.J. (1969) : Availibility of mud phosphorus for the growth of algae. Verh. Int. Limnol., *17* : 467-479.

Gonzalves, E.A. and Joshi, D.B. (1946) : Fresh water algae near Bombay. I. The seasonal succession of the algae in a tank at Bandra. J. Bom. Nat. Hist. Soc., *46* : 154-176.

Hutchinson, G.E. (1957) : A treatise on limnology. Vol. I. Geography, physics and chemistry. John Wiley and Sons Inc., New York, USA.

Jana, B.B. and Sarkar, H.L. (1971) : The limnology of Swetganga-A thermal spring of Bakreswar, West Bengal, India. Hydrobiologia, *37* : 33-47.

Jhingran, V.G. (1982) : Fish and Fisheries of India. 2nd ed. Hindustan Publishing Corporation (India), Delhi.

Kadam, S.P. (1990) : Environmental study of lake Rankala, Jayanti Nala and lake Kotitirsh from Kolhapur city. Environment and Ecology, *8* : 95-97.

Kant, S. and Anand, V.K. (1978) : Inter-relationship of phytoplanktons and physical factors on Mansabal lake Jammu. Indian J. Ecol., *5* : 134-140.

Kant, S. and Kachroo, P. (1979) : A Review of the phytoplanktonic studies. Prof. M. R. Suxena Mem. Volume, Hyderabad.

Kant, S. and Raina, A.K. (1990) : Limnological studies of two ponds in Jammu. II. Physico-chemical parameters. J.Environ. Biol., *11* : 137-144.

Kaushik, S., Agarker, M.S. and Saksena, D.N. (1991b) : Water quality and periodicity of phytoplanktonic algae in Chambal tank Gwalior, Madhya Pradesh. Bionature, *11* : 87-94.

Kaushik, S. and Saksena, D.N. (1989) : Physico-chemical factors and the aquatic insect density of a pond receiving cotton mill effluent at Gwalior. Indian J. Ecol., *16(1)* : 64-67.

Kaushik, S., Saxena, M.N. and Saksena, D.N. (1991c) : Phytoplankton population dynamics in relation to environmental parameters in Matsya Sarovar at Gwalior (M.P.). Acta Botanica Indica, *19* : 113-116.

Kaushik, S., Sharma, S. and Saxena, M.N. (1989) : Aquatic insect of population in relation to physico-chemical characteristics of Chandanpura pond, Gwalior (M.P.). In : Environmental Risk Assessment (Eds. Y.N. Sahai, P.B. Deshmukh, T.A. Mathai and K.S. Pillai). Academy of Environmental Biology, India, Muzaffarnagar, pp. 243-254.

Kaushik, S., Sharma, S. and Saxena, M.N. and Saksena, D.N. (1990) : Abundance of insects in relation to physico-chemical characteristics of pond water at Gwalior (M.P.). Proc. Nat. Acad. Sci. India, *60(b)* : 103-106.

Kaushik, S., Sharma, S. and Saksena, D.N. (1991a) : Ecological studies of certain polluted lentic waters of Gwalior region with reference to aquatic insect communities. In : Current trends in Limnology. I (Ed. N.K. Shastree). Narendra Publishing House, New Delhi, pp. 185-200.

Khan, A.A. and siddqui, A.Q. (1974) : Seasonal change in the limnology of a perennial fish pond at Aligarh. Indian J. Fish., *21* : 463-478.

Khatavkar, S.D., Kulkarni, A.Y. and Goel, P.K. (1990) : Phytoplankton flora of some fresh Waterbodies in South-western Maharashtra. Environment and Ecology, *8* : 267-275.

Khatri, T.C., (1985) : Physico-chemical features of Idukki reservoir, Kerala during Pre-Monsoon Period. Environment and Ecology, *3* : 134-137.

Krishnamoorthy. K.P. and Visweswara, G. (1963) : Hydrobiological studies with reference to the sudden fish mortality. Hydrobiologia, *21* : 275-303.

Kulshrestha,S.K. (1988) : Review of studies on Bhopal lakes. Proc. Nat. Symp. Past, Present and Future Bhopal lakes, pp. 11-22.

Kulshrestha,S.K. (1989) : Degradation of aquatic environment in Madhya Pradesh. Proc. Symp. Environmental Protection. pp. 259-271.

Lee, G.F., Jones, R.A. and Rast, W. (1981) : Alternative approcah to trophic state classification for water quality management. Occasional paper No. 66. Department of Civil Engineering Environmental Engineering Programme, Colorado, State University, Fort Collins Co.

Lehr, J.H., Gass, T.E., Pettyjohn, W.A. and Demarre, J. (1980) : Domestic water treatment. McGraw Hill Book Co., New York.

Mairs, D.F. (1966) : A total alkalinity atlas for marine lake waters. Limnol. Oceanogr., *11* : 68-72.

Mohan, K.S. (1987) : Chemistry of two freshwater lakes of Hyderabad, India, silicates. Poll. Res., *6* : 69-72.

Mohan, K.S. and Reddy, N.S. (1987) : A study of cations in two freshwater lakes of Hyderabad (India). Poll. Res., *6* : 5-8.

Mohan, K.S. and Zafar, A.R. (1986) : Water quality of two drinking water reservoirs of Hyderabad (India). Poll. Res., *5* : 111-122.

Moyle, J.B. (1946) : Some chemical factors influencing the distribution of aquatic plants in Minnesota. Amer. Midl. Natl., *34* : 402-426.

Munawar, M. (1970) : Limnological studies on fresh water ponds of Hyderabad, (India). II. The biocoenose. the distribution of univellular and colonial phytoplankton in the polluted and the unpolluted environemnts. Hydrobiologia, *36* : 105-128.

Nasar, S.A.K. and Dutta Munshi, J. (1974) : Seasonal variation in the physico-chemical and biological properties of a tropical shallow pond. Jap. J. Ecol., *24* : 255-259.

Ohle, W. (1938) : Die Bedeutung der Anstanschvorgange Zwischen Schalamn und wasser fur den Stoffkreislauf der Gewasser, "Vom Wasser".

Olsen, S. (1950) : Aquatic plants and hydrospheric factors. I. Aquatic plants in Swjutland, Denmark Sevensk. Botanisk Tidskriff, 44 : 1-34.

Prasad, B.N. and Manjula, S. (1980) : Ecological study of blue green algae in river Gomti. Indian J. Environ. Hlth., *22* : 151-168.

Pruthi, H.S. (1993) : Studies on the bionomics of fresh water in India. I. Seasonal changes in the physical and chemical condition of the waters of the tank in the Indian Museum Compound. Internat. Rev. Ges. Hydrobiol. Hydrography, *28* : 46-47.

Qadri, M.Y. and Yousuf, A.R. (1978) : Seasonal variations in the physico-chemical factors of a subtropical lake of Kashmir. J. Inland Fish Soc. India, *10*: 89-96.

Qadri, M.Y. and Yousuf, A.R. (1980) : Limnological studies on lake Malpur. Geobios, *7* : 117-119.

Rai, D.N. and Dutta Munshi, J.S. (1979) : Observations on diurnal changes of some physico-chemical factors of threee tropical swamps on Dharbhanga (North Bihar) India. Comp. Physiol. Ecol., *4* : 52-55.

Ramanibai, P.S. and Ravichandran, S. (1987) : Limnology of an urban pond at madras Indian. Poll. Res., *6* : 77-81.

Reid, G.K. and Wood, R.D. (1976) : Ecology of Inland Waters and Estuaries, D. Van Nostran company, New York.

Roy, S.P. (1977) : Limnobiotic investigation of a freshwater tropical shallow pond of Bhagalpur (Bihar), India. J. Bhagalpur Univ., *10* : 110-117.

Sarkar, S.K. (1989) : Phosphorus and aquatic ecosystem. Indian Rev. Life Sci., *9* : 227-251.

Sarkar, H.L. and Rai, H. (1964) : The limnology of the Surajkund, India. Physico-chemical conditons and plankton. Trans. amer. Micr. Soc., *83* : 260-266.

Sawyer, C.H. (1960) : Chemistry for sanitary Engineers. McGraw Hill Book Co., New York.

Schaperclaus, W. (1933) : Text book on pond culture : rearing and keepind of carps, trout and allied fishes (Ed. F. Hund). Fish Leaflet, Washington, *11* : 260.

Seenayya, G. (1971) : Ecological studies in the plankton of certain fresh water ponds of Hyderabad, (India). II. Phytoplankton. Hydrobiologia, *37* : 55-88.

Shardendu and Ambasht, R.S. (1988) : Limnological studies of a rural and an urban tropical aquatic ecosystem : Oxygen forms and ionic strength. Tropical Ecology, 29 : 98-109.

Shastree, N.K., Islam, M.S. Pathak, S. and Afshan, M. (1991) : Studies on the physico-chemical dimensions of the lentic hydrosphere of Ravindra Sarovar (Gaya).

In : Current Trends in Limnology. Vol. I. (Ed. N.K. Shastree). Narendra Publishing House, Delhi, pp. 133-152.

Siddiqui, E.N., Ahmad, M.S. and Bazmi, S.H. (1990) : Limnology of ponds at Darbhanga, Bihar. Environment and Ecology, *8* : 1147-1152.

Singhal, R.N., Jeet, S. and Davies, R.W. (1986) : The physico-chemical environment and the plankton of managed ponds in Haryana, India. Proc. Indian Acad. Sci. India, *95(B)* : 353-363.

Spence, D.H.N. (1964) : The macrophytic vegetation of lochs, swamps and associated fens. In : The vegetation of Scotland (Ed. J.H. Burnett). Oliver and Boyd, Edinburgh, 306-425.

Spence, D.H.N. (1967) : Factors controlling the distribution of freshwater macrophytes with particular reference to Scottish lochs. J. Ecol., *55*: 147-170.

Sreenivasan, A. (1968) : Limnology of tropical impoundments of two hard water reservoir in Madras state. Arch. Hydrobiol., *65* : 205-222.

Sreenivasan, A. (1970) : Limnology of tropical impoundments, A comparative study of the major reservoirs in Madras state (India). Hydrobiologia, *36* : 443-469.

Sreenivasan, A. (1972) : Limnology of a seepage type impoundment, Odathurai tank. J. Inland fish. Soc. India, *4* : 162-168.

Sreenivasan, A. (1974) : Limnological factors of a tropical impoundment Bhavani sagar reservoir, Tamil Nadu, India. Int. Revue ges. Hydrobiol., *59* : 327-342.

Sreenivasan, A. (1976) : Limnological studies of primary production in three temple pond ecosystem. Hydrobiologia, *48* : 117-123.

Sreenivasan, A. (1991) : Time series changes in the ecology of south Indian reservoirs, J. Inalnd Fish Soc. India, *23(2)* : 69-79.

Sreenivasan, A. Sounderaraj, R. and Antony, F. (1964) : Limnological studies of tropical impoundments. II. Hydrological features and plankton of Bhavanisagar Reservoir (Madras) for 1961-61. Proc. Nat. Acad. Sci. India, *59(B)* : 53-71.

Sukumaran, P.K. Raghvan, S.L. and Rahman, M.F. (1984) : Observations on physico-chemical conditions of water and plankton of two tanks in Malanand region of Karnataka. J. Inland Fish Soc. India, 16 : 53-55.

Toetz, D.W. (1976) : Mineral cycling in reservoirs. Ann. Okla. Acad. Sci., *5* : 21-28.

Tripathi, B.D. and Srikandar, M. (1989) : Ecology of lakes and ponds in India. In : Water pollution (Eds. A.K. Sinha, P.N. Vishwanathan and R. Booji). Gyanodaya Prakashan, Nainital, pp. 51-64.

Trivedy, R.K. and Goel, P.K. (1986) : Chemical and biological methods for water pollution studies. Environmental publications Karad (India).

Unni, K.S. (1972) : An ecological study of the macrophytic vegetation of Doodhadhari lake, Raipur (M.P.). 3. Chemical factors. Hydrobiologia, *40* : 25-36.

Unni, K.S. (1983) : Comparative water chemistry of a plankton dominated and a macrophytic dominated lake in Chhindwara, M.P. Proc. Nat. Acad. Sci. India. Part B., *53(b)* : 81-88.

Valecha, V. and Bhatnagar, G.P. (1988) : Seasonal changes of phytoplankton in relation to some physico-chemical factors in lower lake of Bhopal. Geobios, *15* : 170-173.

Valecha, V., Trivedy, R. and Bhatnagar, G.P. (1987), Biological assessment of trophic status of lower lake of Bhopal. Poll. Res., *6* : 91-93.

Venkateswarlu, V. (1983) : Taxonomy and ecology of algae in the river Moosi, Hyderabad, India. II Bacillariophyceae. Bibliotheca Phycologica. Editor J. Cramer, *66* : 1-44.

Verma, M.N. (1969) : Hydrobiological study of a tropical impoundment Tekanpur reservoir Gwalior, India; with special reference to the breeding of Indian carp. Hydrobiologia, *34*: 358-368.

Verma, S.R. and Shukla, G.R. (1968) : Hydrobiological studies of a temple tank Devi Kund in Deoband (U.P.), India. Indian J. Environ. Hlth., *10* : 177-188.

Verma, S.R. and Shukla, G.R. (1970) : The physico-chemical conditions of Kamla Nehru Tank Muzzafar Nagar in relation to biological productivity. Env. Hlth., *12(2)* : 110-128.

Vollenweider, R.A. and Frei, M. (1953) : Vertikaleundzeithliche Vertellung der Leitfahig Keit in einn eutrophen Gewasser Wazrend der sommer stagnation. Hydrobiologia, *15* : 158-167.

Vyas, L.N. and Kumar, H.D. (1968), Studies on the phytoplankton and other algae of Indrasagar tank, Udaipur, India. Hydrobiologia, *31* : 421-434.

Webber, W.J. Jr. and Stumm (1963), Mechanism of hydrogen ion buffering in natural waters. J. Amer. Wat. Works Assoc., *155* : 1553.

Welch, P.S. (1952), Limnology, McGraw Hill Book Company, New York.

Wetzel, R.G. (1975), Primary production river ecology. Blackwell Scientific Publication, Oxford.

World Health Organization (1984), Guidelines for drinking water quality. Vol. 2. Health Criteria and other supporting information. Geneva. ISBN 924-154169-5, pp. 327.

Yousuf, A.R. and Shah, G.M. (1988), Comparative limnology of some freshwater habitats of Kashmir. Geobios, *7* : 58-61.

Zafar, A.R. (1964) : On the ecology of algae in certain fish ponds of Hyderabad, India. I. Physico-chemical complexes. Hydrobiologia, *23* : 179-195.

Zafar, A.R. (1966) : Limnology of th Husain Sagar Lake - Hyderabad, India. Phykos, *5(1-2)* : 115-126.

Zutshi, D.P. and Vass, K.K. (1978), Limnological studies on Dal Lake. Chemical features. Indian J. Ecol., *5* : 90-97.

2

Limnological Studies on Chandrampalli Reservoir, Gulbarga

K. Vijaykumar, Devendra Holkar and Kuldeep Kaur

Water is one of the most essential item needed by man, plants and other living beings for their survival. Water maintains an ecological balance between various groups of living organisms and their environment. Water is the most precious resources on the earth, covering almost seventy per cent of the earth. India in particular has a great potential of water resources, both fresh and marine, and a great variation in environmental conditions provide a wide range of habitats and contrasts with its varied physical and meteorological features. Indeed the economics of all life processes and living communities are hinged on this most vital element. Even from the narrower practical point of view, a body of freshwater which it is abused and polluted. There is no need, therefore to justify the efforts to gain knowledge and acquire requisite skills to ensure the preservation of this vital substance. A proper understanding of the freshwater contents and conservation of this meagerly available precious resources of our country is need of the hour.

Freshwater biology, till recent times has never been a popular subject amongst the students of natural science. This is because of its local nature. The findings of one waterbody may not hold good for another; what is true for the Waterbodies of temperate regions may not be so for the tropical waters. Considerable processes has been achieved in our knowledge of the Waterbodies of temperate region, as the developed countries come under this category. Little is known about the Waterbodies of the tropical countries which are either underdeveloped or underdeveloping. Thus there is void in our knowledge about the tropical inland waters. Even the basic floristic and faunastic studies have not been carried out in a planned way. The Indian subcontinent is no exception to this situation. The literature on freshwater fauna and physico-chemical parameters, being too scattered and spread over past many years, has become almost inaccessible. Because of this the standing fresh Waterbodies have always been of comparatively more interest with investigators than the other types of freshwater systems probably because of their continental existence. Limnology has come a long way since the time Forel (1892) in understanding the dynamics of a standing Waterbodies subsequently, limnology was studied with reference to the organisms, specially, plankton (Hensen, 1887, Fristsch, 1907) which led to the so called branch - "hydrobiology". The long held idea that a suitable environment is necessary for any organisms since life depends upon the cortinuance of proper exchange of essential substances and energies between the organism and its surrounding environmental led to extensive work of several investigators *viz.*, West and West (1912), Delf (1915), Round (1953), Kemmere and Narhold (1969), Morgon and Margaret (1970), Mori (1971), Lin (1973), Miller *et al.*, (1973), Tilman (1977), Russo (1978), Soundar Guard and Jensen (1979) and several others.

The humble beginning dating back to the late half of the century of freshwater studies by workers such as Sewell (1934), Iyenger (1939), Ganapathi (1940) and such others come a long way through the development of scientific temperament and concerted efforts not only in freshwater biology but in field of science in our country. Initially such studies were directed mostly towards gathering baseline information on water chemistry, planktonic organisms to be applied are used in fisheries practices.

In India, a remarkable contribution is made in this field and different trends of study are observed. Some important contributions to field of limnology have been made by various investigators from the different parts of the country, since last six decades. However, until 1931 no significant contributions were made. The works of Iyenger (1939), Gonzalves and Joshi (1946), Chacko and Krishnamurthy (1954), Rao (1955), Phillipose (1959) described on the periodicity of algae of some muddy rain water pools from Chennai gained much attentions, many of others also have contributed towards specific aspects of freshwater studies *viz.*, Zafar (1959), Singh (1960), Sreenivasan (1963 and 1964), Munawar and Zafar (1967) whose works deals with the aspects like thermal stratification of certain reservoirs and annual heat budget of Waterbodies, to mention few. Munawar (1970 and 1974); Rao (1975) studied the ecology of freshwater pond in and around Hyderabad. Patil *et al.* (1975), have investigated on phytoplankton and zooplankton succession during prestabilization and stabilization of domestic sewage in facultative pond of Dharwad, Bharati and Kori (1975), worked on limnological studies in ponds and lakes of Dharwad. Ali and Khan (1976 and 1979), Misra and Yadav (1978) have made comparative study of hydrobiological features of a perennial freshwater body of Central India. Ayyappan and Gupta (1980, 1981 and 1985) while studying a perennial pond at Coastal Karnataka and traced out a significant co-relation between physical and chemical parameters and plankton. Gaddad *et al.* (1983) who mainly dealt with sewage pollution aspects of Jagath Tank, Gulbarga. Hegde and Bharati (1985) while describing the physico-chemical conditions of two shallow Waterbodies near Dharwad indicated a clear inter-relationship between phytoplankton activity. Rao (1985) observed the hydrobiological conditions of a tank near Bangalore, Karnataka and opined that the soil quality affects the nature of productivity of waterbody. Khan and Khan (1985) while studying limnology of shallow waterbody near Aligarh, described seasonal fluctuations of various physico-chemical parameters. Yousuf *et al.* (1986) made an attempt to understand the physico-chemical conditions of lake near Srinagar and suggested that the planktonic population density seems to be limited when compared to the larger Waterbodies. Yousuf and Shah (1988) reported on a comparative limnological assessment of seven shallow Waterbodies in and around Kashmir. Khatavakar *et al.* (1989) has reported on diel variations in some physico-chemical parameters of polluted Waterbodies of Kolhapur, Maharashtra State and were found to be highly significant between N, P and phytoplankton taxa. Roy (1989) dealt with the productivity of aquatic insects in a freshwater pond of Bhagalpur.

In the early 1990, the work on limnological aspects documented by various workers *viz.*, Adholia *et al.* (1990 and 1991) studied on abundance, distribution and total population of macrozoobenthos in relation with limno-chemical parameters of Mansarovar Reservoir. More and Gajjar (1990) investigated on the seasonal effects on the flora of Palan pond in Gujarat. Mukherjee *et al.* (1990) worked on carbondioxide dynamics and speciation of polluted aquatic body at Ranchi. Vijaykumar and Paul (1990) studied on hydrobiological features of Bhosga Reservoir, Gulbarga. Tamot *et al.* (1991) worked on physico-chemical and biological characteristics of upper lake, Bhopal during the Ganesh and Durga immersion before and after such activity. Johnson (1991) worked on total nitrogen and phosphate content of two freshwater lakes of Hyderabad. Belsare *et al.* (1992) studied the status of free dissolved enzymes, their inter-relationship with bacterial content and certain physical and chemical parameters in Bhopal lakes. Subbamma (1992) and Subbamma and Sharma (1994) worked on monthly variation in the abundance of plankton of temple pond near Machilipatanam, Andhra Pradesh. Vijaykumar and Basalingappa (1992) studied on influence of physico-chemical parameters on zooplankton populations in freshwater body of Dharwad. Panda and Dash (1993) worked on water quality and pollution load of three community ponds of Bargarh town, Orissa. Chakrabarty and Saha (1993) carried a pilot study in three artificially impounded water sources in Darjeeling town during rainy season to check an account of hydro-biological and physico-chemical features with special reference to their nutrients. Dutta *et al.* (1993) studied on the diurnal variations of certain physico-chemical parameters of perennial pond at Guwahati. Pati and Sahu (1993) worked on biomonitoring of water pollution in Rengali Reservoir, Orissa. Ahmed and Singh (1993) studied on diurnal variations in physico-chemical parameters of water and zooplankton in a freshwater tank at Patna during summer, monsoon and winter season. Gupta and Sharma (1993) on diel variations of shallow pond of Udaipur. Sinha and Sinha (1993) studied on seasonal variation of zooplanktonic communities in an ancient freshwater stagnant fish pond of Munger, Bihar. Rana and Nirmalkumar (1993) prepared an arbitrary composite tropic index from the average values of various physical, chemical and biological parameters of the four study ponds. Mansoori *et al.* (1993) carried out on limnological aspects of Lakshmital lake, Jhansi during rainy season. Dhamija and Jain (1994 and 1995) investigated for some physico-chemical parameters of freshwater body of Jabalpur. Mukherjee *et al.* (1994) developed a system which defines the assessment of detergent input in freshwater from the influx of the total alkalinity and the resulting material balance of carbon, as well from the increase in the Phosphate load. Rajkumar *et al.* (1994) studied on environment relationships in urban and rural habitat in water of the Chennai City. Patnaik and Padhi (1994) worked on algal bioassay aspects of local ponds of Berhampur (Orissa). Subbamma (1994) studied on hydrography of a fish in the Kolleru Lake area Andhra Pradesh. Devi (1994) worked on seasonal variation and population density of rotatorian in three lakes of Hyderabad. Kumar (1995) worked on limnological aspects of the freshwater tropical wetland of Santhal Pargana (Bihar). Rao and Mahmood (1995) studied on physico-chemical and biological parameters of Habsiguda pond, Hyderabad. Mukesh *et al.* (1995) investigated on plankton diversity of Sadatpur Reservoir, Maharashtra. Choudhary (1996) has made an attempt to assess the impact of limnological conditions of two perennial ponds at Bhagalpur. Raju and Durani (1996) estimated the nano-planktonic primary production in three temple ponds of Bhubaneshwar. Birsal (1996) studied on the primary productivity of the Supa Reservoir during its filling stage, Karnataka. Alam and Khan (1996) worked on plankton population in four freshwater ponds at Aligarh.

Primary production is one of the most important source of energy input in freshwater ecosystem; the production of organic matter in aquatic environment has held the attention of the hydrobiologists since long. The synthesis of basic food *i.e.*, transformation of inorganic substances into organic form is similar to that of terrestrial environment. Some of the studies on the primary productivity of Indian freshwaters include those of Sreenivasan (1963, 1964a, 1964b and 1972) from tanks of Tamil Nadu. Khan and Siddiqui (1971) opined that seasonal fluctuations in gross and net production values were quite apparent in surface waters of Aligarh fish pond. Nasar and Nasar (1976); Gopal *et al.* (1978), have added to primary production studies of different types of ponds at Bhagalpur. Seenayya and Zafar (1979) reported the phytoplankton production in Mir Alam lake of Hyderabad. Ayyappan and Gupta (1985) have studied on the primary production in Ramasamudra at Mangalore and described that the primary production was controlled by several hydrobiological factors. Siddiqui *et al.* (1987), worked on the phytoplankton population of Bhulki pond at Darbhanga, Bihar. Velecha and Bhatnagar (1989) studied on the production of phytoplankton in an eutrophic lower lake in Bhopal. Shukla and Bais (1990) observed co-relation of phytoplankton biomass and productivities with physical properties of Bila Reservoir during winter season. Vijaykumar and Paul (1990, 1991 and 1994), Vijaykumar *et al.* (1991) and Vijaykumar (1992 and 1996) reported the primary production of different Waterbodies of Dharwad and Gulbarga.

Planktons act both as predators and consumers; play an important role in transformation of energy from one tropic level to the next highest, ultimately leading to fish production which is final product of the aquatic environment. The zooplankton populations in a small Waterbodies are subjected to extreme fluctuations, the cause of which is not adequately understood even though exhaustive literature on plankton studies are available. The Indian notable contributions to the knowledge of zooplankton are of Arora (1931); Sewell (1934) who studied on planktonic rotifers. Gouder and Joseph (1961) documented a detailed information on copepods. George (1961) has worked on the distribution of zooplankton in pond and lake. Govind (1963) investigated on the relation between copepodes and physico-chemical parameters in Tungabhadra Reservoir, Hospet, Karnataka. Michael (1968) worked on several aspects such as distribution and abundance of zooplanktons in different Waterbodies near Chennai. Rajendra (1973) observed the occurrence and abundance of copepods of lentic Waterbodies in Tamil Nadu. Patil *et al.* (1975) revealed the importance of cladocerans in the seasonal succession of zooplankton in sewage stabilization pond at Dharwad. Notable contribution on rotifers fauna in different Waterbodies from West Bengal has been made by Sharma (1976, 1978a, 1978b, 1979a and 1979b). Rao and Mohan (1976, 1977, 1980 and 1982) carried out a detailed investigations on the rotifers of Andhra Pradesh. Ayyappan and Gupta (1980) carried an investigation on the ecological studies of zooplankton of tank in Dakshina Kannada, Karnataka. Patil and Gouder (1982) carried out a detailed study on the systematics of rotifers and cladocerans and also the ecological aspects of several ponds of Dharwad. Laal (1984) studied on planktonic rotifers of Bihar State. Masood *et al.* (1987) documented the seasonal variations of copepods in freshwater ecosystem of Kashmir. Datta *et al.* (1987) worked on abundance of zooplanktons in freshWaterbodies of Calcutta, while Shaha and Pandit (1988) compared the density of zooplankton between the lotic and lentic environment. Vijaykumar (1992, 1994 and 1996) also studied on the distribution on the zooplankton in different ponds at Gulbarga area.

Information on limnology and its related aspects pertaining to innumerable standing Waterbodies in the northern parts of Karnataka State of peninsular India especially from Gulbarga is lacking except works of Gaddad *et al.* (1983) while attempting the aspects pertaining to sewage pollution which extended over a period of one year made a cursory mention of some physico-chemical parameters of Moat at Gulbarga. Vijaykumar (1992) who has contributed much on hydrobiological and its related aspects from three fresh Waterbodies *i.e.*, Jagath Tank, Bhosga Reservoir and Pala Tank. Pappa (1995) added the information on limnology from Gobbur tank of Gulbarga. Their contribution to the limnology from this area is a finite resources to the young hydrobiologists of this area. Apart from these works, there is practically no information of the limnological aspects of standing Waterbodies of this area.

Hence the present account is an attempt to accumulate information pertaining to various aspects of hydrobiology of standing Waterbodies from the part of peninsular India.

The present investigation has been carried out on Chandrampalli Reservoir located on Sarnala (Krishna Basin), near Chandrampalli Village in Chincholi Taluk of Gulbarga District which falls under 17^0 21' 0" N latitude and longitude 77^0 25' 0" E and located about 467 m above the sea level.

The preliminary study of different water parameters were carried for a period of 10 months from June 1996 to March 1997. The investigations encompassed, collection of various data pertaining to the meteorological, hydrography, primary production and zooplankton studies.

The Study Area

Chandrampalli Reservoir is a small irrigation waterbody having length of about 3,023 ft, catchment area is about 170 sq. miles, height is 94 ft and gross storage capacity of about 1,200 MCFT (Fig. 1).

Chandrampalli Reservoir presents a typical climatic condition of peninsular south India with semi-arid conditions. The temperature ranging 14^{0}C to 47^{0}C (however during the study period it ranged from 10 to 46^{0}C). The average rainfall being 118.32

mm per annum. This area falls under maidan zones as described by David *et al.* (1974) and typically has undulating contour, thus making scope for depressions and catchment areas.

The Chandrampalli reservoir get its main source during the south west and north east monsoon seasons through the Sarnala and its surrounding areas.

Meteorology

The information pertaining to various climatic factors such as total rainfall was obtained from the Asst. Ex. Engineer, Chandapur Sub Division, Chincholi, Gulbarga District for the period of June 1996 to March 1997 during study period.

Field Studies

Surface water samples were collected from eight different station in the reservoir every month using thoroughly washed bottles for the study of various physico-chemical parameters. All collections and observations were made between 9 A.M. to 2 P.M. throughout the period of the study including primary productivity.

Hydrography

All the estimations were followed according to APHA, AWWA, WWCF, 1985, 16th Edition.

Temperature

Atmospheric and water temperature at each stations were recorded using an ordinary mercury thermometer to the nearest 0.1 ^{0}C. Water temperature, however was recorded at depth of about 6 cm below the surface.

Relative Humidity

The relative humidity was measured with hygrometer and expressed in the percentage (%).

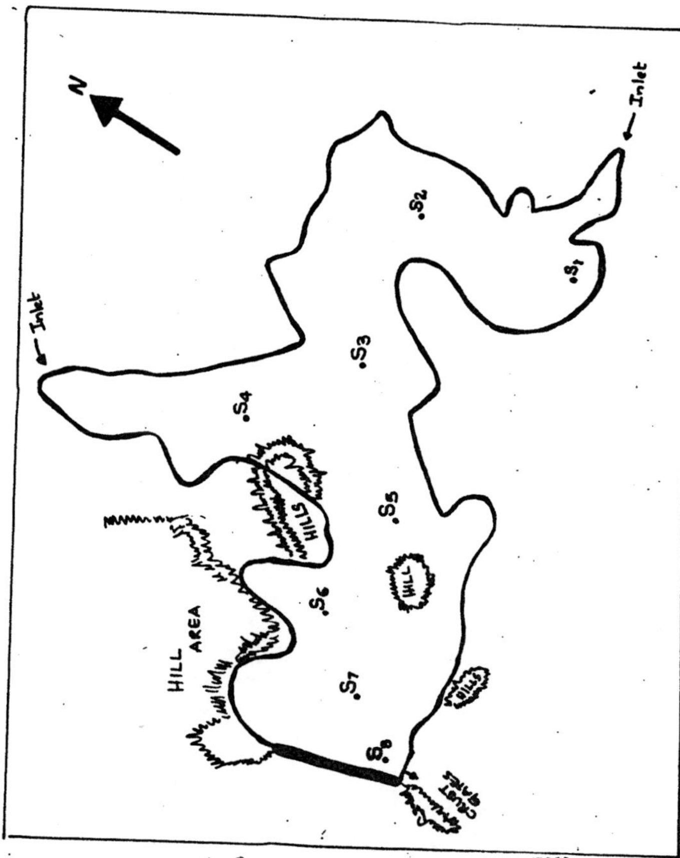

FIG: 1. SKETCH MAP OF CHANDRAMPALLI RESERVOIR

Fig. 1 : Sketch Map of Chandrampalli Reservoir

Hydrogen-Ion-Concentration

pH recordings were made at the different stations using a digital portable pH meter (Elico model Li - 120).

Transparency

Transparency or light penetration of water was measured by immersing Secchi disc and observing its visuality.

Dissolved Oxygen

Water samples were carefully drawn into 250 ml reagent bottles avoiding any kind disturbance before commencing any other estimations, the separated samples were fixed immediately and carried to the laboratory for further estimations.

Free Carbondioxide

Samples were collected in 250 ml of reagent bottles for estimation of free carbondioxide in the field and as well as in the laboratory for accurate results from the different stations of the Reservoir.

Primary Production

Primary production was estimated by the much preferred Gaarder and Grans (1927) "Light and Dark" bottle method. A set of light and dark bottles were kept in the field with water samples and suspended into the surface waters using floats at each stations so as to maintain the bottles 10 to 15 cm below the water surface. The dark bottles including stopper painted with black emulsion paint. The primary production was estimated during 9 A.M to 1P.M with 4 hours incubation period.

Laboratory Studies

Water samples already collected in 250 ml reagent bottles for the purpose of estimation of various parameters were brought to the laboratory and subjected to the analysis immediately to get accurate results. APHA, AWWA, WWCF, 1985, 16th edition (Standard methods for the examination of waste waters). Trivedy and Goel (1980) was referred for the estimation of various parameters *viz.*, Dissolved oxygen, free carbondioxide, pH, alkalinity, chloride, Hardness, Phosphorus-phosphate, and ammonical nitrogen.

Dissolved Oxygen

Dissolved oxygen content of water was estimated by modified Winkler's method. And the results were expressed in mg/1.

Free Carbondioxide

Free carbondioxide content was estimated by the titrimetric method using phenolpthalein indicator and sodium hydroxide (0.05 N). The values were presented in mg/1.

Total Alkalinity

The phenolpthalein and methyl orange alkalinity were determined by titrimetric method using hydrochloric acid and thus carbonate and bicarbonate alkalinity were calculated. The values were presented as mg/1.

Hardness

The total hardness of water is estimated by titrating the water samples with standard EDTA titrate (0.01 N) using buffer solution and Erichrome Black-T indicator and the values were presented in mg/1.

Chloride

Chloride content was estimated by titrating the water samples with silver nitrate and using potassium chromate as indicator, values were presented in mg/1.

Phosphate Phosphorous

Stannous chloride method was employed to determine the orthophosphate content of water samples. The resultant blue colour intensities were measured at 690 nm range using spectrophotometer (Elico-model CL-24). The values were presented in mg/1.

Ammonical Nitrogen

Ammonical nitrogen concentration was measured by using Nessler's reagent method to 20 ml of sample, 4 drops $ZnSo_4$, 6 N NaOH added and filtered, and then 0.2 ml of Nessler's reagent was added and incubated for 5 minutes at room temperature. A Blank solution containing 10 ml of double distilled water was treated in same manner, after keeping five minutes the colour developed was read in a spectrophotometer (Elico-model CL-24). At 420 nm after adjusting the instrument to 100 per cent transmittance with blank. This amount of ammonical nitrogen present in sample was calculated referring to standard graph. And the calculated values were expressed as mg/1.

Primary Production

The primary production in situ was estimated by much preferred Gardner and Grans "Light and Dark" bottle method which is subjected to incubation for 4 hours in surface waters and the samples were fixed using $MnSO_4$ and alkali iodide and brought to the laboratory, while the initial dissolved oxygen content was determined using the sample from the first set of the bottles following the procedure as already mentioned for dissolved oxygen. The oxygen content of dark and light bottles were estimated in the field and as well as in laboratory for accurate results and the results were expressed as in the form of gross production, Net production, and community respiration ($gC/m^3/hr$).

Gross Production : $gC/m^3/hr = \frac{DL - DI}{hr} \times 0.375$

Net Production : $gC/m^3/hr = \frac{DL-DD}{hr} \times 0.375$

Community Respiration : $gC/m^3/hr = \frac{DI-DD}{hr} \times 0.375$

Where,

DI = Dissolved oxygen in the initial bottle : in mg/1.
DL = Dissolved oxygen in the light bottle : in mg/1.
DD = Dissolved oxygen in the dark bottle : in mg/1.
hr = Duration of exposure.

where 0.375 is respiratory quotient.

The present research work carried out at Chandrampalli Reservoir, Chincholi Taluk, Gulbarga District and divided into three distinct seasons *i.e.*, Southwest monsoon season, Northeast monsoon season and Summer season. Southwest monsoon season in this area extends from June to September while Northeast monsoon extends from October to January and Summer from February to May. But in present study, the results were presented upto March 1997.

Total Rainfall

The total rainfall during different months are presented in Table 1.

The rainfall during southwest monsoon period was recorded high with a maximum of 269.00 mm. during August month and while during Northeast monsoon period the rainfall was recorded high (264.00 mm) during October, 1996 and during summer months of 1997 there was no record of rainfall in this area.

Atmospheric Temperature

The monthly atmospheric temperature for different seasons are given in Table 2.

The atmospheric temperature during Southwest monsoon season at Stations I, II, III, IV, V, VI, VII and VIII varied from 25.0 to 29.0°C; 25.0 to 29.5°C; 25.2 to 29.6°C; 24.9 to 29.6°C; 24.9 to 29.8°C; 25.0 to 30.0°C; 25.2 to 29.8°C and 25.3 to 29.8°C respectively.

The atmospheric temperature during Northeast monsoon season at stations I, II, III, IV, V, VI, VII and VIII varied from 27.0

Table 1 : Total rainfall recorded at Chandrampalli Reservoir during 1996-97 period

Months	*Rainfall (mm)*
June 1996	52.00
July	168.22
August	269.00
September	230.00
October	264.00
November	20.00
December	00.00
January 1997	00.00
February	00.00
March	00.00

to 29.0°C; 27.0 to 29.2°C; 27.5 to 29.2°C; 27.6 to 29.5°C; 27.5 to 29.6°C; 27.6 to 29.5°C; 27.8 to 29.7 °C ; 27.8 to 29.9°C respectively.

The atmospheric temperature during summer months of 1997 varied at eight stations from 28.0 to 30.0°C; 28.1 to 31.0°C; 28.2 to 31.5°C; 29.0 to 31.5°C; 29.1 to 31.6°C; 29.4 to 31.6°C; 29.4 to 31.7°C and 29.5 to 31.7°C respectively.

Water Temperature

The fluctuations of water temperature at different stations during different seasons are presented in Table 3.

The water temperature during Southwest monsoon of 1996 varied from 20.0 to 24.0°C; 20.0 to 24.1°C; 21.1 to 24.1°C; 20.1 to 24.4°C; 20.1 to 24.5°C; 20.2 to 24.6°C; 20.1 to 24.6°C and 20.1 to 24.9°C from stations I to VIII respectively.

During the Northeast monsoon season of 1996 and January 1997, the Water temperature recorded from stations I, II, III, IV, V, VI, VII and VIII ranged between 22.0 to 25.0°C; 22.0 to 25.1°C; 22.5 to 25.2°C; 22.6 to 25.5°C; 22.5 to 25.6°C; 22.6 to 25.5°C; 22.8 to 25.6°C and 22.8 to 25.9°C respectively.

During early Summer months of 1997, the water temperature ranged between 22.0 to 25.0 °C

Relative Humidity

Seasonal variations of Relative Humidity (RH) of Chandrampalli Reservoir at different stations are presented in Table 4.

The fluctuations of Relative Humidity during Southwest monsoon season ranged between 38 to 55 per cent; 38 to 56 per cent; 38 to 52 per cent; 39 to 52 per cent; 41 to 52 per cent; 43 to 58 per cent; 42 to 58 per cent; 42 to 53 per cent and 42 to 53 per cent at stations I to VIII respectively.

During Northeast monsoon period the Relative Humidity varied from 35 to 45 per cent at station I; 35 to 46 per cent at station II, 38 to 44 per cent at station III; 38 to 46 per cent at station IV; 39 to 43 per cent at station V; 39 to 42 per cent at station VI; 40 to 42 per cent at station VI and 40 to 43 per cent at station VIII respectively.

During the early Summer months of 1997 the Relative Humidity of reservoir varied from 35 to 39 per cent; 35 to 40 per cent; 35 to 42 per cent; 36 to 41 per cent; 36 to 42 per cent; 37 to 42 per cent; 35 to 43 per cent and 36 to 41 per cent at Stations I to VIII respectively.

Transparency

The seasonal fluctuations of transparency of Chandrampalli reservoir are presented in Table 5.

During Southwest monsoon seasons of 1996 the Transparency values ranged between 20 to 30 cm; 20 to 31 cm; 20 to 33 cm; 21 to 32 cm; 21 to 33 cm; 21 to 30 cm; 22 to 32 cm and 22 to 31 cm at stations I, II, III, IV, V, VI, VII and VIII respectively.

During Northeast monsoon season the Transparency values varied between 24 to 45 cm at Station I, 24 to 46 cm at station II, 23 to 47 cm at station III, 22 to 49 cm at station IV, 24 to 50 cm at station V, 24 to 46 cm at station VI, 24 to 41 cm at station VII and 25 to 40 cm at station VIII respectively.

During early Summer season of 1997, the Transparency values varied from 35 to 57 cm respectively.

Hydrogen-Ion-Concentration

The Hydrogen-ion-concentration of Chandrampalli reservoir for different seasons are presented in Table 6.

During Southwest monsoon season, the pH values at stations I to VIII fluctuated from 8.3 to 9.0; 8.3 to 8.9; 8.0 to 9.1; 8.2 to 9.2; 8.1 to 9.0; 8.3 to 9.2; 8.2 to 9.0 and 8.4 to 9.1 respectively.

Table 2 : Atomospheric temperature ^{0}C at eight selected stations of Chandrampalli Reservoir

Months	*Stations*							
	I	*II*	*III*	*IV*	*V*	*VI*	*VII*	*VIII*
June 1996	25.0	25.0	25.2	24.9	25.0	25.3	25.5	25.6
July	29.0	29.5	29.6	29.6	29.8	30.0	29.8	29.7
August	28.0	28.0	28.1	28.4	28.5	28.6	28.6	28.9
September	25.0	25.1	25.3	25.4	24.9	25.0	25.2	25.3
October	27.0	27.0	27.5	27.6	27.5	27.6	27.8	27.8
November	28.0	28.1	28.2	29.0	29.0	29.1	29.4	29.5
December	29.0	29.2	29.0	29.5	29.6	29.5	29.6	29.7
January 1997	29.0	29.1	29.2	29.4	29.6	29.5	29.7	29.9
February	28.0	28.1	28.2	29.0	29.1	29.4	29.4	29.5
March	30.0	31.0	31.5	31.5	31.6	31.6	31.7	31.7

Table 3 : Water Temperature °C at eight selected stations of Chandrampalli Reservoir

Months	*Stations*							
	I	*II*	*III*	*IV*	*V*	*VI*	*VII*	*VIII*
June 1996	20.0	20.0	20.1	20.1	20.1	20.2	20.1	20.1
July	24.0	24.1	24.0	24.0	24.2	24.2	24.3	24.3
August	24.0	24.0	24.1	24.4	24.5	24.6	24.6	24.9
September	20.0	20.1	20.3	20.4	20.9	21.0	21.2	21.3
October	22.0	22.0	22.5	22.6	22.5	22.6	22.8	22.8
November	23.0	23.1	23.2	24.0	24.1	24.4	24.5	24.5
December	25.0	25.1	25.2	25.5	25.6	25.5	25.6	25.9
January 1997	24.0	24.1	24.2	24.4	24.6	24.5	24.7	24.9
February	22.0	22.0	22.2	23.0	23.1	23.1	23.4	23.5
March	24.0	24.1	24.3	24.6	24.7	24.8	24.9	25.0

Table. 4 : Relative Humidity (%) at eight selected stations of Chandrampalli Reservoir

Months	*Stations*							
	I	*II*	*III*	*IV*	*V*	*VI*	*VII*	*VIII*
June 1996	45.0	45.0	46.0	48.0	50.0	45.0	42.0	42.0
July	55.0	56.0	52.0	52.0	50.0	58.0	53.0	52.0
August	45.0	39.0	42.0	43.0	42.0	43.0	42.0	42.0
September	38.0	38.0	38.0	39.0	41.0	43.0	44.0	42.0
October	40.0	40.0	40.0	40.0	41.0	41.0	42.0	43.0
November	35.0	35.0	38.0	38.0	39.0	39.0	40.0	40.0
December	45.0	46.0	44.0	46.0	43.0	42.0	42.0	40.0
January 1997	40.0	39.0	41.0	42.0	42.0	40.0	41.0	42.0
February	35.0	35.0	35.0	36.0	36.0	37.0	35.0	36.0
March	39.0	40.0	42.0	41.0	42.0	42.0	43.0	41.0

Table 5 : Transparency (cm) at eight selected stations of Chandrampalli Reservoir

Months	*Stations*							
	I	*II*	*III*	*IV*	*V*	*VI*	*VII*	*VIII*
June 1996	25.0	24.0	25.0	25.0	25.0	26.0	24.0	25.0
July	20.0	20.0	20.0	21.0	21.0	21.0	20.0	22.0
August	30.0	31.0	33.0	32.0	33.0	31.0	32.0	31.0
September	25.0	25.0	25.0	26.0	25.0	26.0	25.0	27.0
October	24.0	24.0	23.0	22.0	24.0	24.0	24.0	25.0
November	35.0	35.0	36.0	36.0	38.0	36.0	39.0	40.0
December	40.0	41.0	44.0	45.0	41.0	46.0	41.0	40.0
January 1997	45.0	46.0	47.0	49.0	50.0	45.0	48.0	49.0
February	35.0	38.0	39.0	40.0	42.0	43.0	45.0	45.0
March	45.0	49.0	50.0	52.0	53.0	54.0	56.0	57.0

During the Northeast monsoon season, the pH values at I to VIII, fluctuated for 8.0 to 8.8; 8.0 to 8.9; 7.9 to 8.6; 7.8 to 8.7; 7.9 to 8.4; 8.0 to 8.8; 8.1 to 8.6 and 7.9 to 8.8 respectively.

While during the early Summer season of 1997, the Hydrogen-ion-concentration fluctuated between 8.0 to 8.2; 8.1 to 8.2; 8.1 to 8.2; 8.3 to 8.4; 8.2 to 8.3; 8.1 to 8.1; 8.1 to 8.3; 8.1 to 8.4 and 8.0 to 8.2 respectively.

Dissolved Oxygen

Dissolved Oxygen recorded at eight stations of Chandrampalli Reservoir during different season are presented in Table 7.

Dissolved Oxygen concentration during Southwest monsoon season of 1996, recorded at stations I to VIII were ranged between 2.5 to 4.4 mg/l; 2.4 to 4.5 mg/l; 2.3 to 4.1 mg/l; 2.8 to 4.0 mg/l; 2.1 to 3.9 mg/l; 2.4 to 4.2 mg/l; 2.3 to 4.1 mg/l and 2.5 to 4.3 mg/l respectively.

During the Northeast monsoon seasons the Dissolved oxygen values recorded at stations I to VIII were varied between 2.2 to 5.6 mg/l; 2.2 to 5.7mg/l; 2.1 to 5.9 mg/l; 2.0 to 5.2 mg/l;2.2 to 5.9 mg/l; 2.4 to 5.9 mg/l; 2.4 to 5.8 mg/l and 2.5 to 5.9 mg/l respectively.

While during the early Summer season of 1997, the Dissolved Oxygen concentration ranged between 4.9 to 8.2 mg/l; 4.5 to 8.2 mg/l; 4.1 to 7.9mg/l; 4.1 to 7.3 mg/l; 4.2 to 8.1 mg/l; 4.1 to 7.9 mg/l; 4.2 to 7.8 mg/l and 4.4 to 8.0 mg/l respectively.

Free Carbondioxide

The Free Carbondioxide concentration of Chandrampalli reservoir during different season are present in Table 8.

During the Southwest monsoon the free carbondioxide concentration of the water ranged from 2.2 to 9.9 mg/l; Traces to 9.9 mg/l; Traces to 8.8 mg/l; Traces to 9.9 mg/l; Traces to 4.4 mg/l; 2.2 to 9.9 mg/l; Traces to 7.7 mg/l and Traces to 8.8 mg/l respectively.

During the Northeast monsoon season the free carbondioxide concentration the water varied between 5.0 to 5.5 mg/l at Station I; 5.5 to 6.6 mg/l at Station II; Traces to 6.6 mg/l at Station III; and IV; Traces at V Station; Traces to 5.5 mg/l at Station VI; Traces to 3.3 mg/l at Station VII and Traces to 1.1 mg/l at Station VIII respectively.

While during the early Summer season, the Free Carbondioxide ranged between 2.2 to 3.3 mg/l at station I; 1.1 to 2.2 mg/l at Station II and III; Traces to 1.1 mg/l at Station IV, V and VIII; Traces at VI and VII respectively.

Alkalinity

Variation of Total Alkalinity of Chandrampalli Reservoir during different season of 1996-97 are presented in Table 9.

During the Southwest season the values of Total Alkalinity varied from 125 to 160 mg/l at Station I; 124 to 160 mg/l at Station II; 125 to 165 mg/l at Station III; 125 to 166 mg/l at Station IV; 130 to 185 mg/l at Station V; 125 to 185 mg/l at Station VI; 117 to 166 mg/l at Station VII and 125 to 170 mg/l at Station VIII respectively. During the Northeast monsoon season the values of Total Alkalinity varied from 150 to 185 mg/l; 155 to 180 mg/l; 156 to 175 mg/l; 157 to 175 mg/l; 165 to 185 mg/l; 175 to 190 mg/l; 150 to 185 mg/l and 155 to 188 mg/l and Station I to VIII respectively.

While during the early Summer season of 1997, the Total Alkalinity concentration varied from 175 to 185 mg/l at Station I; 180 to 185 mg/l at Station II; 160 to 175 mg/l at Station III; 160 to 175 mg/l at Station IV; 175 to 180 mg/l at Station V; 185 to 187 mg/l at Station VI; 165 to 180 mg/l at Station VII and 175 to 185 mg/l at Station VIII respectively.

Chloride

The concentration of chloride in Chandrampalli Reservoir during different season are presented in Table 10.

During the Southwest monsoon season the values of chloride concentration varied from 34.68 to 56.80 mg/l; 42.60 to 62.48 mg/l; 42.60 to 63.90 mg/l; 34.68 to 63.90 mg/l; 42.6 to 58.22 mg/l; 42.60 to 62.48 mg/l; 42.60 to 59.64 mg/l and 34.60 to 63.90 mg/l at Stations I to VIII respectively.

During Northeast monsoon season the values of chloride concentration varied from 42.28 to 56.80 mg/l at Station I; 39.76 to

Table 6 : Hydrogen-ion-concentration at eight selected stations of Chandrampalli Reservoir

Months	*Stations*							
	I	*II*	*III*	*IV*	*V*	*VI*	*VII*	*VIII*
June 1996	8.9	8.8	8.8	8.6	8.6	8.3	8.5	8.6
July	9.0	8.9	9.1	9.2	9.0	9.2	9.0	9.1
August	8.6	8.4	8.0	8.2	8.1	8.3	8.2	8.4
September	8.3	8.3	8.4	8.6	8.5	8.3	8.2	8.6
October	8.0	8.0	7.9	7.8	7.9	8.0	8.1	7.9
November	8.8	8.9	8.6	8.7	8.4	8.8	8.6	8.8
December	8.2	8.0	8.2	8.0	8.1	8.2	8.1	8.0
January 1997	8.1	8.1	8.2	8.0	8.0	8.1	8.2	8.3
February	8.0	8.1	8.3	8.2	8.2	8.1	8.4	8.0
March	8.2	8.2	8.4	8.3	8.1	8.3	8.1	8.2

Table 7 : Dissolved Oxygen (mg/l) at eight selected stations of Chandrampalli Reservoir

Months	*Stations*							
	I	*II*	*III*	*IV*	*V*	*VI*	*VII*	*VIII*
June 1996	4.4	4.5	4.1	4.0	3.9	4.2	4.1	4.3
July	2.6	2.7	2.7	2.9	2.9	3.0	3.0	3.1
August	2.5	2.4	2.3	2.8	2.1	2.4	2.3	2.5
September	2.9	3.0	3.0	3.2	3.2	2.8	2.9	3.0
October	2.2	2.2	2.1	2.0	2.2	2.2	2.4	2.5
November	2.9	3.0	3.0	3.1	2.9	3.2	3.2	3.1
December	3.1	3.1	3.2	3.4	3.3	3.5	3.4	3.5
January 1997	5.6	5.7	5.9	5.2	5.4	5.9	5.8	5.9
February	8.2	8.2	7.9	7.3	8.1	7.9	7.8	8.0
March	4.9	4.5	4.1	4.1	4.2	4.1	4.2	4.4

Table 8 : Free Carbondioxide (mg/l) at eight selected stations of Chandrampalli Reservoir

Months	Stations							
	I	*II*	*III*	*IV*	*V*	*VI*	*VII*	*VIII*
June 1996	2.2	2.2	*	1.1	2.2	2.2	1.1	*
July	4.4	3.3	1.1	*	4.4	4.4	4.8	0.8
August	9.9	9.9	8.8	9.9	*	9.9	*	8.8
September	8.8	*	*	6.6	*	6.6	7.7	6.6
October	5.5	5.5	*	6.6	*	5.5	*	*
November	5.0	5.5	3.3	6.6	*	*	3.3	*
December	5.2	6.6	6.6	2.2	*	*	*	1.1
January 1997	5.6	5.5	4.4	*	*	*	*	1.1
February	2.2	1.1	1.1	1.1	*	*	*	*
March	3.3	2.2	2.2	*	1.1	*	*	1.1

* Traces.

Table 9 : Total Alkalinity (mg/l) at eight selected stations of Chandrampalli Reservoir

Months	*Stations*							
	I	*II*	*III*	*IV*	*V*	*VI*	*VII*	*VIII*
June 1996	130	140	125	160	130	130	117	140
July	125	124	133	125	157	125	124	125
August	160	160	165	166	180	185	166	170
September	140	140	130	156	167	180	157	117
October	175	160	165	166	180	175	175	165
November	180	180	175	160	180	185	165	170
December	185	175	160	175	185	180	185	185
January 1997	150	155	156	157	165	190	150	155
February	175	185	160	175	175	185	180	185
March	185	180	175	160	180	187	165	175

55.38 mg/l at Station II; 44.02 to 44.86 mg/l at Station III; 46.86 to 49.70 mg/l at Station IV; 48.28 to 58.22 mg/l at Station V; 42.28 to 59.64 mg/l at Station VI; 53.96 to 59.64 mg/l at Station VII and 48.28 to 58.22 mg/l at Station VIII respectively.

While during the early Summer season of 1997, the Chloride values varied from 26.98 to 49.70 mg/l; 25.56 to 49.70 mg/l; 34.68 to 42.02 mg/l; 34.68 to 49.70 mg/l; 42.60 to 49.70 mg/l; 42.28 to 42.60 mg/l; 42.60 to 53.96 mg/l and 34.60 to 48.28 mg/l at Stations I, II, III, IV, V, VI, VII and VIII respectiveiy.

Hardness

The Total Hardness of Chandrampalli Reservoir during different season are presented in Table 11.

The Hardness values during the Southwest monsoon season varied from 62 to 124 mg/l at Station I; 70 to 124 mg/l at Station II; 69 to 130 mg/l at Station III; 68 to 134 mg/l at Station IV; 72 to 130 mg/l at Station V; 68 to 124 mg/l at Station VI; 72 to 134 mg/l at Station VII and 68 to 130 mg/l at Station VIII respectively.

During the Northeast monsoon season the values varied from 130 to 150 mg/l; 125 to 140 mg/l; 106 to 150 mg/l; 106 to 130 mg/l; 92 to 130 mg/l; 106 to 134 mg/l; 96 to 134 mg/l and 109 to 134 mg/l at Station I to VIII respectively.

While during the early Summer of 1997, the values varied between 124 to 150 mg/l; 124 to 140 mg/l; 134 to 140 mg/l; 106 to 130 mg/l; 106 to 135 mg/l; 130 to 134 mg/l; 106 to 135 mg/l and 96 to 124 mg/l at Stations I, II, III, IV, V, VI, VII and VIII respectively.

Phosphate-Phosphorus

The Orthophosphate concentration of Chandrampalli Reservoir during different seasons are presented in Table 12.

During the Southwest monsoon season the values of Orthophosphate ranged from 0.11 to 0.18 mg/l; 0.12 to 0.17 mg/l; 0.12 to 0.18 mg/l; 0.12 to 0.17 mg/l; 0.13 to 0.16 mg/l; 0.12 to 0.18 mg/l and 0.13 to 0.17 and 0.12 to 0.16 mg/l at Stations I to VIII respectively.

During the Northeast monsoon season the values of Orthophosphate varied from 0.06 to 0.13 mg/l; 0.08 to 0.14 mg/l;

0.08 to 0.14 mg/l; 0.09 to 0.15 mg/l; 0.09 to 0.13 mg/l; 0.10 to 0.14 mg/l and 0.10 to 0.16 mg/l; and 0.11 to 0.12 mg/l at Stations I to VIII respectively.

While during the early Summer months of 1997, the values varied form 0.07 to 0.08 mg/l; 0.08 to 0.09 mg/l; 0.07 to 0.08 mg/l; 0.09 to 0.10 mg/l; 0.11 to 0.12 mg/l; 0.11 to 0.13 mg/l; 0.11 to 0.12 mg/l and 0.13 to 0.14 mg/l at Stations I to VIII respectively.

Ammonical Nitrogen

The Ammonical Nitrogen concentration of Chandrampalli Reservoir for different season are shown in Table 13.

During the Southwest monsoon season the concentration of ammonical nitrogen is varied from 0.23 to 0.26 mg/l; 0.20 to 0.30 mg/l; 0.17 to 0.64 mg/l; 0.20 to 0.58 mg/l; 0.21 to 0.61 mg/l; 0.27 to 0.78 mg/l; 0.26 to 0.67 mg/l and 0.27 to 0.67 mg/l at Stations I to VIII respectively.

During the Northeast monsoon season the values ranged from 0.20 to 0.50 mg/l; 0.30 to 0.55 mg/l; 0.35 to 0.64 mg/l; 0.33 to 0.60 mg/l; 0.25 to 0.67 mg/l; 0.28 to 0.68 mg/l; 0.26 to 0.69 mg/l and 0.29 to 0.68 mg/l at Stations I, II, III, IV, V, VI, VII and VIII respectively.

While during early Summer of 1997, the values are fluctuated from 0.29 to 0.30 mg/l at Station I; 0.30 to 0.32 mg/l at Stations II and III; o.30 to 0.58 mg/l at Station IV; 0.58 to 0.59 mg/l at Station V; 0.61 to 0.67 mg/l at Station VI; 0.61 to 0.69 mg/l at Station VII and 0.69 to 0.70 mg/l at Station VIII respectively.

Gross Primary Productivity

The variations of Gross Primary Productivity of Chandrampalli Reservoir during different seasons is expressed in Table 14.

During the Southwest monsoon season the Gross Primary productivity values varied from 0.215 to 0.281 gC/m^3/hr at Station I; 0.215 to 0.300 gC/m^3/hr at Station II; 0.206 to 0.309 gC/m^3/hr at Station III; 0.253 to 0.337 gC/m^3/hr at Station IV; 0.159 to 0.337 gC/m^3/hr at Station V; 0.290 to 0. 365 gC/m^3/hr at Station VI; 0.271 to 0.375 gC/m^3/hr at Station VII and 0.253 to 0.403 gC/m^3/hr at Station VIII respectively.

Table 10 : Chloride (mg/l) at eight selected stations of Chandrampalli Reservoir

Months	*Stations*							
	I	*II*	*III*	*IV*	*V*	*VI*	*VII*	*VIII*
June 1996	34.68	42.60	42.60	34.68	42.60	42.60	42.60	34.60
July	56.80	62.48	63.90	63.90	58.22	59.64	65.32	63.90
August	55.38	45.44	42.60	42.60	58.22	62.48	59.64	62.48
September	45.44	46.86	46.86	49.70	48.28	48.28	53.96	59.64
October	45.44	46.86	46.86	42.60	58.22	48.28	59.64	59.64
November	46.86	45.44	46.86	49.70	48.28	48.28	53.96	58.22
December	42.28	39.76	44.02	46.86	48.28	42.28	53.96	48.28
January 1997	56.80	55.38	46.86	49.70	58.22	59.64	59.64	58.22
February	49.70	49.70	44.02	49.70	49.70	42.28	53.96	48.28
March	26.98	25.56	34.68	34.68	42.60	42.60	42.60	34.60

Table 11 : Hardness ($MgCaCO_3/l$) at eight selected stations of Chandrampalli Reservoir

Months	*Stations*							
	I	*II*	*III*	*IV*	*V*	*VI*	*VII*	*VIII*
June 1996	62	70	69	68	72	68	72	68
July	124	125	124	124	124	124	134	130
August	106	92	96	92	92	106	92	92
September	106	106	130	134	130	106	92	106
October	130	134	106	130	106	106	96	109
November	134	125	124	106	92	134	134	134
December	130	130	125	124	98	130	125	124
January 1997	150	140	150	130	130	130	125	109
February	150	140	140	130	135	130	135	124
March	124	124	134	106	106	134	106	96

Table 12 : Phosphate-Phosphorus (mg/l) at eight selected stations of Chandrampalli Reservoir

Months	*Stations*							
	I	*II*	*III*	*IV*	*V*	*VI*	*VII*	*VIII*
June 1996	0.11	0.12	0.12	0.12	0.14	0.14	0.13	0.12
July	0.16	0.14	0.16	0.16	0.14	0.16	0.16	0.14
August	0.15	0.13	0.12	0.14	0.13	0.12	0.14	0.12
September	0.18	0.17	0.18	0.17	0.16	0.18	0.17	0.16
October	0.13	0.13	0.14	0.15	0.13	0.14	0.16	0.17
November	0.07	0.08	0.09	0.10	0.12	0.13	0.10	0.11
December	0.06	0.08	0.08	0.09	0.09	0.10	0.10	0.11
January 1997	0.08	0.08	0.09	0.10	0.11	0.12	0.11	0.12
February	0.08	0.09	0.08	0.10	0.12	0.13	0.11	0.13
March	0.07	0.08	0.07	0.09	0.11	0.11	0.12	0.14

Table 13 : Ammonical Nitrogen (mg/l) at eight selected stations of Chandrampalli Reservoir

Months	*Stations*							
	I	*II*	*III*	*IV*	*V*	*VI*	*VII*	*VIII*
June 1996	0.26	0.28	0.26	0.28	0.26	0.27	0.26	0.27
July	0.23	0.29	0.17	0.20	0.21	0.64	0.67	0.67
August	0.26	0.20	0.93	0.58	0.61	0.78	0.62	0.61
September	0.25	0.30	0.64	0.58	0.61	0.78	0.61	0.61
October	0.20	0.30	0.35	0.33	0.25	0.28	0.26	0.29
November	0.40	0.42	0.43	0.60	0.67	0.68	0.69	0.68
December	0.50	0.55	0.58	0.59	0.56	0.60	0.69	0.67
January 1997	0.28	0.30	0.64	0.58	0.61	0.68	0.61	0.61
February	0.30	0.32	0.30	0.58	0.59	0.67	0.69	0.70
March	0.29	0.30	0.32	0.39	0.58	0.61	0.61	0.69

Table 14 : Gross Primary Productivity (gC/m^3/hr) at eight selected stations of Chandrampalli Reservoir

Months	*Stations*							
	I	*II*	*III*	*IV*	*V*	*VI*	*VII*	*VIII*
June 1996	0.225	0.225	0.206	0.253	0.159	0.365	0.375	0.403
July	0.215	0.215	0.234	0.271	0.281	0.300	0.309	0.309
August	0.253	0.300	0.309	0.290	0.290	0.318	0.196	0.253
September	0.281	0.300	0.300	0.337	0.337	0.290	0.271	0.300
October	0.253	0.243	0.262	0.234	0.253	0.225	0.225	0.225
November	0.281	0.318	0.300	0.309	0.337	0.318	0.328	0.290
December	0.309	0.309	0.337	0.318	0.337	0.318	0.318	0.318
January 97	0.412	0.412	0.431	0.487	0.487	0.421	0.412	0.384
February	0.656	0.731	0.665	0.721	0.815	0.648	0.665	0.675
March	0.187	0.215	0.262	0.262	0.243	0.243	0.243	0.234

Table 15 : Net Primary Productivity (gC/m^3/hr) at eight selected stations of Chandrampalli Reservoir

Months	*Stations*							
	I	*II*	*III*	*IV*	*V*	*VI*	*VII*	*VIII*
June 1996	0.459	0.468	0.412	0.431	0.337	0.571	0.562	0.618
July	0.281	0.337	0.337	0.375	0.384	0.393	0.421	0.421
August	0.375	0.403	0.412	0.393	0.337	0.431	0.412	0.318
September	0.459	0.468	0.468	0.506	0.506	0.440	0.450	0.468
October	0.365	0.346	0.365	0.328	0.365	0.346	0.356	0.356
November	0.446	0.506	0.496	0.487	0.515	0.515	0.525	0.478
December	0.496	0.496	0.534	0.534	0.553	0.506	0.496	0.506
January 97	0.740	0.740	0.787	0.778	0.787	0.759	0.712	0.712
February	1.031	1.019	1.050	1.019	1.181	1.012	1.096	1.301
March	0.450	0.450	0.459	0.459	0.431	0.421	0.440	0.450

During the Northeast monsoon season the Gross Primary Productivity values varied form 0.253 to 0.412 gC/m^3/hr; 0.243 to 0.318 gC/m^3/hr; 0.262 to 0. 431 gC/m^3/hr; 0.234 to 0.487 gC/m^3/hr; 0.253 to 0.487 gC/m^3/hr; 0.225 to 0.421 gC/m^3/hr; 0.225 to 0.412 gC/m^3/hr and 0.225 to 0.384 gC/m^3/hr at Stations I to VIII respectively.

While during early Summer months of 1997, the values of Gross Primary Productivity varied from 0.187 to 0.656 gC/m^3/hr at Station I; 0.215 to 0.731 gC/m^3/hr at Station II; 0.262 to 0.665 gC/m^3/hr at Station III; 0.262 to 0.721 gC/m^3/hr at Station IV; 0.243 to 0.815 gC/m^3/hr at Station V; 0.243 to 0.648 gC/m^3/hr at Station VI; 0.243 to 0.665 gC/m^3/hr at Station VII and 0.234 to 0.675 gC/m^3/hr at Station VIII respectively.

Net Primary Productivity

The net primary productivity values of Chandrampalli Reservoir during different seasons are presented in Table 15.

The net primary productivity values during the Southwest monsoon season varied from 0.281 to 0.459 gC/m^3/hr at Station I; 0.337 to 0.468 gC/m^3/hr at Station II and III; 0.375 to 0.506 gC/m^3/hr at Station IV; 0.337 to 0.506 gC/m^3/hr at Station V; 0.393 to 0.579 gC/m^3/hr at Station VI; 0.412 to 0.562 gC/m^3/hr at Station VII and 0.318 to 0.618 gC/m^3/hr at Station VIII respectively.

During the Northeast monsoon season the values of net primary productivity ranged from 0.365 to 0.740 gC/m^3/hr; 0.346 to 0.740 gC/m^3/hr; 0.365 to 0.787 gC/m^3/hr; 0.328 to 0.778 gC/m^3/hr; 0.365 to 0.787 gC/m^3/hr; 0.346 to 0.759 gC/m^3/hr; 0.356 to 0.712 gC/m^3/hr and 0.356 to 712 gC/m^3/hr at Stations I to VIII respectively.

While during Summer months of 1997, the values of net primary productivity ranged from 0.450 to 1.031 gC/m^3/hr; 0.450 to 1.019 gC/m^3/hr; 0.459 to 1.050 gC/m^3/hr; 0.459 to 1.019 gC/m^3/hr; 0.431 to 1.181 gC/m^3/hr; 0.421 to 1.012 gC/m^3/hr; 0.440 to 1.096 gC/m^3/hr and 0.450 to 1.301 gC/m^3/hr at Stations I to VIII respectively.

Community Respiration

The community respiration values of Chandrampalli Reservoir during the different seasons are presented in Table 16.

Table 16 : Community Respiration ($gC/m^3/hr$) at eight selected stations of Chandrampalli Reservoir

Months	Stations							
	I	*II*	*III*	*IV*	*V*	*VI*	*VII*	*VIII*
June 1996	0.234	0.243	0.196	0.178	0.178	0.206	0.187	0.215
July	0.131	0.121	0.103	0.103	0.103	0.093	0.112	0.112
August	0.121	0.103	0.103	0.103	0.046	0.112	0.121	0.065
September	0.178	0.187	0.168	0.168	0.168	0.150	0.178	0.168
October	0.112	0.103	0.103	0.093	0.112	0.121	0.131	0.131
November	0.159	0.187	0.196	0.178	0.178	0.196	0.196	0.187
December	0.187	0.187	0.206	0.215	0.121	0.187	0.178	0.187
January 97	0.309	0.328	0.356	0.290	0.300	0.337	0.300	0.328
February	0.375	0.365	0.384	0.328	0.365	0.365	0.421	0.356
March	0.262	0.234	0.196	0.196	0.187	0.178	0.196	0.215

The Community Respiration values during the Southwest monsoon season ranged from 0.121 to 0.234 gC/m^3/hr; 0.103 to 0.243 gC/m^3/hr; 0.103 to 0.196 gC/m^3/hr; 0.103 to 0.178 gC/m^3/hr; 0.046 to 0.178 gC/m^3/hr; 0.093 to 0.206 gC/m^3/hr; 0.112 to 0.187 gC/m^3/hr and 0.065 to 0.215 gC/m^3/hr at Stations I to VIII respectively.

During the Northeast monsoon season the values of Community Respiration ranged form 0.112 to 0.309 gC/m^3/hr; 0.103 to 238 gC/m^3/hr; 0.103 to 356 gC/m^3/hr; 0.093 to 0.290 gC/m^3/hr; 0.112 to 0.300 gC/m^3/hr; 0.121 to 0.337 gC/m^3/hr; 0.131 to 0.300 gC/m^3/hr; 0.131 to 0.328 gC/m^3/hr at Stations I to VIII respectively.

While during early Summer months of 1997, the values of Community Respiration varied from 0.262 to 0.375 gC/m^3/hr; 0.234 to 0.365 gC/m^3/hr; 0.196 to 0.384 gC/m^3/hr; 0.196 to 0.328 gC/m^3/hr; 0.187 to 0.365 gC/m^3/hr; 0.178 to 0.365 gC/m^3/hr; 0.196 to 0.421 gC/m^3/hr and 0.215 to 0.356 gC/m^3/hr at Stations I, II, III, IV, V, VI, VII and VIII respectively.

For the first time, the investigations on the Chandrampalli Reservoir near Chincholi Taluk, Gulbarga District have studied to reveal the limnological parameters.

Though some work has been carried out on the limnological study of the lakes, tanks, small ponds and reservoirs in India. It has been found always difficult to compare between two Waterbodies. The influence of abiotic factors on waterbody and its inter-relationship with plankton productivity. This is mainly because of the the differences in water masses, their size, shape, depth, climatic conditions and region to which they belong. Therefore the information about important physical, chemical and biological characteristics of the reservoirs and their seasonal changes were studied for a period of 10 months. And the work will be carried out for further to study the in detail of macro flora, fauna and pollution problems in this reservoir.

This reservoir is perennial waterbody (mean depth 94 feets) with about two inlets bringing surplus water from the Sarnala river and its surrounding areas. The water from this reservoir is mainly used for irrigation and drinking purposes.

Rainfall in this region is mainly Southwest monsoon. The annual average rainfall of this region is about 1000 mm. During the period of this study, the total rainfall was 1003.22 mm, during rainy season the reservoir was filled during August month and the gross capacity of reservoir is about 1,200 MCFT and the total catchment area is 170 sq. miles. The total irrigation command area is 12,900 acres.

The total rainfall recorded during Southwest monsoon was 719.22 mm and the highest recorded during August month (Table.1)

The heat transmitted with light is responsible for establishing the thermal stratification in Waterbodies, and it influences the respiration of both plants and animals. The atmospheric temperature fluctuated from season to season in the range of 25.0 to 31.7^0 C (Table. 2)

While the surface water temperature, the average maximum at all stations was 24.9^0 C during Southwest monsoon and 25.9^0 C during Northeast monsoon season, the average maximum and minimum was 22.0 to 25.9^0 C , meanwhile for Summer season the maximum and minimum temperature was 22.0 to 25.0^0 C.

Ayyappan and Gupta (1981), Vijaykumar (1992), while working on the limnology of Ramasamudra Tank and Jagath Tank has observed its close follow with that of air temperature. Many other workers have also observed the similar trend in different Waterbodies (Gonzalves and Joshi, 1946; Singh and Swarup, 1979; George *et al.*, 1986).

The present investigations though the similar trends observed from June 1996 to March 1997. Goldman and Horne (1983) are of the opinion that the structure of aquatic ecosystems is directly related to solar radiations responsible for light and heat penetration in the water. Since photosynthetic activity depends on solar radiation, however, the transparency of the water is directly dependent on the turbidity of the water and transparency is being influenced not only by suspended particles, but by the planktonic organisms (Ayyappan and Gupta, 1981). Therefore the information pertaining to the transparency of the water in any aquatic ecosystem gains considerable significance.

The relative humidity was varied from 35 to 55 per cent all over the period (Table 4). The relative humidity was going hand in hand with the atmospheric and the water temperature which will have greater influence on evaporation of water. Relative humidity may not influence directly to the Waterbodies or aquatic organisms but may have some influence on the aquatic insects which are inhabiting the aquatic Waterbodies.

It is well known fact that changes in the pH concentration of water will bring about subtle changes in the structural and functional variations in the organisms of the waterbody. Measurements of hydrogen-ion-concentration is an important means of understanding the nature of chemical conditions which prevail in an aquatic ecosystem (Welch, 1992).

In the present study, the pH concentration has been recorded from June, 1996 to March, 1997. During Southwest monsoon the pH of water ranged from 8.0 to 9.1 and during Northeast monsoon the pH values fluctuated from 7.8 to 8.9, while during the Summer season the pH concentration ranged from 8.0 to 8.4 (Table 6).

It is observed that the pH values of Chandrampalli reservoir followed a specific seasonal trend from June to March. The pH at all stations remained more or less constant, while in the month of July and November a distinct peaks were observed.

In this study, it is interesting to note that the pH values were more or less towards alkaline throughout the period of the study and almost similar range of variation in pH has been recorded by Das and Srivastav, (1956), Vijaykumar (1992), Pappa (1995) found a direct co-relation between water temperature and pH; such a relation could also will drawn in the present work.

It is discernible from the above observations that the pH values in semi-arid tropical standing Waterbodies are dependent on local conditions.

Dissolved oxygen is one of the most important parameters of the water quality, directly effecting survival and distribution of flora and fauna in an ecosystem. The two main sources of dissolved oxygen are diffusion from air and photosynthesis, while major factors responsible for its depletion are biochemical oxidation and respiration by flora and fauna. The quantity of dissolved oxygen in water is directly or indirectly dependent on water temperature, partial pressure of oxygen in the air, concentration of dissolved salts, amount of chlorophyll content etc., and temperature and hydrogen-ion-concentration (Welch, 1952 and Wetzel, 1975).

In the present investigations, the dissolved oxygen in the Chandrampalli Reservoir on an average fluctuated between 2.1 to 4.5 mg/l; 2.0 to 5.9 mg/l and 4.1 to 8.2 mg/l during Southwest, Northwest monsoon seasons and Summer seasons respectively (Table 7).

The dissolved oxygen concentration was recorded high during June and attained its peak in January, February 1997. This increase in gross primary production which released more oxygen after photosynthesis and this enhancement is due to increase in transparency during this particular period.

Rao (1975), Bharati and Kori (1975), Sreenivasan (1976), Mishra and Yadav (1978), Singhal *et al.* (1985), Saxena and Mishra (1991) while working on different Waterbodies from various parts of India, reported low values of dissolved oxygen during the high temperature period. The higher values of oxygen observed during summer in the present investigations substantiate the views of Ayyappan and Gupta (1981), Bhat and Negi (1985), Shukla and Bais (1990), and Vijaykumar (1992) who also reported the higher values during the summer and they attributed this to increased solar radiation and thus a considerable good standing crop of phytoplankton.

The amount of free carbondioxide depends on the decomposition of top soil and chemical nature of the underlying rocks. By far, it is relatively abundant in natural waters and its importance in photosynthetic activity requires no further explanation, since daily seasonal carbon flow in the system forms the very base of the photopyramid (Goldman and Harne, 1983).

In the present investigations the free carbondioxide values were considerably more in the month of August and September than during the latter. (Table 8).

It is rather interesting to note that the increasing trend of free carbondioxide values are compared with the decreased trend in the dissolved oxygen from all stations during the study period. Further it is of interest to note that the observations reveals, period of low values of free carbondioxide which attributed slight increase in pH values as also noted by Jhingran (1975), Singhal *et.al* (1985 and 1986) reported that free carbondioxide and water temperature varied independently with low values of free carbondioxide when aquatic vegetation was more abundant and high values of free carbondioxide when water inflow to the reservoir was greatest. In the present study the low values of the free carbondioxide was observed during all seasons in different stations.

The role of alkalinity in the determination of the productive capacity of aquatic environment has been described by Phillipose (1959) according to whom inland waters are categorized into three types, further according to the same author, water with low alkalinity were comparatively low production.

In the present study, it could be observed that the total alkalinity values in selected stations of Chandrampalli reservoir more or less indicated a similar pattern of fluctuations though with minor differences. The values tend to increase during Southwest monsoon season of 1996 with distinct peak during August. There after almost all same trend towards Northeast monsoon and summer season was observed.

The present study reveals that the comparative higher values of total alkalinity in the Chandrampalli reservoir which reaches upto 190 mg/*l* of $CaCo_3$ during Northeast monsoon season of 1996 (Table 9).

Singhal *et al.* (1986), while describing the physico-chemical environment of managed ponds in Haryana reports no clear seasonal trends of alkalinity though with a maximum of 146.00 ppm recorded during June and further they opined that the decline during rainy months probably be due to dilution.

Natural water normally contains low chlorides than bicarbonates and sulphates. Large contents of chlorides in freshwater is an indication of organic pollution (Thrash *et al.* (1994). Though chloride level as high as 250 mg/l is safe for human consumption, a level above this imparts salty taste to potable water.

During the present investigations, the chloride concentration exhibited an increasing trend during Southwest monsoon and a sharp decline in chloride content could be noticed during March (Table 10 and Fig. 11) Gonzalves and Joshi (1946), Singh (1960), Mishra and Yadav (1979), Ali and Khan (1979), Vijaykumar (1992) Pappa (1995), Kumar (1995) and Raju and Durani (1996) have also noticed that the chloride concentration increases in Waterbodies.

Hardness to water is imparted by alkaline earth metal cations mainly calcium and magnesium present in it. Ecologically temporary hardness plays a key role in buffering capacity thus, neutralizing an offset in pH, due to addition of acidic products. This has great effect on biotic diversity of an ecosystem. Hardness of water also restricts water users. In the present study, the hardness ranges from 70 to 150 mg $CaCO_3$/l in the month of June, 1996 and

January and February 1997 respectively (Table 11). High values of hardness In January and February, 1997 is due to concentration of carbonates and bicarbonates through runoff from rain water.

Waterbodies having hardness 15 ppm or above are satisfactory for the growth of fishes while hardness less than 5 ppm gives slow growth, distress and leads to death (Swingle, 1967). Sreenivasan (1964) also co-related the hardness of water with organic productivity from the present study on hardness, it is clear that the water is more suitable for aquaculture practices.

The phosphates (Po_4) in natural waters occur in very small quantities. It an important nutrient in the maintenance of the fertility of pond. In an aquatic ecosystem phosphorus occurs both in inorganic and organic form of the two, the inorganic phosphorus plays and important role as a nutrient along with nitrates. In Chandrampalli reservoir phosphate (phosphorus) content recorded was minimum (0.06 mg/l) in the month of December, 1996 and maximum (0.18 mg/l) in the month of September (Table 12). Though the phosphate content in waters at eight stations were low during the study period. The phosphate more than 2 mg/l in open water gives a sign of organic pollution (Pomeroy, *et al.* 1965). As a nutrient, excess of phosphate stimulates development of algal blooms, such blooms of blue green algae like Microcystis are toxic in nature. But in the present investigation, because of low phosphate content no algal bloom were observed.

Nitrogen occurring in the freshwater is in the form of an element of nitrogenous compounds. Nitrogen is mostly derived from atmosphere and denitryfying bacteria giving nitrates, nitrites and ammonia. Nitrogenous compounds may be derived from outside sources like rain, surface runoff and seepage. Nitrogen is generally known to show no seasonal or depth variations except these resulting from temperature changes, as observed by Pearsall (1921 and 1930), Gonzalves and Joshi (1946), Singh (1960), Rajendra (1973), Rao (1975), Sreenivasan (1976), Seenayya and Zafar (1979), Valecha and Bhatnagar (1989), Vijaykumar (1992 and 1996).

In the present investigations during Southwest monsoon season, the ammonical nitrogen varied from 0.17 to 0.93 mg/l at station III. The highest concentration was recorded 0.93 mg/l at station III. The values of ammonical nitrogen content during Northeast monsoon season of 1996 observed at stations I to VIII ranged between 0.20 to 0.69 mg/l. The lower values were recorded during summer months of 1997 during the study period (Table 13). The increase may be attributed to increasing temperature and the decline might be due to the stripping action of methane bubbles released by anoxic decay in the sediments and also may be due to direct utilization of ammonical nitrogen. By the phytoplankton as pointed out by Vijaykumar (1992) and Pappa (1995) while describing the nitrogen content of Jagath Tank, Bhogsha reservoir and Pala Tank during 1985 - 1988 and Gobbur Tank during 1994 to 1995, observed the seasonal variations of ammonical nitrogen with higher values during summer. In the present study increasing concentration of ammonical nitrogen may substitute the present view that influx of nitrogenous nutrients from agricultural lands might be influencing the nitrogen budget of this waterbody. However, the concentration recorded by Kanbur (1990), at Moat Gulbarga varied between 36.7 to 45.5 mg/l seems to be quite higher than the present study. However, it could be observed that higher gross primary production during the study period may be due to increase in the levels of nutrients including that of ammonical nitrogen.

It is thus describable that the nutrient contents in the protected Waterbodies are more governed by local influence, always with change in the seasonal influence rather than flowing set pattern.

Gross primary production of Chandrampalli reservoir followed more or less similar trends of fluctuations at all the stations and also exhibited an increasing trend through the Southwest monsoon to Northeast monsoon with a distinct peaks. The higher values being remained with steep increase in the month of February, 1997, which was highest recorded during the present study (0.815 $gC/m^3/hr$) Table 14.

Net primary production estimations of the Chandrampalli reservoir followed rather similar trends of oscillation at all the stations, recording lower values during Southwest monsoon of 1996 and exhibited an increasing trend towards Northeast monsoon season and summer (Table 15).

Community respiration observation reveals that an increase trend towards Northeast monsoon season of January 1997 and summer months during period of the study. [Though there were low values were observed in all stations, but higher values were observed during summer of 1997].

From India, there are quite good number of workers on primary production of the standing Waterbodies. Sreenivasan (1963) reported on the primary production in the three upland lakes from Nilagiries in Tamil Nadu with one of the lake in Ooty recording high primary production values. Even though these observations pertaining to water comparatively large dimensions, such co-relation could also be observed in the present study. The fact is well substantiated by the extensive contributions of the same author, Sreenivasan (1964a, 1964b, 1972 and 1976), from ponds and tanks distributed in and around Tamil Nadu where he reported that varying ranges could be attributed to different factors *i.e.* low oxygen, free carbondioxide and processes such as self shading. Khan and Siddiqui (1971) observed high gross primary productivity between the period from March to May and lower values for the rest of the periods from tropical fish ponds at Aligarh and further opined that there is direct relationship between primary production, transparency, temperature and phosphate-phosphorus among nutrients. Ayyappan and Gupta (1985) recorded gross primary production from Ramasamudra tank in Coastal Karnataka which ranged upto 264.27 $gC/m^3/hr$ and they are of the view that the higher rates of primary production during March to July coincided with very low pH, high free carbondioxide, low total alkalinity, high specific conductivity.

In an aquatic ecosystem, zooplanktons play an important role not only in converting plant food into animal food but also provides an important food sources for other higher organisms including fish. Keeping this idea, an attempt has been made to study the availability of the zooplankton species. Among zooplankton *Keratella tropica, Asplanchna pilinia, Brachious falcatus* and *B. caudatus* among rotifers and cyclops *Bicuspidatus thomasi, Neodiaptomus,* cyclopoid sp. among copepodes and among cladocerns, *Daphnia magna, D. pulex, Moina sp, Bosnia sp.* and *Ceridophinia sp.* have exhibited high population during the study period.

Among macro-organisms, *Macrobrachium lammeri* and *Barytelphusa sp.* among Crustaceans. In Insects (aquatic), 14 species were recorded *i.e.* Naids of May flies, naids of Dragon flies, *Hydrometra vittata, Lemnometra fluviorum, Microvetia sp., Laectotrephes maculatus, Diplonychus rusticum, Notonecta sp., Rantra elongata, Belostuma indica, Carixa sp, Cybister sp, Hydoticus sp* and *Hydrophilus sp* were recorded in high number during the study period.

However, clear study of the zooplanktons, crustaceans, Insecta, Molluscans, Fishes, Amphibians and birds in the Chandrampalli reservoir will be under taken in future course for Ph.D. work to study their role in aquatic ecosystem of Chandrampalli reservoir.

References

Adholia, U.N., Chakrabarty, A., Srivastava, V. and Vyas, A : 1990. Community studies on Macrobenthos with reference to limno-chemistry of Manasarovar, Bhopal. *J. Natcon.*3(2): 139 - 154.

Adholia, U.N. 1991: Phytoplankton community in relation to limno-chemistry of Manasarovar Reservoir, Bhopal. *J.Nat.Con.* 3(2): 155 - 186.

Ahmad, S.H. and Singh, A.K. 1993: Correlation between physico-chemical factors and Zooplankton during diurnal variations in a freshwater tank at Dholi, Bihar, India. *J.Environ.Biol.* 14(2). 95 - 105.

Ahmad, S.H. and Singh, A.K. 1993: Diurnal variations in physico-chemical factors and zooplanktons in the surface Niche of a perennial tank of Patna (Bihar), India. *J.Ecobiol.* 5(2) 111 - 120.

Alam, A and Khan. A.A. 1996: Dynamics of plankton communities in four freshwater lentic ecosystems in relation to varying dominant biota. *Poll.Res.* 15(3): 289 - 291.

Ali, M. and Khan, A.A. 1976 : Diurnal variations in three ponds of Aligarh. Indian.*J. of Zoology.* 4 : 341 - 355.

Ali, M. and Khan, A.A. 1979 : Limnological studies on a Sewage fed pond of Aligarh. *Sci and Env.* 1 (2) : 185 - 191.

APHA, AWWA and WPCF. 1985 : Standard methods for the examination of water and waste water. 16th edition. pp. 1268.

Arora, G.L. : 1931: Fauna of Lahore, II. Entomostraca Water fleas of Lahore. Bull. Dept of Zoology. Punjab Univ. I (10) : 62 - 100.

Ayyappan, S. and Gupta, T.R.C. 1980 : Limnology of Ramasamudra Tank : *J. Inland Fish Soc. India.* 12(2) : 1 - 12.

Ayyappan, S. and Gupta, T.R.C. 1980b : Limnology of Ramasamudra Tank : bottom sediments. *Mysore. J. Agri. Sci.*14 : 408 - 420.

Ayyappan, S. and Gupta, T.R.C. 1981: Limnology of Ramasamudra Tank : Hydrography. *Mysore. J. Agri. Sci.* 15: 305 - 312.

Ayyappan, S. and Gupta, T.R.C. 1985: Limnology of Ramasamudra Tank : Primary production *Bull. Bot. Soc.* Sagar. 32 : 82 - 88.

Belsare, D.K., Prasad, D.Y., Belsare, S.D. and Gupta, S. N. 1992 : limnological studies on Bhopal lake : 5 Evaluation of Bacterial load by enzyme activity in Bhopal lakes. *Proc. of National Academy of Sci.*, India. Vol. LXII.

Birsal, N.R. 1996 : The primary productivity of the Supa reservoir during its filling phase. *J.Environ. Biol.* 17 (2) : 127 - 132.

Bharati, S.G. and Kori, A.P. 1975 : Limnological studies in ponds and lakes of Dharwad : Diurnal variation in Saydapur Keri. Part.I. *Karnataka Univ. J. Sci.*20 : 157 - 167.

Bhat, S.D. and Negi, U. 1985 : Physico-chemical features and phytoplankton in a subtropical pond. *Comp. Physiol and Ecol* 10(2) 85 - 88.

Chacko, P.M. and Krishnamurthy, P.I. 1954 : On the plankton of three freshwater fish ponds in Madras City, India. *Indo Pacific. Fish. Coun* (UNESCO) : 103 - 107.

Chakrabarty, D. and Saha, G.K. 1963 : The physico-chemical and hydrobiological features of three stagnant waters sources in Darjeeling Hills. *Geobios.* 20 (2): 121-123.

Choudhary, S.K., Nayak,M.R. and Verma, P.K. 1996 : Impact of eutrophication on the biochemical properties of Eichhornia crasspes Mart. Growing in two perennial ponds at Bhagalpur. *Envi and Ecol.* 14 (1) : 48-54.

Das, S.M. and Srivastav, K. 1956 : Quantitative studies on the freshwater plankton of the fish tank in Lucknow, India. *Proc. Nat. Acad. Sci.* 26 : 82-89.

Datta, N.C., Mandal, N. and Bandopadhy, B.K. 1987 : Seasonal abundance of rotifers in a perennial freshwater pond in Calcutta. *J.Environ.Biol* 8(1) : 63 - 71.

David, A., Rao, N.G.S. and Ray, P. 1974 : Tank fishery resources of Karnataka. *Bull. Cen. Inland. Fish. Res. Inst.* Barrackpore. 20 : 1 - 87.

Delf, E.M. 1915 : The algal vegetation of some ponds on hampstead Health. *New Phytol.* 14 : 63 - 80.

Devi, M.J. 1994 : Seasonal variation and population density of rotifers in three lakes of Hyderabad. *J.Aqua.Biol.* 9(1 and 2) : 41 - 44.

Dhamija, S.K. and Jain Yatish. 1994 : Variations in the physico-chemical characteristics of lentic waterbody of Jabalpur (M.P.) *J.Envi and Poll.* 1(3 and 4) : 125 - 128.

Dhamija, S.K. and Jain Yatish. 1995 : Studies on the water quality index of a lentic waterbody at Jablapur (M.P.) *Poll.Res.* 14(3) : 341 - 346.

Dutta, A. and Kalita, J., Nuvaid, M., Sheik.M.S. and Sharma, D.D. 1993 : Diurnal variations of certain physico-chemical parameters of two standing Waterbodies of in the vicinity. *Envi and Ecol.* 11(4) : 820 - 824.

Forel, F.A. 1892 : Le. Leman : Monographic limnologique.F. Rogue, Lausanne.

Fritsch, F.E. 1907 : The subarial and freshwater algae flora of the tropics. *Ann.Bot.* 21 : 235 - 275.

Gaddad, M.S. and Nimbargi, P.M. 1983 : Ecological studies of two polluted Waterbodies of Gulbarga. *Poll.Res.* 2(2) : 49 - 52.

Gaarder, T. and Gran, H.H. 1927 : Investigation of the production of plankton in the Oscofiord. Rapp. P.V. *Revn Commnit. Explor. Scient. Mer. Mediterr.* 42. 1 - 48.

Ganapati, S.V. 1940: An ecological study of garden Pond containing abundant zooplankton. *Proc. Indian. Acad. Sci* (Bombay) 17(2) : 41-58.

George, J.P., Venugopal, G. and Venkateshvaran. 1986 : Anthropogenic eutrophication in a perennial tank : effect on the growth of *Cyprinus Carpio communis. Indian. J. Environ. Health.* 28(4) : 303-313.

George, M.G. 1961 : Observations on the rotifers from shallow ponds in Delhi. *Curr.Sci.* 30(7) : 268 - 269.

Goldman, C.R. and Horne, A.T. 1983 : Limnology. McGraw Hill company pp 464.

Gonzalves, E.A. and Joshi, D.R. 1946 : Freshwater Algae near Bombay. *J. Bom.Nat.Hist.Soc.* 46 : 154 - 176.

Gopal, B., Sharma, K.P. and Trivedi, R.K. 1978 : Studies on ecology and production in Indian freshwater ecosystem at primary producer level with emphasis on macrophytes in : J.S. Singh and Gopal, B. (Eds) Glimpses of ecology. Int. Sci. Pub., Jaipur : 349 - 376.

Goudar, B.Y.M. and Joseph, K.J. 1961 : On the correlation between the natural populations of freshwater zooplankton, cladocera, copepoda and rotifera and some ecological factors. *J. Karnatak. Univ. Sci.* 6 : 89 -96.

Govind, B.V. 1963 : Preliminary studies on plankton of the Tungabhadra Reservoir. *Indian.J.Fish.* 10(1) : 148 - 158.

Gupta, M.C. and Sharma, L.L. 1993 : Dial variations in selected water quality parameters and zooplankton in shallow pond of Udaipur, Rajasthan. *J.Ecobiol.* 5(2) : 139 - 142.

Hensen, V. 1887 : Vber Die Bestimmung des plnktos oder des in meere treibenden materials on pflanzen. Und. *Thieren. Ber. Kammu. Wiss. Unters. Dt. Meera.* 5 : 1 - 109.

Hegde, G.R. and Bharati, S.G. 1985 : Comparative phytoplankton ecology of freshwater pond and lakes of Dharwad, Karnataka State, India. *Bull. Bot. Soc. Sagar.* 32 : 24 - 29.

Iyenger, M.O.P. 1939 : Algal problem peculia to the tropics with special reference to India. Proc. 25th *Indian.Sci.Con.* 25(4) : 141 - 149.

Johnson, M.E.C. 1991 : Nitrogen content of two freshwater lakes. *J.Aqua.Biol.* 6(1 and 2) : 19 - 23.

Johnson, M.E.C. 1991 : Phosphate content of two freshwater lakes. *J.Aqua.Biol.* 6(1 and 2) : 24 - 26.

Jhingran, V.G. 1975 : Fish and Fisheries of India. Hindustan Publ. Crop. (India). pp 954.

Kanbur, V.V. 1990 : Ecological influence on nitrogen metabolism in sewage near Jagath Tank at Gulbarga. *Trends in Ecotoxicology :* 117 - 122.

Kemmerere, A.J. and Narhold, M.N. 1969 : A method for gross primary productivity measurement. *Limnol. Oceanogr.* 14 (4) : 607-610.

Khan, A.I. and Khan, A.A. 1985 : Physico-chemical conditions in Seikha Jheel at Aligarh. *Envi* and *Eco.* 3 : 269-274.

Khan, A.A and Siddiqui, A.R. 1971 : Primary production in tropical fish pond at Aligarh, India, *Hydrobiologia.* 37 : 447-456.

Khatavakar, S.D., Kulkarni, A.Y. and Goel, P.K. 1989: Observations on the diurnal cycles of phytoplankton and some nutrients during summer in the surface waters of shallow mesotrophic lake. *Geobios.* 16 : 210-214.

Kumar, A. 1995 : Some limnological aspects of the freshwater tropical wetland of Santhal Pargana, (Bihar), India. *J.Envi* and *Poli.* 2 (3) : 137-141.

Laal, A.K. 1984 : Ecology of planktonic rotifers in tropical freshwater pond in a Patna, Bihar. Indian. *J. Anim. Sci.* 54 (3) : 291-294.

Lin, C.K. 1973 : phytoplankton succession in a eutrophic lake with special reference to blue green algal bloom. *Hydrobiologia.* 39(3) : 321 - 334.

Mansoori, H.A., Tiwari, J.P. and Gurudeo, A.S. 1993 : Limnological study of Lakshmital lake, Jhanshi, India : during rainy season. *Geobios.* 20 : 120 - 122.

Masood, H.B., Yousuf, A.R. and Qadri, M.Y. 1987 : Hydrobiology of Anchor Kashmir. *Com.Physical.Ecol* 12 : 131 - 139.

Michael, R.G. 1968 :Studies on the zooplankton of a tropical fish pond, India. *Hydrobiologia.* 32 (1 - 2) : 47 - 68.

Miller, W.E., Maloney, T.E. and Greene, J.E. 1973 : Algal productivity in 19 lake waters as determined by algal assays. *Wat.Res.* 8 : 667 - 679.

Mishra, G.P. and Yadav, A.K. 1978 : A comparative study of physico-chemical characteristics of rivers and lake water in Central India. *Hydrobiologia.* 59(3) : 275 - 278.

More, P.G. and Gajjar, H.J. 1990 : Hydrobioligical study of Palan pond in Valsad, Gujarat. *J.Ecobiol.* 2(2) 128 - 135.

Morgan, A. and Margaret, K. 1970 : Seasonal changes in the waters of lake Chilwa (Malvi) in a drying phase 1968-69, *ibid* 31(1) : 105 - 128.

Mori, S. 1971 : Fourth report of regular limnological survey of lake Biwa (1968 - 1970).III : Phytoplankton. Memo. of the Sci.Kyoto. Univ. *Series.Biol.* 5(1) : 35 - 36.

Mukesh, S.S., Patil, R.P. and Kulkarni, S.D. 1995 : Plankton diversity of Sadatput reservoir. *J.Aqua.Biol.* 10(1) : 24 - 25.

Mukherjee, B. Tetarway, S., Pati, M. Prasad, T., Mukherjee, M. and Dhar, S. 1990 : Carbondioxide dynamics and metabolic rythm in polluted aquatic body at Ranchi. *J.Eco. Biol.* 2(2) : 145 - 154.

Mukherjee, B., Pankajakshi,G.V.N., Bose, P., Kumar. R., Kumar P., Kumar A., Kumari R., Singh S. and Kanogia, S. 1994 : The phosphate load from detergent inputs and its effect in freshwater. *J.Envi.Biol.* 15(1) : 27 - 39.

Munawar, M. and Zafar, A.R. 1967 : On the ecology of algae in certain fish ponds of Hyderabad. India. III, the periodicity. *Hyderobiologia.* 30(1) : 96 - 112.

Munawar, M. 1970 : Limnological studies on fresh water ponds of Hyderabad, India. II. The Biocoenose, distribution of unicellular and colonical phytoplankton in the polluted and unpolluted environments. Hydrobiologia, *36* : 105-128.

Munawar, M. 1974 : Limnological studies of freshwater ponds of Hyderabad, India. IV. The Biocoenose periodicity of species, compositions of unicellular and colonical phytoplankton in polluted and unpolluted environments.. *Hydrobiologia, 45 (1)* : 1- 32.

Nasar, S.A.K. and Nasar, S.K.T. 1976 : An unusual record of primary production of freshwater ponds. *Proc.Indian. Sci. Congr.* 62 : 144.

Pappa, A. 1995 : Preliminary studies on limnology of Gobbur Tank. M.Phil Dissertation, Gulbarga University, Gulbarga. 60pp.

Panda, S.K. and Dash, M.C. 1993 : Water quality of some community ponds of Bargarh Town *Poll.Res.* 12(1) : 35 - 47.

Patil, H.S. Doddakundi, G.B. and Rodgi, G.S. 1975 : Succession in zoo and phytoplankton in a sewage stabilization pond. *Hydrobiologia.* 47(2) 253 - 264.

Pati, S. and Sahu, B.K. 1993 : Biomonitering of water pollution in Rengali reservoir, Orissa. *Envi* and *Ecol.* 11 (4) : 982-984.

Patil, C.S. and Gouder, B.Y.M. 1982 : Freshwater fauna of Dharwad, Karnataka State, India : Cladocera. *J. Karnatak Univ. Sci.* 27 : 115-126.

Patnaik, S. and Padhi, S. 1994 : Algal bioassay studies as environmental indicators of local ponds of Behrampur (Orissa). *J. Envi and Poll.* 1 (1) : 7-10.

Phillipose, M.T. 1959: Freshwater phytoplankton of Inland fisheries. *Proc. Symp. Algology.* ICAR : 272-291.

Pearsall, W.H. 1921: A suggestion on to the factors influencing the distribution of free floating vegetation. *J. Ecol.* 9 : 241-253.

Pearsall, W.H. 1930: Phytoplankton in the English lakes -1: The production in the water of Dissolved substances of biological importance. *J. Ecol.* 18: 306-320.

Pomeroy, L.R., Smith, E.E. and Grant, C.M. 1965: The exchange of phophate between estuarine water and sediments. *Limnol. Oceanoor.* 10 (2) : 167-172.

Rajkumar, R., Ramanibai, R. and Niranjali, D. 1994: Observation of species (Plankton) environmental relationships in urban aquatic ecosystems. *J. Envi. Biol.* 15 (3) : 177-183.

Rajendra, M. 1973: Copepoda. R.G. Michael (Ed). A Guide to the study of freshwater organisms. *J. Madurai Uni.* Suppl. 1 : 103-151.

Raju, K.R. and Durani, P.K. 1996 : Nano-planktonic production in temple ponds of Bhubaneswar. *J. Envi. Biol.* 17 (4) 317-320.

Rana, B.C. and Nirmal Kumar, J.I. 1993 : A composite rating of tropic status of certain ponds of Gujarat, India. *J. Envi. Biol.* 14 (2) : 113-120.

Rao, C.B. 1955 : On the distribution of algae in a group of six small ponds. II algal periodicity. *J. Ecol.* 43: 291-308.

Rao, V.S. 1975 : An ecological study of three freshwater ponds of Hyderabad, India. III. The phytoplankton. *Hydrobiologia.* 47 : 319 - 337.

Rao, K.R. and Mohan, C.P. 1976 : On the occurence of rotifers *Asplanchnella seiboldii* (levdig) in Indian waters. *Curr. Sci.*45 (6) : 234 - 235.

Rao, K.R. and Mohan, C.P. 1977 : Rotifers as indicators of pollution. *Curr.Sci.* 46(6) :190.

Rao, K.R. and Mohan, C.P. 1980 : Morphometric studies of *Brachionus ferficula* Wierzeski. *Geobios.* 7 : 89 - 90.

Rao, K.R. and Mohan, C.P. 1982 : : Studies on freshwater rotifers from Vishakhapatnam, India. *Rec. Zool. Surr. India* 79 : 397 - 403.

Rao K.D.S. 1985 : Hydrobiological studies of an Irrigation tank in Bangalore. *Mysore . J. Agri. Sci.*19 : 194 - 197.

Rao, V.N. and Mahmood, S.K. 1995 : Nutrient status and biological characteristics of Hubsiguda Pond. *J.Envi. and Poll.* 2(1) : 31 - 34.

Round, F.E. 1953 : An investigation of two benthic algal communities in Mahan Tran Yorkshiva, *J.Ecol.* 41 : 174 - 197.

Roy, D.N. and Sharma, U.P. 1989 : Zooplankton periodicity in weed wet lands of north Bihar. *Freshwater. Biol.* 1(2) : 161 - 166.

Russo, A.R. 1978 : Some ecological observation on a permanent pond in southern England, primary production of planktonic seasonal succession. *Hydrobiologia.* I (60) : 33 to 48.

Saxena, D.N. and Mishra, B.R. 1991 : The water quality index and self purification capacity of sewage collecting Channel, Morar (Kalpi) River at Gwalior, Madhya Pradesh, *J. Fresh. Water. Biol.* 3 (2) : 169-175.

Seenayya, G. and Zafar, A.R. 1979 : An ecological study of the Mir Alam lake, Hyderabad, India. *Ind. J. Bot.* 2 (2) 214-220.

Sewell, R.B.S. 1934 : Fauna of chilka lake. Copepoda. *Mem. Ind. Mus.* 5 : 771-857.

Shaha, L.L and Pandit, B. 1988: Density of zooplankton between lentic and lotic ecosystem. *Geobios News Report.* 7 : 31-35.

Sharma, B.K. 1976 : Rotifers collected from north west India, *News. Zool. Sur. India* 2 (6) : 255-258.

Sharma, B.K. 1978a : Contribution to the rotifera fauna of West Bengal-I family lecanidae. *Hydrobiologia.* 58 (1) : 81-86.

Sharma, B.K. 1978b : Contributions to the rotifer fauna of West Bengal - II, Genus Lepadella Bory de. St. Vincent. *Hydrobiologia.* 58 (1) : 86-88.

Sharma, B.K. 1979a : Rotifers from West Bengal - III : Further studies on the Eurotatoria. *Hydrobiologia.* 64 (3) : 239-250.

Sharma, B.K. 1979b : Rotifers from West Bengal - IV. Further contributions to the eurotatoria. *Hydrobiologia.* 65 (1) : 39-47.

Shukla, S.N and Bais, V.S. 1990 : Changes in Physics - chemical profile of the Bila reservoir: during winter season. *Trends in Ecotoxicology.* 291 - 293.

Siddiqui, E.N. Ahmad. M.S. and Bazmi, S.H. 1987 : Phytoplanktonic production of Bulki pond at Darbhanga, Bihar. *Envi.and Ecol.* 5 (2) : 291 - 293.

Singhal. R.N., Jeet, S and Devies, R.W. 1985 : The relationship among Physical, chemical and plankton characteristics of unregulated rural ponds in Haryana, India. *Trop.Ecol.* 26 : 43-53.

Singh, V.P. 1960 : Phytoplankton ecology of the Inland waters of Uttar Pradesh. *Proc.Symp. Algol.* ICAR., New Delhi. 243 - 271.

Singh, S.R. and Swarup. 1979 : Limnological studies of Suraha lake (Ballia) II. The periodicity of phytoplankton. *J.Indian.Bot.Soc.* 58 : 319 - 329.

Singhal. R.N., Jeet, S and Devies, R.W. 1986 : The physical chemical environment and plankton of managed rural ponds in Haryna, India. *Proc.Indian.Acad.Sci.* (Anim. Sci.) 95 (3) : 353 - 363.

Sinha, K.K. and Sinha, D.K. 1993 : Seasonal trends in physico-chemical factors and zooplankton in a freshwater pond of Munger, Bihar. *J.Ecobiol.* 5(4) : 299 - 302.

Sreenivasan, A. 1963 : Primary production in three upland lakes of Madras State, India. *Curr. Sci.*32 : 130 - 131.

Sreenivasan, A. 1964a : Limnological features and primary production in a polluted moat at Vellore, Madras State. *Envi.Health.* 6 : 237 - 245.

Sreenivasan, A. 1964b : The limnology of primary production and fish production in tropical pond. *Limnol.Oceanogr.* 9 (3) : 391 - 396.

Sreenivasan, A. 1972 :Limnology of seepage type impoundment Odathurai tank. *J.InlandFish.Soc.India.*4 : 117 - 132.

Sreenivasan, A. 1976 : Limnological studies on primary production in three temple pond ecosystem, *Hydrobiologia.* 48 : 117 - 123.

Sounderguard, M and Sand-Jensen, K. 1979 : Physico-chemical environment, phytoplankton biomas and production in oligotrophic soft water lake Kalgoard, Denmark. *Hydrobiologia.* 63 (3) : 241 - 253.

Subbamma, D.V. and Sarma, R.D.V. 1992 : Studies on the water quality characteristics of a temple pond. *J.Aqua.Biol.* 7(1 and 2) : 22 - 27.

Subbamma, D.V. 1992 : Plankton of a temple pond near Machilipatnam, Andra Pradesh. *J.Aqua.Biol.* 7(1 and 2) 17 - 21.

Subbamma, D.V. and Rama Sarma, D.V. 1994 : Hydrography of a fish pond in the Kolleru Area, Andhra Pradesh, India *J.Aqua. Biol.* 9(1 and 2) : 27- 29.

Swingle, H.S. 1967 : Standardization of chemical analyses for water and pond muds. *FAO. Fish. Rep.* 44 (4) : 397 - 421.

Tamot, P., Khare, S., Shrivastava, P. 1991 : Limnological studies of upper lake Bhopal during certain religious activities. *J.Natcon.* 3 (2) 141 - 146.

Tilman, D. 1977 : Resources competition between planktonic algae. An experimental and theoretical approach. *Ecology.* 58 (2) : 338 - 348.

Trivedi, R.G. and Goel, P. 1980 : Chemical and biological methods for water pollution studies. Enviro Media Publishers. 1 - 250.

Thrash, I.C., Suchking, E.V. and Bealc, J.P. 1944 : The examination of water suppliers (ed) E.W. Taylor.

Valech, V and Bhatnagar, G.P. 1989 : Primary productivity of phytoplankton in eutrophic lower lake Bhopal, India. *Envi and Ecol.* 7(1) : 202 - 205.

Vijaykumar, K. and Paul, R. 1990 : Physico-chemical studies on the Bhosga Reservoir in Gulbarga, Karnataka. *Eco biol.* 2(4) : 332 -335.

Vijaykumar, K. and Paul, R. 1991 : Primary production of Nuggikeri tank at Dharwad. *J.Nat.Con.* 3 (1) : 33 - 41.

Vijaykumar, K. and Paul, R., Kadadevaru, G.1991 : Physico-chemical features of Attikolla pond during Pre-monsoon period. *Envi and Ecol.* 9 (2) : 393 - 395.

Vijaykumar, K. and Basalingappa, S. 1992 : Effect of physico-chemical factors on seasonal abundance of cladocera in Devikoppa tank. *J.Karnatak Univ. Sci.* 34. 135 - 140.

Vijaykumar, K. 1992 : Limnological studies of perennial and seasonal standing Waterbodies of Gulbarga area. Ph.D. Thesis. Gulbarga University, Gulbarga - 160 pp.

Vijaykumar, K. and Paul, R. 1994, Effect of nutrient enrichment on phytoplankton production. *Proc. Acad. Envi. Biol.* 3(2) : 171 - 175.

Vijaykumar, K. 1994, Seasonal variations in the primary productivity of a tropical pond. *J.Eco.Biol.* 6(3) : 207 - 211.

Vijaykumar, K. 1996, Rectifer fauna of Devikoppa tank, Dharwad, Karnataka, India. 14th chapter. Advances in Fish and Wildlife Ecology and Biology (ed) Kaul.B.L. Daya Publishing House. 187 - 192.

Welch, P.S. 1952, Limnology, McGraw Hill Book Company. New York, 2nd edition. 1 - 538.

West, S.A. and West, S.S. 1912, On the periodicity of phytoplankton of some Irish lakes. *Trans Roy. Irish Acad.* 33b : 77 - 116.

Wetzel, R.G. Limnology. W.B. Saunders Company, Toranto, 1 - 743.

Yousuf, A.R., Shah, M.G. and Qadri M.V. 1986, Limnological aspects of Mirgund wetland. *Geobios News Report* 5 : 27 - 30.

Yousuf, A.R., Shah, M.G. 1988, Comparative limnology of some freshwater habitat of Kashmir. *Geobios News Report* 7 : 58 - 61.

Zafar, A.R. 1959, Taxonomy of lakes. *Hydrobiologia* 13 (3) : 287 - 299.

3

Limnological Studies of Shiroli Reservoir – A Case Study

D.V. Muley and P.T. Gaikad

Introduction

At present limnology plays an important role in the decision making processes for problems of dam construction, pollution control, fish enhancement and aquaculture practices. Applied limnology has great scope in healthy existence of natural and manmade Waterbodies and to harvest the natural resources at sustainable level (Goldman and Horne, 1983).

In order to utilise a freshwater body successfully for fish production it is very important to study the abiotic and biotic factors, influencing the biological productivity of the said waterbody. Such investigations involves physical factors like temperature, humidity, rainfall, turbidity and transparency of water; chemical factors like pH, dissolved oxygen, free CO_2, Chlorinity, hardness of water and important nutrients like phosphates and nitrates and effects of pollution if any. In biological investigation study of micro and macro flora and fauna always provides the clear picture of the ecological relationship existing in the waterbody.

In India some hydrobiological work on historic shallow Waterbodies like temple tanks and village ponds in south India have done by Chacko (1949, 1950); Ganapati (1940, 1957, 1959, 1962), Krishnamurthi (1965), Michael (1968) and Vijayraghavan (1971, 1973). Recent studies, on hydrobiology of the freshwater lentic habitats is well documented by Sawant (1983), Angadi (1985), Subbamma and Rama Sarma (1992) and Pandey *et al.* (1993).

According to Jhingran (1975), the total man made reservoir area in India is about 10,94,960.616 hect. Out of which 40 per cent is formed by small reservoirs and tanks. In Maharashtra State out of 1,51,114.71 hect. area of total freshWaterbodies about 89 per cent area is constituted by small reservoirs and tanks. Such small and shallower freshWaterbodies have been found to be much more productive than large impoundments (Holts, 1966).

Around the Kolhapur city, about 10 tanks of varying sizes are present, which are used for minor irrigation, washing and bathing purpose. As a representative of these tanks "Shiroli reservoir" was selected for the limnological studies. The present study is aimed to investigate some of the important physical and chemical parameters along with the flora and fauna of the reservoir. Similarly by studying the phyto and zooplankton qualitatively to find out what type of exotic fishes can be introduced in the tank in future so as to utilize the waterbody successfully for fish production.

Materials and Methods

The study are 'Shiroli reservoir' (11.61 hect.) is situated at about 8 km. North West of Kolhapur city (Latitude - 16^0 14' 15" North, Longitude - 74^0 13' 7" East) at the height of 634 meters from mean sea level. This waterbody is covered with earthen impoundments on the southern and western sides.

The investigation on physical, chemical and biological parameters were carried out during August 1994 to July 1995. The water samples were collected from five selected sites (A to E) around the tank by considering the shape and size of the tank, human activities at different sites, vegetation and position of inlets and outlets of the tank.

Fortnightly water samples were collected from five stations during hours (7.30 to 8.30 a.m.) and brought to the laboratory for further analysis. Physical parameters like water temperature, water level, turbidity and transparency were studied according to Welch (1948) and Lind (1974). The data about relative humidity, aerial temperature and rainfall was obtained from meteorology section, college of Agriculture, Kolhapur. During the chemical investigation seven major parameters namely pH, dissolved oxygen, free carbondioxide, chlorinity, $CaCo_3$ hardness (APHA *et al.*, 1985) and important nutrients like phosphates (Murphy and Riley, 1962) and nitrates (Charles *et al.*, 1945) from different stations were studied.

Attempts were also made to study the aquatic flora and fauna of the waterbody qualitatively during monsoon (Aug.-Sept.) winter (Dec.-Jan.), and summer (April-May) seasons of the year 1994-95. In biological limnology maun emphasis was given to phytoplankton and zooplankton as it forms basic trophic level in the aquatic food chain. The plankton samples were collected by using standard nylon plankton net having mesh size 98/sq. inch. The planktons were preserved in 4 per cent formalin and identified using standard manuals by Ward and Whipple (1959), Nidham and Nidham (1962) and Tonapi (1980). Similarly the animal species were identified and preserved in 4 per cent formalin solution and enlisted in the tabular form.

Results and Discussion

The investigation on the Shiroli reservoir, near Kolhapur, have for the first time studied to reveal the limnological status of the representative small waterbody.

Though some work has been done on the limnological study of lakes, tanks, reservoirs and small tanks in India. It has been found always difficult to compare between two Waterbodies. The influence of abiotic factors on waterbody and its interrelationship with plankton productivity. This is mainly because of the differences in water masses, their size, shape, depth climatic conditions and the region to which they belong. Macroflora, fauna and pollution problems in these waters also play an important role, making it difficult to compare the characteristic of individual waterbody with another. Therefore, the information about important physical, Chemical and Biological characteristics of the reservoir and their seasonal changes were studied.

This tank is a perennial waterbody (mean depth 2.85 meters) with about four inlets bringing surplus water from the catchment area. The water from this reservoir is mainly used for washing purpose. The water level can be maintained by the water lift system from Panchaganga river during summer season when water level drops down, while the excess water flows out through spill way, during monsoon.

The heat transmitted with light is responsible for establishing the thermal stratification in Waterbodies, and it influences the respiration of both, plants and animals. Phytoplankton exhibits higher respiration and hence, slightly lower efficiency at high temperature. The air temperature fluctuates from season to season in the range of 16.65°C to 31.8°C, while in monsoon, it was moderate (Table 1). The fluctuations in surface water temperature were in the range of 21°C to 31°C. The increasing water temperature during summer months were observed (Table 2). After the monsoon, temperature drops down in both air and water.

Table 1 : The table showing fortnightly readings of air temperature in °C for the period of August, 1994 to July, 1995

Date	*Air Temperature*		
	Maximum	*Minimum*	*Mean*
16-08-94	26.7	21.0	23.85
30-08-94	25.2	20.9	23.05
13-09-94	28.4	18.8	23.60
27-09-94	32.2	18.4	25.30
11-10-94	32.1	20.9	26.50
25-10-94	29.2	19.1	24.15
08-11-94	30.5	18.4	24.45
22-11-94	28.3	14.1	21.20
06-12-94	26.7	13.0	19.85
20-12-94	23.5	9.8	16.65
03-01-95	28.6	11.4	20.00
17-01-95	26.7	11.6	19.15
31-01-95	30.4	14.6	22.50
14-02-95	32.9	15.2	24.05
28-02-95	34.4	14.8	24.60
14-03-95	35.5	17.7	26.60
28-03-95	34.4	17.0	25.70
11-04-95	37.2	19.5	28.35
25-04-95	39.6	24.0	31.80
09-05-95	33.8	20.5	27.15
24-05-95	35.3	22.5	28.90
06-06-95	34.5	23.6	29.05
20-06-95	30.8	22.6	26.70
04-07-95	30.7	22.7	26.70
18-07-95	26.1	21.5	23.80

The percentage of relative humidity was studied from August, 1994 to July, 1995. The lowest relative humidity was recorded on 14th March 1995 while, highest relative humidity was recorded on 11th October, 1994. The relative humidity has direct

Table 2 : Showing fortnightly readings of water temperature in °C at stations A to E for the period of August 1994 to July, 1995

Date	*A*	*B*	*C*	*D*	*E*
16-08-94	26	26	25	26	26
30-08-94	26	26	26	26	26
13-09-94	27	27	28	27	26
27-09-94	28	28	29	29	27
11-10-94	29	30	29	29	29
25-10-94	27	28	28	28	27
08-11-94	26	26	26	26	26
22-11-94	25	25	24	25	24
06-12-94	25	24	23	23	23
20-12-94	23	22	22	22	23
03-01-95	21	22	21	21	22
17-01-95	23	23	23	23	23
31-01-95	23	24	23	23	24
14-02-95	23	24	23	23	24
28-02-95	23	23	23	23	24
14-03-95	26	26	25	25	25
28-03-95	27	26	26	26	26
11-04-95	27	27	27	27	27
25-04-95	28	28	27	28	28
09-05-95	29	29	29	29	29
24-05-95	31	30	31	30	30
06-06-95	28	29	28	28	28
20-06-95	29	29	28	28	28
04-07-95	26	26	26	26	25
18-07-95	26	26	26	26	25

correlation with the monsoon rainfall. The relative humidity values though decreased in September, it reaches to the highest in the month of October 1994 and again it decline subsequently and reach to the lowest in March, 1995, from where is again slowly increased considerably till July 1995 (Table 3). Low humidity in the summer months, causes a great loss to surface water due to evaporation.

Rainfall in this area is mainly from South-West monsoon. The annual average rainfall of this region is above 1000 mm. During the period of this study, total rainfall was 814.8 mm.

The water level fluctuations coincides with the seasonal pattern of temperature. The gradual fall in the water level in the post monsoon period and reaches to minimum (1.7 meters) level in the month of June-July, 1995 (Table 4). Pre-monsoon rains cause little increase in water level. Wirth (1970) have dealt with the

Table 3 : Showing fortnightly readings of per cent Relative Humidity for the period of August, 1994 to July 1995.

Date	*Percentage of Relative Humidity*		
	Maximum	*Minimum*	*Mean*
16-08-94	92.0	88.4	90.20
30-08-94	95.1	94.1	94.60
13-09-94	89.8	79.5	84.65
27-09-94	90.5	64.1	77.30
11-10-94	92.7	97.7	95.20
25-10-94	89.0	92.0	90.50
08-11-94	88.0	63.0	75.50
22-11-94	80.0	51.0	65.50
06-12-94	99.0	44.0	71.50
20-12-94	89.0	45.0	67.00
03-01-95	87.4	41.4	64.40
17-01-95	89.4	48.4	68.90
31-01-95	88.8	37.7	63.25
14-02-95	88.7	28.8	58.75
28-02-95	58.1	25.1	55.10
14-03-95	63.7	28.5	46.10
28-03-95	81.8	24.5	53.15
11-04-95	79.5	26.1	52.80
25-04-95	70.5	27.5	49.00
09-05-95	79.2	54.1	66.65
24-05-95	84.7	43.8	64.25
06-06-95	84.4	49.8	67.10
20-06-95	84.1	65.1	74.70
04-07-95	87.8	64.0	75.90
18-07-95	91.7	82.2	86.95

different aspects of level fluctuations in freshwater bodies in India. Pantulu *et al.* (1966) tried to find out its relation with fisheries in Damodar Valley reservoir. During the summer months (April - May) the water level being decreased considerably in tanks, makes the fish harvesting easier in less man efforts to get maximum harvest. This season also coincides with the maturity period of the weed fishes from the tank and they can be eradicate before the release of their eggs.

The turbidity of natural water may be due to the suspended inorganic substances such as silt, clay or due to any planktonic organisms. It is important limiting factor in the productivity of reservoirs, tanks and ponds. Turbidity also depends upon the type of the bottom, if it is rocky with sand or gravel, then there will be low turbidity (Jhingran, 1983). Turbidity plays an important role in productivity of the reservoir. The data on the turbidity was comparatively higher than other three stations. This may be due to human activities in this part of the reservoir. Turbidity was maximum at all stations in the months of July, 1995, whereas it was minimum in August 1994 (Table 5).

The turbidity and transparency has shown negative co-relation, as the water transparency drops gradually after September, 1994; and again it gradually increased and reached to maximum in May, 1995 (Table 6). This might be due to wave action, algal blooms, and fishing activity in the tank, made the water more disturbed.

Table 4 : Showing fortnightly readings of water level in meters at station A for the period of August, 1994 to July, 1995

Date	*Water level in Meters*
16-08-94	3.40
30-08-94	3.50
13-09-94	3.30
27-09-94	3.00
11-10-94	3.45
25-10-94	3.40
08-11-94	3.40
22-11-94	3.35
06-12-94	3.20
20-12-94	3.15
03-01-95	3.15
17-01-95	3.20
31-01-95	3.10
14-02-95	3.00
28-02-95	2.85
14-03-95	2.72
28-03-95	2.60
11-04-95	2.47
25-04-95	2.30
09-05-95	2.25
24-05-95	2.17
06-06-95	2.00
20-06-95	1.90
04-07-95	1.70
18-07-95	2.75

In chemical analysis parameters like pH, dissolved oxygen, free carbondioxide, chlorinity, hardness phosphates and nitrates from the surface water were studied.

pH values recorded at all five stations showed almost similar pattern of fluctuations. Minimum pH was recorded in the month of march, 1995 at all the stations, whereas it was maximum at the end of January, 1995 except E, where it was maximum in July, 1995 (Table 7)

Table 5 : Showing fortnightly readings of Turbidity of water in Jackson units at stations A to E for the period of August, 1994 to July, 1995.

Date	*A*	*B*	*C*	*D*	*E*
16-08-94	40	44	42	48	52
30-08-94	55	59	60	65	78
13-09-94	58	63	67	68	82
27-09-94	60	66	69	70	85
11-10-94	80	82	85	91	95
25-10-94	70	75	81	86	90
08-11-94	75	78	76	80	86
22-11-94	74	76	74	78	82
06-12-94	70	72	69	71	78
20-12-94	62	68	62	65	71
03-01-95	61	66	60	62	70
17-01-95	65	70	66	68	75
31-01-95	71	76	70	73	79
14-02-95	69	74	66	70	76
28-02-95	65	70	60	64	71
14-03-95	68	74	66	69	76
28-03-95	70	78	71	73	80
11-04-95	71	77	70	72	81
25-04-95	73	79	72	75	84
09-05-95	77	84	78	79	88
24-05-95	81	86	82	83	91
06-06-95	80	85	81	82	90
20-06-95	84	87	85	87	95
04-07-95	87	90	92	93	100
18-07-95	90	98	98	105	108

Every water body shows the seasonal and diurnal variation in pH. According to Jhingran (1983), pH is alkaline in the afternoon and acidic in the dawn. The pH of the water samples from Shiroli reservoir ranged between 6.3 to 8.7 during the period of investigation. According to Swingle (1967) water having pH range

Table 6 : Showing fortnightly readings of Transparency in Cms at stations A to E for the period of August, 1994 to July, 1995.

Date	*A*	*B*	*C*	*D*	*E*
16-08-94	68	62	61	60	58
30-08-94	47	43	42	42	40
13-09-94	50	45	44	45	42
27-09-94	52	48	46	47	44
11-10-94	53	49	46	47	45
25-10-94	48	44	42	46	42
08-11-94	50	46	44	48	44
22-11-94	54	50	48	52	48
06-12-94	60	54	50	53	50
20-12-94	60	55	53	53	50
03-01-95	60	55	53	53	50
17-01-95	58	52	50	50	48
31-01-95	58	52	50	50	50
14-02-95	55	50	48	46	44
28-02-95	55	50	48	46	44
14-03-95	59	54	50	49	46
28-03-95	62	60	56	54	50
11-04-95	65	63	60	59	52
25-04-95	70	65	65	64	55
09-05-95	70	68	65	65	55
24-05-95	70	65	62	64	55
06-06-95	69	64	59	62	52
20-06-95	65	62	56	60	50
04-07-95	65	62	55	58	50
18-07-95	45	40	38	40	42

of 6.5 to 9.0 are most suitable for pond culture. In low pH values, fishes get prone to attacks of parasites and diseases. As the pH increases above 9.5, there will be absence of CO_2, which is required for the growth of phytoplankton.

Dissolved oxygen (D.O.) is one of the most important parameters of the water quality, affecting survival and distributing of flora and fauna in an ecosystem. The two main sources of dissolved oxygen are diffusion from air and photosynthesis, while major factors responsible for its depletion are biochemical oxidation and respiration by flora and fauna. The quantity of dissolved oxygen in water is directly or indirectly dependent on water temperature, partial pressure of oxygen in the air, concentration of dissolved salts, amount of chlorophyll content etc. Welch (1952), and temperature and hydrogen ion concentration Wetzel (1975).

In the Shiroli reservoir the entire basin is partially eutrophicated due to which the euphotic zone has been reduced. When there is low temperature, solubility of oxygen in water increases considerably, which further reduce when water temperature becomes high. Oxygen concentration in water samples of Shiroli reservoir shows fluctuations between 0.56 to 4.83 ml/l (Table 8). Lowest dissolved oxygen in the month of Sept. - Oct. 1994, might be due to short photo period, and low wave action.

Though there was increase in the dissolved oxygen values during winter, it gradually decrease during summer, which may be attributed to high air and water temperature which prevents diffusion of oxygen from air.

Table 7 : Showing fortnightly readings of pH (Hydrogen ion Concentration) at stations A to E for the period of August, 1994 to July, 1995

Date	*A*	*B*	*C*	*D*	*E*
16-08-94	7.6	7.6	7.8	7.7	7.8
30-08-94	7.6	7.7	7.9	7.7	7.8
13-09-94	7.5	7.8	8.0	7.8	7.9
27-09-94	7.7	7.8	8.0	7.8	7.8
11-10-94	7.8	7.7	7.9	7.7	7.7
25-10-94	7.8	7.8	7.8	7.6	7.6
08-11-94	7.8	7.9	8.0	7.8	7.7
22-11-94	8.0	8.1	8.0	8.1	8.0
06-12-94	8.0	7.8	7.9	7.9	7.5
20-12-94	7.8	7.7	8.0	7.8	7.6
03-01-95	7.9	7.8	8.0	7.9	7.8
17-01-95	7.8	7.4	8.0	7.9	7.6
31-01-95	8.4	8.3	8.1	8.3	7.7
14-02-95	8.0	7.7	7.9	7.8	7.5
28-02-95	6.6	7.0	7.3	6.9	7.3
14-03-95	6.5	6.3	6.9	6.4	6.3
28-03-95	6.5	6.3	6.8	6.4	6.3
11-04-95	6.6	6.5	6.8	6.5	6.4
25-04-95	6.7	6.5	6.9	6.6	6.5
09-05-95	7.3	7.1	7.2	7.1	7.1
24-05-95	7.8	7.8	7.6	7.7	7.4
06-06-95	8.0	7.7	7.7	7.5	7.5
20-06-95	7.6	7.7	6.9	7.2	7.1
04-07-95	8.2	7.8	7.0	7.2	7.3
18-07-95	7.4	7.0	7.5	7.8	8.7

Table 8 : Showing fortnightly readings of dissolved oxygen in ml/l at stations A to E for the period of August, 1994 to July, 1995.

Date	*A*	*B*	*C*	*D*	*E*
16-08-94	1.12	1.26	2.52	2.45	2.24
30-08-94	0.51	0.84	1.12	1.02	1.12
13-09-94	1.35	0.56	1.26	0.84	0.70
27-09-94	0.91	1.12	1.16	0.92	1.06
11-10-94	0.56	0.98	1.12	0.70	0.56
25-10-94	0.51	1.12	1.26	0.84	0.64
08-11-94	1.26	1.68	2.00	1.96	1.19
22-11-94	0.56	1.54	2.24	1.96	1.26
06-12-94	1.68	1.96	2.84	2.00	0.98
20-12-94	1.26	1.68	2.66	1.40	0.84
03-01-95	1.75	2.59	3.50	2.10	1.96
17-01-95	1.82	2.80	3.08	2.10	1.75
31-01-95	2.10	2.52	2.52	2.10	2.24
14-02-95	1.35	2.00	3.08	1.75	1.19
28-02-95	1.82	1.82	2.10	1.82	1.26
14-03-95	3.64	2.24	2.66	2.80	1.26
28-03-95	3.36	2.24	2.59	2.52	1.54
11-04-95	3.22	2.10	2.38	2.24	1.40
25-04-95	3.08	2.10	2.14	2.10	1.30
09-05-95	2.66	2.00	2.10	1.96	1.26
24-05-95	1.96	1.68	1.54	1.61	0.98
06-06-95	0.70	2.84	3.08	3.08	2.24
20-06-95	0.70	2.66	4.74	2.52	2.24
04-07-95	1.26	0.98	3.78	2.38	3.92
18-07-95	2.52	2.94	4.74	4.60	4.83

During the period of investigation free carbon dioxide values ranged from 1.65 to 43.94 mg/l. Maximum value was recorded at station A, while minimum value at station C. Though there was great fluctuations in carbon dioxide in different seasons, the pattern of changes was almost uniform (Table 9). In general, the free carbondioxide content at five stations was in the order of A, B, D, E and C. Due to the unstable state of free Carbondioxide, it escapes as the saturation level is attained. Therefore no correlation with other abiotic or biotic factor could be established.

The amount of carbon dioxide depends on the decomposition of top soil and chemical nature of the underlying rocks. In the day time it is utilized by plants for photosynthesis. Hence in the early morning it will be in maximum quantity than that of the evening (Mann, 1958).

Table 9 : Showing fortnightly readings of free Carbon dioxide in mgl at stations A to E for the period of August, 1994 to July, 1995

Date	A	B	C	D	E
16-08-94	34.55	25.96	12.58	13.98	11.18
30-08-94	37.95	24.97	18.97	15.98	12.98
13-09-94	13.38	15.98	9.38	12.58	11.98
27-09-94	11.98	11.98	8.58	9.98	11.98
11-10-94	17.97	15.98	5.99	15.98	17.97
25-10-94	19.97	12.98	7.59	11.98	20.97
08-11-94	17.97	15.98	9.98	17.97	20.57
22-11-94	19.97	17.97	17.97	29.96	23.97
06-12-94	25.96	20.57	15.98	24.57	25.56
20-12-94	29.96	21.97	13.98	25.96	29.96
03-01-95	25.96	28.56	20.57	30.56	27.96
17-01-95	27.96	27.96	24.57	25.96	24.97
31-01-95	19.97	23.97	21.97	23.97	25.96
14-02-95	23.97	28.56	19.97	25.36	25.96
28-02-95	29.96	30.56	21.57	27.96	21.97
14-03-95	21.97	29.96	7.99	25.96	29.96
28-03-95	21.97	25.96	7.99	25.96	27.96
11-04-95	26.56	25.96	8.58	21.97	32.56
25-04-95	27.36	25.96	5.99	23.97	29.96
09-05-95	29.96	27.96	4.59	25.96	24.97
24-05-95	29.96	29.96	1.65	27.96	24.57
06-06-95	30.56	30.56	3.99	29.96	22.57
20-06-95	43.94	37.95	7.99	31.96	27.96
04-07-95	33.95	29.96	9.98	27.96	29.96
18-07-95	30.56	26.96	6.59	7.99	13.98

Natural water normally contains low chloride than bicarbonates and sulphates. Large contents of chloride in fresh water is an indication of organic pollution (Thrash *et al.*, 1944). Though chloride level as high as 250 mg/l is safe for human consumption, a level above this imparts salty taste to potable water. In the present study, the chloride content was minimum in the month of August, 1944, while is was maximum in June-July 1995 at all stations. The highest chloride content was 308.0 mg/l at station D in June, 1995, whereas the lowest value recorded was 16.5 mg/l at station B in August, 1994. Though there are fluctuations in chloride content, the pattern of changes was almost uniform (Table 10).

Hardness to water is imparted by alkaline earth metal cations mainly calcium and magnesium present in it. Ecologically

Table 10 : Showing fortnightly readings of Chlorinity in mg/l at stations A to E for the period of August, 1994 to July, 1995

Date	*A*	*B*	*C*	*D*	*E*
16-08-94	59.0	54.0	50.0	45.0	48.0
30-08-94	18.0	16.5	17.0	20.75	17.25
13-09-94	37.15	34.3	28.5	40.5	34.5
27-09-94	44.5	41.0	41.0	42.0	41.5
11-10-94	43.0	40.5	34.0	43.0	45.0
25-10-94	44.0	42.0	37.0	46.0	42.0
08-11-94	45.5	44.5	45.5	47.5	44.0
22-11-94	47.5	46.0	46.5	52.0	45.0
06-12-94	49.0	49.0	48.5	50.5	52.5
20-12-94	51.0	52.0	53.0	54.0	53.0
03-01-95	54.5	56.0	52.5	55.0	56.5
17-01-95	56.0	61.5	58.0	59.0	58.0
31-01-95	71.0	69.5	68.5	69.0	69.0
14-02-95	75.0	72.0	70.0	70.5	72.0
28-02-95	71.0	68.0	67.5	64.0	65.5
14-03-95	61.0	62.0	62.0	63.0	65.0
28-03-95	60.0	58.0	59.5	59.0	61.5
11-04-95	61.5	62.5	62.5	64.0	65.0
25-04-95	63.5	66.0	66.0	72.0	74.0
09-05-95	69.5	69.5	85.5	88.0	84.0
24-05-95	82.0	82.5	120.0	129.0	90.0
06-06-95	95.0	97.0	156.0	176.0	104.0
20-06-95	133.5	142.0	275.0	308.0	136.5
04-07-95	117.5	121.5	292.5	141.0	121.5
18-07-95	124.0	118.0	170.0	87.0	135.0

temporary hardness plays a key role in buffering capacity thus, neutralizing an offset in pH, due to addition of acidic products. This has great effect on biotic diversity of an ecosystem. Hardness of water also restricts water uses. In the present study, the hardness ranges from 51.21 to 120.65 $CaCO_3$/l at station E, in the month of November, 1994 and July, 1995 respectively. High values of hardness in July, 1995 is due to the concentration of carbonates and bicarbonates through run off from rain water. Though there were fluctuations in the hardness from March to July, 1995, the pattern of changes was almost similar, particularly from August, 1994 to February, 1995 (Table 11).

Pond or tank waters having hardness of 15 ppm or above are satisfactory for the growth of fishes. While hardness less than

5 ppm gives slow growth, distress and leads to death (Swingle, 1967). Sreenivasan (1964) also correlated the hardness of water with organic productivity. From the present study on hardness it is clear that the water is more suitable for aquaculture practices.

Table 11 : Showing fortnightly readings of Hardness as $CaCO_3$ in mg/l at stations A to E for the period of August, 1994 to July, 1995

Date	*A*	*B*	*C*	*D*	*E*
16-08-94	71.17	62.49	66.83	69.44	82.46
30-08-94	52.94	52.94	56.42	65.96	52.94
13-09-94	53.55	54.68	62.06	70.30	57.28
27-09-94	57.28	56.42	54.68	59.89	60.76
11-10-94	57.28	53.81	49.47	60.76	56.42
25-10-94	57.28	54.68	52.94	60.76	57.28
08-11-94	56.42	52.94	54.68	61.62	57.28
22-11-94	52.94	52.08	52.08	61.62	51.21
06-12-94	53.81	55.55	52.94	59.89	55.55
20-12-94	55.55	57.28	52.08	58.15	53.81
03-01-95	57.28	58.15	54.68	55.55	57.28
17-01-95	57.28	57.28	53.81	54.68	56.42
31-01-95	58.15	57.28	51.21	53.81	56.42
14-02-95	57.28	58.15	51.21	55.55	58.15
28-02-95	60.76	64.23	54.68	59.89	65.96
14-03-95	75.51	81.59	56.42	70.30	77.25
28-03-95	78.12	85.93	61.62	75.51	85.93
11-04-95	81.59	86.80	69.44	78.98	86.80
25-04-95	84.19	87.66	72.91	81.59	88.53
09-05-95	82.46	81.59	72.04	81.59	84.19
24-05-95	77.25	72.91	59.02	78.12	78.98
06-06-95	67.70	72.04	55.55	96.34	69.44
20-06-95	85.93	86.80	85.93	110.23	89.40
04-07-95	88.53	90.27	86.80	88.53	88.53
18-07-95	88.53	80.72	80.72	82.46	120.65

The phosphates (PO_4) in natural waters occur in very small quantities. It is an important nutrient in the maintenance of the fertility of pond. In an aquatic ecosystem phosphorus occurs both in inorganic and organic forms. Of the two, the inorganic phosphorus plays an important role as a nutrient along with nitrates. In Shiroli reservoir phosphate (Phosphorus) content recorded was minimum (0.48 mg/l) in the month of October, 1994 and maximum (2.94 mg/l) in May 1995. Though phosphate content in water at five stations were low from August to November, 1994, there was remarkable increase in the content from December, 1994 onwards and reached to the maximum in May 1995 and again decline slowly till July, 1995 (Table-12). The phosphate more than 2 mg/l in open water gives a sign of organic pollution (Pomeroy *et al.*,1965). As a nutrient, excess of phosphate stimulates development of algal blooms, such blooms of blue green algae like Microcystis are toxic in nature.

Table 12 : Showing fortnightly readings of phosphate (phosphorus) in mg/l at stations A to E for the period of August, 1994 to July, 1995

Date	*A*	*B*	*C*	*D*	*E*
16-08-94	0.96	0.96	0.93	0.84	0.75
30-08-94	0.84	0.84	0.96	0.60	0.66
13-09-94	0.48	0.72	0.84	0.54	0.66
27-09-94	0.75	0.60	0.63	0.69	0.63
11-10-94	0.60	0.48	0.72	0.63	0.75
25-10-94	0.66	0.63	0.69	0.63	0.69
08-11-94	0.54	0.51	0.54	0.51	0.51
22-11-94	0.51	0.48	0.60	0.45	0.54
06-12-94	0.51	0.51	0.81	0.75	0.96
20-12-94	1.89	1.95	1.92	1.86	1.95
03-01-95	1.62	1.92	1.71	1.71	1.92
17-01-95	1.74	1.83	1.86	1.80	1.74
31-01-95	1.80	1.74	1.77	1.83	1.80
14-02-95	1.89	1.77	1.71	1.95	1.83
28-02-95	1.83	1.95	1.80	1.80	1.92
14-03-95	1.83	1.98	1.68	1.77	1.95
28-03-95	1.86	1.95	1.80	1.80	1.95
11-04-95	2.10	2.01	2.10	2.10	2.34
25-04-95	2.40	2.46	2.40	2.52	2.52
09-05-95	2.94	2.94	2.94	2.94	2.94
24-05-95	2.01	2.22	2.40	2.46	2.34
06-06-95	1.56	1.44	1.56	1.44	1.47
20-06-95	1.56	1.62	2.28	1.68	1.74
04-07-95	1.59	1.62	2.37	1.62	1.68
18-07-95	1.62	1.68	1.95	1.62	1.86

Nitrogen occurring in the fresh water is in the form of an element of nitrogenous compounds. Nitrogen is mostly derived from atmosphere and denitrifying bacteria giving nitrates, nitrites and ammonia. Nitrogenous compounds may be derived from outside sources like rain, surface run off and seepage. In Shiroli reservoir, nitrate values ranged between 0.4 mg/l in August, 1994 to 3.6 mg/l in May, 1995. Nitrate values increases considerably from August to October, 1994 and decline slowly till January 1995, but again it increased remarkably and reached to its maximum in May, 1995, giving a peak value and then decreased in preceding months. (Table 13).

Table 13 : Showing fortnightly readings of Nitrates in mg/l at stations A to E for the period of August, 1994 to July, 1995

Date	*A*	*B*	*C*	*D*	*E*
16-08-94	1.7	1.4	1.9	1.9	1.6
30-08-94	1.6	1.0	1.6	0.4	0.6
13-09-94	1.6	1.5	2.5	2.7	2.2
27-09-94	1.6	2.2	1.7	3.0	2.4
11-10-94	2.4	2.4	2.9	2.3	2.2
25-10-94	2.5	2.8	2.8	2.9	3.0
08-11-94	1.9	2.4	2.9	2.2	2.4
22-11-94	2.0	2.1	2.0	2.4	2.5
06-12-94	1.4	2.8	1.6	2.0	2.0
20-12-94	1.1	1.2	1.3	1.1	1.3
03-01-95	1.0	1.2	1.0	1.0	1.2
17-01-95	1.2	1.3	1.1	1.1	1.2
31-01-95	0.9	1.0	0.85	0.8	0.85
14-02-95	1.0	1.2	1.1	1.2	1.1
28-02-95	1.5	1.6	1.4	1.7	1.3
14-03-95	1.8	2.1	1.8	1.8	1.8
28-03-95	1.9	2.2	2.0	2.1	2.1
11-04-95	2.5	2.7	2.7	2.8	2.6
25-04-95	2.8	2.9	2.8	2.8	2.9
09-05-95	3.6	3.6	3.6	3.4	3.4
24-05-95	3.0	3.0	3.1	3.1	3.0
06-06-95	1.9	1.9	2.0	1.8	2.0
20-06-95	1.9	2.2	2.4	1.9	1.9
04-07-95	1.8	1.6	2.3	2.2	1.9
18-07-95	1.7	1.8	2.2	2.1	2.1

Increasing levels of nitrates and phosphates leads to eutrophication, a phenomenon linked with nutrient enrichment, leading to initial higher productivity, abundance of undesirable plants and animal species and finally destruction of an ecosystem. The values of phosphates and nitrates in Shiroli reservoir showed correlation and almost similar trend in the periodic fluctuations.

The high values of phosphates and nitrates during May, 1995, was perhaps due to high temperature, enhancing degradation of organic matter from the reservoir.

Among biotic communities phytoplankton constitutes the first stage in trophic level by virtue of their capacity to fix radiant energy into biological energy through photosynthesis, referred as 'primary productivity'. The plankton plays a vital role in regulating the atmospheric level of O_2 and CO_2 gases and biogeochemical cycles.

Biological Studies

In present study, the biota of the reservoir was surveyed during the period of investigation and listed in table. The phytoplankton was represented by 35 species whereas, zooplankton consists of 29 species. In the macro-organism II aquatic plant species and 42 animal species were recorded, including fishes, amphibians and aves. The species of *Nymphaea* and *Nelumbo* occupied the southern half of the reservoir. Plankton studies were carried out for qualitative estimation.

Flora and Fauna

The flora and fauna of the reservoir is rich qualitatively. Phytoplankton consists of 4 families and their 35 species-Bacillariophyceae-13 species, Chlorophyceae-10 species, Cyanophycea-7 species and Desmidiaceac-5 species.

In Zooplankton samples about 29 species were recorded. These species were observed throughout the period of investigations with seasonal fluctuation. In Zooplankton samples, three species of protozoans, 9 species of rotifers, 10 species of copepods and 7 species of cladocerans were identified. The macrofauna of the reservoir was represented by 42 species and their composition was Crustaceans-2, Insects-14, Molluscs-6, Pisces-14, Amphibians-1 and Aves-5. The list of the organisms found in the Shiroli reservoir during the study are showed in the list.

PHYTOPLANKTON

Bacillariophyceae

1. *Diatom vulgare*
2. *Cymbella affinis*
3. *Tabellaria affinis*
4. *Teracyclus lawstris*
5. *Diatomella balfouriana*
6. *Pinnularia nobilis*
7. *Navicula sp*
8. *Fragilaria capucina*
9. *Synedraulina* sp
10. *Amphora ovalis*
11. *Melosria granulata*
12. *Fragilaria crotonensis*
13. *Synedra affinis*

Chlorophyceae (Green Algae)

1. *Spirogyra* sp.
2. *Nephrocytium quadriseta*
3. *Characium hookeri*
4. *Chlorosarcina minor*
5. *Pediastrum simplex*
6. *Spino closterium curvatum*
7. *Hydroclicyton reticulatum*
8. *Spondylosium moniliforme*
9. *Desmidiom sphaericum*
10. *Sphaerocystis schroeteri*

Cynophyceae (Blue-green Algae)

1. *Microcystis aeruginosa*
2. *Coelosphareium dibium*
3. *Spirulina piatensis*
4. *Anabaenopsis circularis*
5. *Merismopedia punctata*
6. *Anabaena* sp.
7. *Nostoc linckia*

Demidiaceae

1. *Desmid staurastrum*
2. *Staurastrum leptocladium*
3. *Coelastrum microporum*
4. *Staurastrum chaetoceras*
5. *Cosmarium granatum*

Macrophytes (Aquatic)

1. *Nymphaea nouchali*
2. *Nymphaea pubescens*
3. *Nelumbo nucifera*
4. *Ipomoea aquatica*
5. *Limnophila heterophylla*
6. *Ceratophyllum demersum*
7. *Hydrilla verticillata*
8. *Najas minor*
9. *Azolla pinnata*
10. *Blyxa echinosperma*
11. *Ottelia alismoides*

ZOOPLANKTON

Protozoans

1. *Amoeba* sp.
2. *Volvox* sp.
3. *Ceratum* sp.

Rotifers

1. *Branchiour falcatus*
2. *Keratella tourocephala*
3. *Asplanchna pilinia*
4. *Notholcaocum minota*
5. *Filinia terminalis*
6. *Cephalodella megalo cephalo*
7. *Branchionus caudatus*
8. *Kellicottia* sp.
9. *Fillinia branchiata*

Copepods

1. *Diaptomus nauplius* II (Development stage)
2. *Heliodiaptomus* sp.
3. *Neodiaptomus diaphorus*
4. *Cletocampus albuquerquensis*
5. *Rhinediaptomus indicus*
6. *Cyclopoid* sp.
7. *Cyclops bisuspidatus thomasi*
8. *Ergasilus chautaughensis*
9. *Senecella calconoides*

Cladocera

1. *Daphnia pulex*
2. *Diaphanosoma brachyarum*
3. *Ceriodaphnia parthenogenetic*
4. *Simocephalus* sp.
5. *Moina* sp.
6. *Lynceus* sp.
7. *Bosmina* sp.

MACRO-ORGANISMS

Crustaceans

1. *Macrobrachium kistepsis* (Prawn)
2. *Barytelphus* sp.(Crab)

Insects (Aquatic)

1. Naids of may flies
2. Naids of Oragon and damsel flies
3. *Hydrometra vittata*
4. *Lemnometra fluviorum*
5. *Microvetia* sp.
6. *Laecotrephes maculatus*
7. *Diplonychus rusticum*
8. *Notonecta* sp.
9. *Ranatra elongata*
10. *Belostoma indica*
11. *Carixa* sp.
12. *Cybister* sp.
13. *Hydoticus* sp.
14. *Hydrophilus* sp.

Molluscs (Gastropoda)
1. *Viviparus bengalensis* (Bellamya)
2. *Lymnea stagnalis*
3. *Diagoniostoma pulchella*
4. *Planorbis erustus*

Pelecypoda (Bivalvia)
1. *Lamellidens corriganus*
2. *Lamellidens marginalis*

Fishes
1. *Labeo fimbriatus*
2. *Catla catla*
3. *Cyprinus carpio*
4. *Tilapia mossambica*
5. *Hypophthalmichthys molitrix*
6. *Mystius malabaricus*
7. *Garra mullya*
8. *Rasbora daniconius*
9. *Walago attu*
10. *Channa punctata*
11. *Punctius ticto*
12. *Punctius serana*
13. *Labeo rohita*
14. *Cirrhina mrigala*

Amphibians
1. *Rana tigrina*

Birds
1. Pond Heron : *Ardeola gravii gravii*
2. Cattle egret : *Bubulcus ibis coromedus*
3. Brahminy Kite : *Haliastur indus*
4. King fisher : *Alceddo atthis taprobana*
5. White Wagtail : *Motacila altra dukunesis*

The average values of nutrients (phosphate and nitrate) and phytoplankton showed inverse correlation, as low values were recorded in the month of May - June, 1995. This may be due to the shallower depth of the reservoir during this period.

Considering the shallowness of the waterbody and its small size this perennial tank seems to have almost all basic requirements for intensive fish culture practices. With the application of advance techniques the inflow of silt and domestic sewage from the nearby area can be checked. The fish production from this reservoir can be increased by manyfold.

Summary

Limnology is the science dealing with internal actions of processes and methods whereby matter and energy are transformed within the lake or pond. Applied limnology has great scope in healthy Waterbodies and to harvest the natural resources at sustainable level. In India there are great inland fresh water resources. In order to utilise these fresh Waterbodies successfully for fish production, it is very important to study the biotic and abiotic factors influencing the biological productivity of fresh waterbody.

To study the hydrobiology of "Shiroli tank" the following parameters have been investigated such as, shape and size, observations of physical parameters like air and water temperature, humidity, rainfall, water level (depth), turbidity and transparency. Chemical parameters like pH, dissolved oxygen, free carbondioxide, chlorinity, hardness of water and important nutrients like, phosphate and nitrate from water and effect of pollution if any. In biological investigations, study of microflora and microfauna showed ecological status of existing waterbody. All physico-Chemical studies have been carried out fortnightly from August 1994 to July 1995.

The shape of the reservoir is almost elongated and spread in 11.61 hect. in the survey No. 503 of Shiroli village near Kolhapur. The water samples were collected from five different stations (A to E) fixed around the tank for analysis. The physical parameters studied by using various materials and methods are as follows.

The aerial and surface water temperature was measured with the help of standard mercury thermometer. The maximum temperature was recorded in the month of May (31.8^{0}C, 29^{0}C) while minimum in the December (16.6^{0}C, 21^{0}C). Per cent relative humidity and rainfall data was obtained from meteorological department of Agriculture College, Kolhapur. Relative humidity was maximum (95.2 per cent) in October and minimum (46.1 per cent) in the month of March, 1995. Rainfall in this area is mainly from South-West monsoon. The average rainfall of this region is above 1000 mm. During the period of investigation total rainfall was 814.8 mm. Water level is generally observed in man made tanks and reservoirs. Maximum water level of the study area (3.5 meters) was recorded in August 1994, while it was minimum in July 95. In natural waters turbidity may be due to suspended inorganic substances like silt, clay or planktonic organisms.

The highest turbidity was recorded in August 95 while lowest in August 94 and was expressed in Jackson's units. Transparency was measured with the help of Secchi disc in cm. Average transparency of the water ranged from 35 to 125 cm. In the month of August, 94 and August 95 respectively. The pattern of changes was uniform with minor differences.

During the present investigation seven major chemical parameters namely pH, dissolved oxygen, free carbondioxide, chlorinity, hardness, phosphates and nitrates from the surface water were studied. The hydrogen ion concentration (pH) was measured with pocket digital pH meter. Average pH ranges from 6.3 in March to 8.7 in July. The dissolved oxygen was measured by standard Winkler's techniques and expressed in ml/l at N.T.P. Highest value was recorded in July (6.86 ml/l) and lowest in August (0.51 ml/l). Free carbondioxide was estimated by titration method in mg/l. Free carbondioxide range from 1.65 in May to 43.94 in June. Chlorinity was estimated in mg/l with the help of titration method using silver nitrate as titrant. It ranges from 16.5 in August to 308.0 in July. Total hardness from surface water samples was estimated with standard EDTA method and expresses in mg - $CaCO_3$/l. Hardness ranges from 5.21 in October and 120.65 in July. The two important nutrients, phosphates and nitrates were studied from water samples. The amount of phosphate phosphorus (PO_4) was estimated with spectronic - 20 and expressed in mg/l. It ranged from 0.48 in September to 2.94 in May, 95. The nitrates (NO_3) were estimated by Brucine method adopted for spectronic-20 and expressed in mg/l. of nitrate. It ranges from 0.4 in August to 3.6 May 95. Above physico-chemical parameters of water samples from all the stations have been studied fortnightly by using standard methods of water analysis.

In biological investigation, study of microflora and fauna in the waterbody have been investigated by collection, identification and preservation of various species in different seasons. The species of microflora and fauna have been listed separately. The information about fish stocking and fisheries of this tank was collected from District fisheries office, Kolhapur. The fishery observation were made in the field. The results of the above data are tabulated and presented in the final form.

Acknowledgement

The authors are thankful to the Prof. and Head, Department of Zoology, Shivaji University, Kolhapur, for providing necessary facilities.

References

Angadi, S.M. (1985), Hydrobiology of Rajaram tank. M.Phil. Dessertation, Shivaji University, Kolhapur.

APHA, AWWA and WPCA, (1985), Standard methods for examination of water and waste water, 16th Ed. American Public Health Association, Washington D.C.

Chacko, P.I. (1949), Krishna river and its fishes Proc. Indian Sci. Congr., 3, 165-166.

Chacko, P.I. (1950), Suggestion for stocking fish ponds in Madras, Madras Agri. J. 37(5) : 1-5.

Charles, A., Noll, W. and Betz. L.D. (1945). Determination of Nitrates in Bioler water by Brucine Indul and Eng. Chem. (Anly. Edn.), 17(7); 426-427.

Ganapati, S.V. (1940), The ecology of a temple tank containing a permanent bloom of *Microcystis acculoginosa* (Kutz), J.Bom. Nat. Hist. Soc. 42; 65-67.

Ganapati, S.V. (1959), The limnology of two minor irrigation reservoirs near Madras. The Errakuppan reservoir. Hydrobiologia 8; 365-379.

Ganapati, S.V. (191957). Limnological studies of two upland waters in Madras state, Arch. Hydrobiol : 56 (30-61).

Ganapati, S.V. (1962), Studies on the source of the Madras city water supply and other waters of Madras state Complied papers submitted for the Degree of D.Sc. to Madras University.

Goldman, C.R. and A.J. Horne (1983), Limnology : International student edition Mc Graw - Hill International Book Company, Tokyo, Japan.

Holt, S.J. (1966), Contribition of freshwater fish production to human nutrition and well being. In ; The biological basis of freshwater fish production. Edn. Gerking 455-467.

Jhingran, V. G. (1975), Fish and Fisheries of India. Hindustan Publishing Corporation (India), Delhi.

Jhingran, V. G. (1983), Fish and Fisheries of India. Hindustan publishing corporation, New Delhi, India pp 54, 86, 88, 167, 261, 283-292.

Krishnamurthy, K.P. (1965), Hydrobiological studies in Gandhisagar (Jumma tank), Diurnal variation in plankton. Hydrobiologia, 25(12); 99-118.

Lind, O.T. (1974), Handbook of common method in Limnology the C.V. Mosby Company, Saint Louis (U.S.A).

Mann, (1958), Annual fluctuations in sulphate and bicarbonate, hardness in pond, Limnol Oceanogr :3: (418-422).

Michael, R.G. (1968), Fluctuations in the relative abundance of the weed fauna of a tropical fresh water fish pond. Hydrobiologia 31(i) - 37-59.

Murphy, J. and J.Rilley, (1962), A modified single solution, method for determination of phosphate in natural waters. Anal. Chem. Acta. 27 (31-36).

Nidham, J.G. and P.R. Nidham (1962), A guide to the study of fresh water biology. Holder, day Inc. San Erancisco, 108.

Pandey, A.K.S.Z. Siddiqi and K.V.Rama Rao (1993), Physico-Chemical and biological characteristics of Hussain Sagar, an Industially polluted lake, Hyderabad. Proc. Acad. Enviorn. Biol. (2), 161-167.

Pantulu, V.R.K. Alaraja, and Bhimachar, B.S. (1966), Fisheries of Damodar Valley in relation to construction of dams. Proc. Nat. Inst. Sci. India. 32 (B) (5 and 6) : (191-207).

Pomeroy, L.R., Smith, E.E and Grant, C.M. (1965), The exchange of phosphate between estuarine water and sediments. Limnol. Oceanogr. 10(2) : (167-172).

Sawant, A.D. (1983), Some Hydrobiological studies on a tropical tank, Motitalav, at Sawantwadi M.Phil dessertation, Shivaji University, Kolhapur.

Sreenivasan, A. (1964), Limnology of tropical impoundments I. Hydrobiological features and fish production in Stanley reservoir, Metturdam. Int. Res. Ges. Hydrobiol. 51 : 295-306).

Subbamma, D.V. and Rama Sarna (1992). Studies on the water quality Characteristics of a Temple pond near Machilipatnam, Andhra Pradesh, J. Aqua. Biol., Vol. 7(1 and 2), 22-27.

Swingle, H.S. (1967), Standardization of Chemical analyses for water and pond muds *F A O Fish. Rep.* 44(4) : (397-421).

Thrash, J.C., E.V. Suchking and J.P.Beale (1944). The examination of water supplies (Ed) E.W. Taylor.

Tonapi, G.T. (1980), Fresh water animals of India : An ecological approach, Oxford and IBH Publishing Co, New Delhi, India.

Vijayaraghavan, S. (1971), Seasonal variation in primary productivity in three tropical ponds *Hydrobiologia*, 38 (3-4), 395-408.

Vijayaraghavan, S. (1937b), Seasonal variation in dissolved Carbohydrates (D) (HO) content in three freshwater ponds, Indian J. Fish., 20(1) : 157-165.

Ward, H.B. and G.C. Whipple (1959), Freshwater Biology. Edn. W.T. Edmondson. John Willey and Sons. Inc. New York, 1248pp.

Welch, P.S. (1948), Limnological Methods, Mc Graw - Hill Book Company, New-York, 381pp.

Welch, P.S. (1952), Limnology, 2nd ed. McGraw-Hill Book Co, New York, 538pp.

Wetzel, R.G. (1975), Limnology, W.B. Saunders Çompany, Taronto, 743pp.

Wirth, T.L. (1970), Manipulation of reservoir water for improved quality and fish population response. Res. Pop. Wis. Dept. Nat. Resour. 62; (23-24).

4

Limnological Studies of Jagat Tank Gulbarga

S.B. Angadi, B. Lingannaiah and Eshwarlal Sedamkar

Introduction

The habits of men and the forms of their social association with water than with the land, from which they earned their bread. All persons who shared rights to a water course were held responsible for its maintenance and cleaning. Hence, whole community was responsible for the care of large water courses. Water serves in many other ways to maintain life, health, vigour and social stability. The living organisms depends on such natural Waterbodies provides an appeal that attracts many people to come out and protect the same.

The role of water in nature is unique not only from the point of human consideration but numerous of those organisms, which make aquatic medium their dwelling place. In other words limnology is the study of aquatic life not only the organisms themselves but also the external influence of the medium that affects the aquatic life. The reserves of water on the earth are immense, but this is mostly salt water which is unfit for drinking or irrigation purposes. The amount of freshwater is huge as well but its distribution over the globe is uneven. The inland waters are a small fraction of the total hydrosphere, but is of immense importance in governing the various life activities of all organisms.

The standing fresh Waterbodies like lakes, ponds, pools, etc., have always been of comparatively more interest with investigators, than the other types of fresh water systems. It is the diversity in unity that gives a peculiar fascination to limnology. A group of lakes confronts the investigator as a series of very complex physical, chemical and biological system. Each waterbody shows its own characteristics. The study of different lakes helps in understanding the variation of geographical and climatic conditions.

Water is one of the very few natural substances, that can exist as a liquid at the surface of the earth. The water balance of lake is obviously expressed by an equation indicating that the rate of change of volume of the lake, is equal to the rate of inflow from all the sources, less the rate of water loss. The sources of income are,

(1) Precipitation falling on lake surface.
(2) Water in surface influents.
(3) Ground water seeping through the floor of the lake.
(4) Ground water entering by discrete springs.

Brig and Juday (1931), emphasized the distinction between drainage lakes with an outlet and seepage lakes into which ground water enter, and from which water leaves by seeping through the wall of the lake basin. Many lakes in semi-arid regions lie in basins without any kind of effluents, loosing water only by evaporations and such lakes may be termed 'Closed' in contradiction to open lakes having an effluent. Limnology has come a long way since the time of Forel (1892), in understanding the dynamics of the standing Waterbodies. Forbes, (1887), put it many years ago, that much of the intellectual significance of limnology.

The Indian sub-continent with its strategic position presents unique and varied physical, geographical and meteorological features. In out country, from the ancient times lakes and impoundments have provided water for domestic uses, recreation, fisheries as well as transportation that was particularly important prior to modern commerce. The lentic habitats present a variety of features, because of their distribution ranging from high altitude to the arid plateau of our country. The organisms of these varieties of lentic habitats have aroused immense interest of many limnologists and hydrobiologists. The humble beginning dating back to the later half of the century, fresh water were carried out by Sewell (1934); Ganapati (1940) and others. Initially such studies were directed mostly towards gathering baseline information on hydrochemistry, planktonic organisms etc.

In the recent years a remarkable contribution was made in this field and different trends of studies were observed towards an systemic approach to understand hydrobiology. The long way is crossed in understand the different aspects of limnology. It is evident from the above mentioned literature, that dynamics of Waterbodies, large and small are governed by the seasonal fluctuations typical to tropical areas, as that of our country.

The problems of increasing fertility of lakes as they age, is a natural phenomenon and proceeds gradually throughout their life times. This process of eutrophication normally takes centuries, though some lakes recently experienced a dramatic increase in this rate creating eutrophic conditions within decades or less.

Freshwater lakes and reservoirs are decaying and dying at an increasing rate. Man's action and inactions are the major causes of their degeneration. Water quality have been declined by a variety of chemical species. The process of eutrophication describes nutrient enrichment of the Waterbodies from an initially nutrient poor (Oligotrophic) condition. Addition of nutrients to the waterbody gradually builds up the in-lake concentrations of nutrients, where quality detriment occurs at which stage the waterbody is termed eutrophic.

Elevated nutrient levels in water allows biomass productivity to increase dramatically, leading to the water quality deterioration, reducing aesthetic values. This accelerated or cultural eutrophication, is caused by an increase in phosphorus and nitrogen exports from the land to Waterbodies arising from man's activities in their catchments. Agricultural intensification *i.e.*, increasing use of nitrogen and phosphorus based fertilizers and urbanisation process have been implicated as the main sources of this nutrient exports. The effects of cultural eutrophication are primarily manifested in lakes, reservoirs, tanks, ponds, pools as their quiescent conditions cause the nutrients present in the water to accumulate.

Although the process of eutrophication is a natural one, the associated time span can be measured in the geological time frame. However, over the last few centuries man has rapidly increased his use of nutrients loading to a lake or reservoir to encourage plant growth (bloom forming), increase in biomass, eventually resulting in a large quantity of dead plant and animal matter sinking to the bottom, and which will be oxidized by bacteria leading to the depletion of oxygen, and releasing volumes of reduced gases like methane, hydrogensulphide etc., This process may spoil the total water system which is an essential one.

The present work is an attempt in the direction of understanding the influence of different physical, chemical parameters on the planktonic life. The increase in the nutrient level from the past have created an immense pressure on Jagat tank. It is of particular interest to note that such aspects in one of the Waterbodies, which fed by sewage waters of the town and interaction of various human activities. Comparison of early works of Gaddad *et al.*, (1983) and Vijaykumar and Paul, (1992), and the present status of the nutrient level and the extent of eutrophication can be assessed. Efforts are made on the whole to contribute hydrobiological information on self containing standing Waterbodies of smaller dimensions with a view to understand their dynamics in the context of the changing mode of man-made disturbances in the present time, where in the name of development, every natural environment is interfered with.

Topography

Gulbarga is a town situated in the Northern part of Karnataka State. ($76^0/41^0$ to 77^0-42^0 longitude and 16^0-12^0 to 17^0-46^0/latitudes), located 452 meters above Mean Sea level. Gulbarga presents a climatic typical of peninsular South India with semiarid conditions. The temperature varies between 16^0C to 45^0C and the average rain fall being 543 millimetre per annum. This area falls under maidan zone as described by David *et al.*, (1974), and typically has an undulating contour, thus making scope for depressions and catchment areas. Thereby numerous small ponds often loosely termed as tanks are common sights in this area. Many of these ponds or tanks are perennial or seasonal, depending upon their morphology and normal rain fall of the region. All Waterbodies are fed by monsoon rains or streams, which serve the inhabitants directly or indirectly for agricultural, irrigation, fishing, recreational purposes and other human activities.

Trophic Status of Waterbody

This tank is situated in the heart of the city, and is also known as Sharana Basaveshwara Tank. The eastern, northern and the southern boundaries are flanked by roads. A part of Western bank is also having a small 10 feet mud road. The tank is having Prosopis,

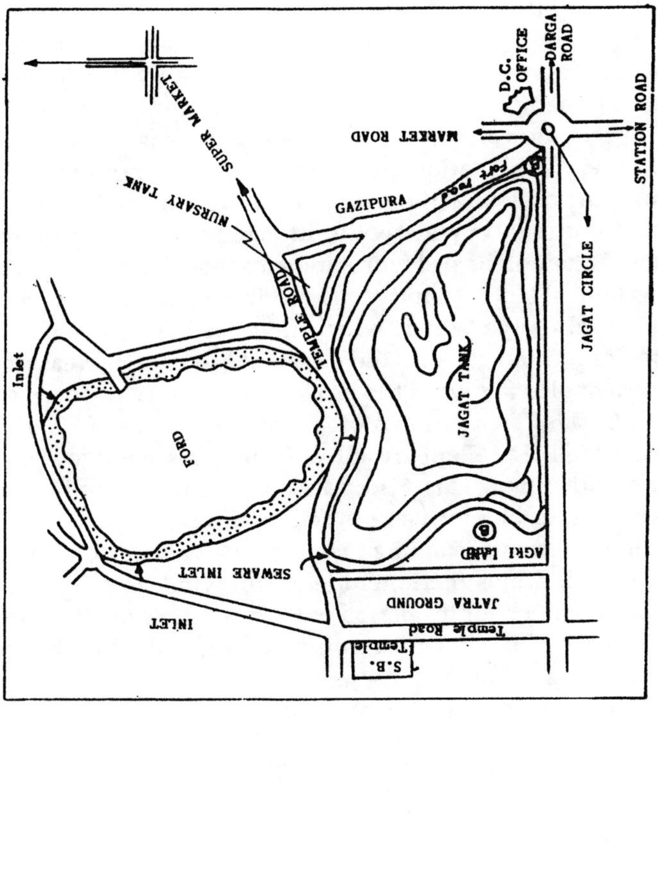

Fig. 1 : Map of Jagat Tank, Gulbarga

Acacia and other vegetation towards the western side. Towards the north western points the tank is linked with a narrow channel with the adjacent moat of the Gulbarga fort, through which a part of the city sewage drains into the tank. The moat is filled with thick growth of *Eichornia crasipis.*

The tank spreads over an area of 0.275 square kilometres, the mean depth of tank is 2.1 metre with a maximum depth of 3.5 metre during the rainy seasons, and the maximum being 1.2 metre during the dry seasons. The tank is fully utilized by Parkithwar fisherment Co-operative Society Ltd., for the fishing activities for the past 28 years the various features mainly drainage of sewage into tank, fishery exploitation and human utilization of the tank for various purposes and the problem of eutrophication, that deteriorates the quality of waterbody, attracted our keen attention to understand the details of Hydrobiological status and nutrient level of this tank. Hence, a study was conducted for assaying the nutritional level of the water, which favoured the growth of planktonic flora, often causing the seasonal blooms. Increasing level of nutrient also favoured the growth of other Angiosperms, like *Ipomea, Cypers, Trapha,*traces of *Eichornia, Pistia and Lemma* growing very well. Excellent growth of some other benthic algae like, *Nitella* and others was also noticed. Generally only the summer and winter blooms were found to appear.

The information pertaining to various climatic factors such as, total rain fall, atmospheric temperature, wind velocity and humidity was obtained for the period of February 1994 to February 1995. However, the data pertaining to rain fall the above information is referred in the present context.

The chemical and physical parameters were analysed according to APHA (1985). Other planktonic analysis were analysed according to standard limnological methods.

RESULTS AND DISCUSSION

Physico Chemical Parameters

Hydrogen ion concentration

In the present investigation it was observed that, pH played an important role in the formation of blooms. It has been observed that higher pH (8.7 to 9.1, Table 4) favoured the growth of blooms. Similarly Nazneen (1980), stated that the bloom formed due to high pH level. In the Jagat Tank the pH was alkaline throughout the study period, mainly due to bicarbonate as evidenced by methyl orange alkalinity. The same was also observed in tropical water by George *et al.*, (1986). Gadded *et al.*, (1983) and Vijaykumar and Paul, (1992), were also documented on the Jagat Tank, the pH ranges from 8 to 9.07 and 7.36 to 9.33 respectively. The alkaline nature of waterbody favoured the maximum growth of mixed bloom dominated by blue green algae and other forms. Many other workers have also correlated the higher pH and the formation of blooms (Philiphose, 1959; Verma *et al.*,1978, Sharma *et al.*1981, George *et al.*,1986 and Prasad, 1990).

Temperature

Since the work was carried out during post monsoon and winter months, the duration of sunshine was lesser and the air temperature and water temperature was low. The tank being shallow responded quickly to these changes and such observations have been made by Verma *et al.*, (1969), Vijayaraghavan, (1971), and George *et al.*, (1986) in tropical water. The same pond showed the water temperature ranging between the maximum of 28°C and minimum of 19°C as recorded by Vijaykumar, (1992).

Seasonal changes in temperature of water varied between 20°C to 28°C and 20°C to 27.5°C in site A and B respectively, (Table 4). The maximum bloom formation observed at 28°C, (October to November) and as the temperature decreased the disappearance of the bloom was noticed. It confirms the importance of temperature in bloom formation. The bloom was mixed and dominated by blue green algae, while less frequency of algal groups. The other mixed bloom dominated by *Cyclotella meneghiniana* and other diatoms was observing during low temperature period (December to February). Many other workers have also observed the importance of temperature in the growth and periodicity of blue green algae, (Fritsch, 1907; Gonzalves and Joshi, 1946; Prescott, 1938; Hutchinson, 1967; Singh and Swarup, 1979 and George *et al.*, 1986).

Free Carbondioxide

In the present investigation rapid fluctuation of free carbon-dioxide has been observed. Highest values of free carbondioxide was observed during November in both the sites A and B (19.58 ppm, Table 4). Similarly the fluctuation of carbon-dioxide level have also been observed by Gonzalves and Joshi, (1946) and Tripathi *et al.*, (1985). In the present study it was also observed that during the maximum bloom period (October and November) there was maximum carbon-dioxide level. It is probably due to photosynthetic activity of blooming algae which has caused free carbon-dioxide level to go down to 1.1ppm during December and correspondingly total phytoplankton reduced (except dominance of *Cyclotella meneghiniana* and other diatoms, Table 2).

Similarly other workers have also been correlated maximum occurrence of phytoplankton with the maximum quantity of dissolved carbon-dioxide (Prasad, 1990; Singh and Swarup, 1979) on the other hand Schindler, 1971 and Cole, 1975 concluded that the free carbon-dioxide supply rarely limits the growth of phytoplankton Alternately the photosynthetic activities of phytoplankton adopts by utilizing bicarbonate as carbon source.

Alkalinity

Phenolpthalin and methyl orange alkalinity contributes the total alkalinity. The phenolpthalin alkalinity is an indication of higher rate of carbon assimilations. It was almost all equal at two sites ranging from 10.1 to 24.1ppm throughout the sampling period (Table 4).

Methyl orange alkalinity showed higher concentration during November which agrees with George *et al.*,(1986); Prasad, (1990) and Tripathi *et al.*, (1985). It can be attributed to the rain occurring in October, which may have brought surrounding sewage and waste water effluents thereby increases the alkalinity. Alikunhi, (1957), states that the highly productive water have to be above 100 ppm alkalinity. Similarly Philiphose, (1959) and Gosh *et al.*, (1974), have correlated the high contents of alkalinity with the high productivity. In the present work it confirms that the water of

Jagat Tank is a hard one and total alkalinity ranges from 144.3 ppm to 321.2 ppm (Table 4). It has also been supported by hardness analysis (Table 4), supports this view. The increase or decrease in bicarbonate may respectively be ascribed to the photosynthetic or respiratory activity of the algae. This observation is in agreement with Ganapati, (1940).

Dissolved Oxygen

In the present study dissolved oxygen varies from 3.0 ppm to 11.4ppm (Table 4). The maximum dissolved oxygen was noticed in the month of October and November and lowest in the month of March and September. Correspondingly the phytoplankton productivity was also similar to that of dissolved oxygen (Table 4). This could be attributed to the intensive photosynthetic activity of phytoplanktons. Similar observations were also made by various workers on dissolved oxygen and productivity (Ganapati, 1960; Gosh *et al.*, 1974; George *et al.*, 1986 and Prasad, 1990).

Chlorides

Chlorides are not harmful beyond 250 ppm, however it imparts a peculiar taste to water rendering it unacceptable from the point of drinking purposes. It has been observed by various workers, thus the high chloride contents was due to pollutions make by animals and human activities (Kenwood, 1920; Thresh *et al.*, 1944 and Ganapati, 1960). However so far no one has given satisfactory explanation of chloride on phytoplankton. In the present work it was observed that maximum chloride content (259.3 ppm, Table 4), was during September. During this period the South-west monsoons are very meagre. However during October when North-East monsoons sets heavy rain occurred and the chloride content starts gradually decreasing, upto January (198.85 ppm). Similarly Vijayakumar, (1992), recorded the maximum chloride content in the same tank (279 ppm) during South-west monsoons. However other workers have reported higher concentration of chlorides during summer in tropical water (Roy, 1955; Zafar, 1964; Venkateswaralu, 1969 and Olania *et al.*, 1976).

Dissolved Organic Matter

In the present study not much significant variation in dissolved organic matter was observed in both site A and B. As the dissolved organic matter is high, ranging from 8.1 to 10.0 ppm (Table 4) indicates the eutrophic nature of the lake. All the other workers who have worked on the eutrophic lake found high dissolved organic matter (Gonzalves and Joshi, 1946; Fritsch and Rich, 1913; Hodgets, 1922; Lind, 1938). In the present investigation dissolved organic matter favoured the growth of mixed bloom dominated by blue green algae and chlorococcales and Euglenoid members (Pearsall, 1932; Rao, 1955). It was also observed that from December to February, gradual decrease in the dissolved organic matters indicating the presence of mixed bloom dominated by *Cyclotella* and other diatoms. Similarly other workers have also observed that, high dissolved organic matter favoured the growth of Euglenoid members and chlorococcales (Hosmani and Bharati, 1974 and 1980a and 1980b; Seenayya, 1971; Singh, 1959 and Munawar, 1970 and 1972).

Nitrites

In the present study the concentration of nitrites fluctuates from season to season. In the month of October to December the nitrites content was high 0.129 ppm to 0.215 ppm (Table 4). It was also observed that during this period the mixed bloom dominated by blue green algae. Further, in the present study it was observed that the high nitrite contents promoted the mixed bloom dominated by *Cyclotella meneghiniana* and other diatoms (Table 2). The gradual decrease in nitrite concentration in the month of January and February was probably due to consumption by mixed bloom dominated by *Cyclotella* and other diatoms. Similarly Bharati and Hosmani, (1973), attributed low level of nitrites may be due to utilization by phytoplanktons. Contrary to our observation George *et al.*, (1986), recorded high level of nitrites during winter months. The observation made in the present study may suggest that nitrite is an intermediatory compound, which has been converted into nitrates or molecular nitrogen. This will be again fixed by nitrogen fixing organism. That may be the result the higher concentration of nitrites during the month of October to December may be due to the activity of mixed bloom dominated by blue green algae and other nitrifying organism in the waterbody.

Phosphates

According to Schindler, (1971), phosphate play an important role on algal population and some times it acts as a primary limiting factor in lentic waters. The main source of phosphate is sewage pollution (Welch, 1952; Hutchinson, 1957 and Horne, 1978). According to Ganapati, (1960) in tropical water, phosphates are always present in sufficient quantity. In the present study it was observed that the phosphate concentration was high during summer (February to March) and gradually decreased during rainy season (September to November). The phosphate level gradually increased during winter (December to February, Table 4). On the other hand as the cyanophycean bloom (Table 2) increased during the month of October, the phosphate level decreased. The fall of phosphate level during the time of bloom may be due to its utilization by the mixed bloom dominated by blue green algae. Further, in the winter period alongwith the gradual increase of phosphate may have favoured the other mixed bloom dominated by *Cyclotella meneghiniana* and other diatoms (Table 2). Other workers have also observed the phosphate level in various Waterbodies as we have observed in the present study (Gonzalves and Joshi, 1946; Singh, 1959; Zafar, 1966 and Tripathi *et al.*, 1985). Out observations more appropriately fits with the results made by Seenayya and Raju, (1972), according to them the phosphates were high just prior to the development of bloom. Their concentration decreased with the multiplication of algae and exhausted when the mixed bloom was dominated by blue green algae attained the climax.

Primary Productivity

The primary production involves Chaemo autotrophic processes, forming the base of energy flow in the ecosystem. An understanding of primary productivity becomes all the more essential, in the evaluation of efficiency of energy utilization at different trophic level. In the present study, Gross primary productivity, Net primary productivity and community respiration are more or less similar, in site A and B. The gross primary productivity and net primary productivity almost similar in the month of February and March. As and when bloom sets, there is a gradual increase in both gross primary productivity and net primary productivity (September to November). Further both were declined from December to February (Table 5). The community respiration values always show variation as it was found in primary production throughout the study period. From the data obtained in the present study with the primary production and physico chemical parameters, it was found that except phosphates, dissolved oxygen, free carbondioxide, carbonates, bicarbonate and nitrites, other parameters may not be much correlated in support of primary production in general. However the carbondioxide, carbonates and bicarbonates were in sufficient quantity during the time of bloom formation alongwith the gradual increase in dissolved oxygen. It was observed that the maximum photosynthetic activity during bloom period can be correlated with the net oxygen produced and carbondioxide fixed (Table 1 and 5).

On the other hand the phosphates and nitrites are deemed to be considered as nutrients are also available in sufficient quantity, in the waterbody. The phosphates were maximum during the months of February to September and it was reduced during the time of climax of the bloom, indicates the role of phosphates for the growth of phytoplankton, in turn primary productivity. The increased nitrite value during bloom period indicates the maximum activity of phytoplankton, the consumption and recycling of nitrite by nitrifying bacteria and blue green algae, which formed the domination in the bloom. The other workers have also correlated primary production with physico chemical factors involved. Among various workers Prasad, (1990), opined that the primary production depends on nitrogen and phosphates which favours maximum growth of phytoplankton. Ayyappan and Gupta, (1985), recorded gross primary productivity from a tank in coastal Karnataka, due to the presence of high level of free carbondioxide. Khan and Siddiqui, (1971) observed high gross primary productivity is due to transparency, temperature and phosphates. Fog and Wolfe,

(1954), Sawyer, (1966) Ketchum, (1969), Schindler and Fee, (1975) opined that among chemicals and physical parameters the and phosphates, other nutrients and radiant energy play an important role in primary production.

Algal Blooms

Some times fertile waterbody show extensive coluration and massive occurrence of algal forms resulting a "Bloom" also called algal blooms or water blooms. The bloom may be pure (dominated with single species) or mixed (dominated with many number of species). The blooms may be permanent (persists for longer periods) or temporary (appears and disappears in shorter periods). Blooms are usually observed in lakes, tanks, ponds and pools, when masses of blue green algae float at the surface on calm days (Round, 1979). Blooming in temperate waterbody usually occurs only in late summer, but in tropical lake it can be permanent feature (Round, 1979). Some blooms do occur the rainy season and the reasons attributed to this may be in the influx of organic matters. The occurrence of bloom in various Waterbodies throughout the country have been extensively studied by lyengar, (1938) and Ganapati, (1940 and 1957). In the present investigation it was observed that the maximum persistence of bloom during the month of October when the rain falls was maximum, and the lake received the maximum influx. During this time it was found that the mixed bloom dominated by blue green algae and other forms. Among blue green algae *Microcystis aeruginosa, Microcystis pulverea, var, incerta, Anabaena spiroidis, Spirulina major, Spirulina princeps, Oscillatoria* species, *Merismopedia minima. Merismopedia glauca* (Table 2) were observed as major bloom forming algae. Besides *Cyclotella meneghiniana*, another form appears to be large in number, the other forms of diatoms. The other chlorococcalean members like *Kirchineriella obesa, Oocystis gigas, Micrasteiris quadridenta, Oocystis gloeocystiformis, Scenedesmus quadricauda, S. dimorphus, S. caudato-acculeatus, S. armatus, Pediastrum duplex.* The Euglenineae members like *Trechlomonas bulla, Euglena minuta, E. elastica, E. acus, and Leponcillis ovum* (Table 2), were also part of the bloom. Similarly other workers have also made an observation on mixed bloom dominated by blue green algae (Ganapati, 1940 and 1959; Huber pestolozzi, 1938).

Table 1 : Physico-chemical parameters of the water during Cynophycean bloom

No.	Parameter	Before bloom (Feb to Sep)	During bloom (Oct to Nov)	After bloom (Dec to Feb)
1.	Temperature			
	(a) Atmosphere	28.12	26.60	22.60
	(b) Water	25.33	25.25	21.80
2.	pH	8.98	8.96	8.85
3.	CO_2	10.90	15.53	0.18
4.	Chlorides	219.44	217.93	204.10
5.	Dissolved Oxygen	4.61	10.10	8.83
6.	Phosphates	0.83	0.28	0.465
7.	Nitrites	0.043	0.124	0.146
8.	Dissoled organic matter	9.71	9.30	8.15
9.	Hardness			
	(A) Total	256.75	282.12	252.90
	(B) Permanent	173.50	167.00	150.16
	(C) Temporary	83.25	115.12	102.74
10.	Alkalinity			
	(A) Phenolpthalin	19.73	13.07	13.13
	(B) Methyl orange	187.80	288.60	207.35
	(C) Total	206.72	301.67	220.48
11.	Primary productivity			
	(A) NPP	3.13	7.35	1.69
	(B) GPP	3.14	7.79	1.95
	(C) CR	0.145	0.172	0.166

The factors, that promote such abundant growth seems to be shallowness, warm temperature, high nitrogen and phosphate contents and a good supply of half bound carbonates (Deshikachary, 1959). Many workers have observed that the temperature is an important factor, that promote the growth of phytoplankton (Fritsch, 1907; Pearsall, 1923; Godward; 1934; Prescott, 1938; Philiphose, 1959; Singh, 1959; Bharati and Hosmani, 1975; Watanbe, *et al.*, 1986). In the present findings we have also observed high temperature alongwith nitrites, phosphates, high alkalinity, rich inorganic matter and chlorides (Table 1) favoured the growth of mixed bloom dominated by blue-green algae. King,(1970), Shapiro, (1973) and Ganapati, (1940) have also made the similar observation as that of us. In the opinion of Fritsch, (1907) and Clarke, (1924), the blue-green

Table 2 : Seasonal occurrence of phytoplanktons per litre

No.	*Name of Organics*	*Before bloom (Feb to Sep)*	*During bloom (Oct to Nov)*	*After bloom (Dec to Feb)*
1.	*Microcystis aeruginosa*	$392X10^3$	$1204X10^3$	$140X10^3$
2.	*Microcystis pulvarias, var incerta*	$736X10^3$	$2660X10^3$	$336X10^3$
3.	*Anabaena spiroides*	$616X10^3$	$2828X10^3$	$224X10^3$
4.	*Oscillatoria*	$84X10^3$	$56X10^3$	$308X10^3$
5.	*Merismopedia glauca*	$364X10^3$	$168X10^3$	$112X10^3$
6.	*Merismopedia minima*	$420X10^3$	$140X10^3$	$112X10^3$
7.	*Spirulina princeps*	$28X10^3$	$56X10^3$	$28X10^3$
8.	*Spirulina major*	$196X10^3$	$392X10^3$	$56X10^3$
9.	*Cyclotella meneghinaina*	$868X10^3$	$84X10^3$	$2800X10^3$
10.	*Navicula rhombodis*	$56X10^3$	$28X10^3$	$308X10^3$
11.	*Gramphonema gracile*	$56X10^3$	$28X10^3$	$224X10^3$
12.	*Synedra ulna*	$28X10^3$	$28X10^3$	$504X10^3$
13.	*Actinastrum hantzschii*	$56X10^3$	$28X10^3$	$112X10^3$
14.	*Kirchineniella obesa*	$56X10^3$	$84X10^3$	$56X10^3$
15.	*Schroederia indica*	$28X10^3$	$140X10^3$	$56X10^3$
16.	*Oocystis gigas*	$168X10^3$	$140X10^3$	$56X10^3$
17.	*Scenedesmus quadricauda*	$252X10^3$	$56X10^3$	$140X10^3$
18.	*Scenedesmus dimurphus*	$112X10^3$	$112X10^3$	$168X10^3$
19.	*Scenedesmus caudato-acileatus*	$112X10^3$	$56X10^3$	$84X10^3$
20.	*Scenedesmus armatus*	$112X10^3$	$56X10^3$	$196X10^3$
21.	*Oocystis glo-ecysformis*	$112X10^3$	$56X10^3$	$140X10^3$
22.	*Pediastrum duplex*	$84X10^3$	$56X10^3$	$196X10^3$

Table 2 : Contd..

No.	Name of Organics	Before bloom (Feb to Sep)	During bloom (Oct to Nov)	After bloom (Dec to Feb)
23.	*Micrasteiris quadridenta*	$112X10^3$	$84X10^3$	$56X10^3$
24.	*Trechlomanas bulla*	$140X10^3$	$84X10^3$	$84X10^3$
25.	*Euglena minuta*	$112X10^3$	$168X10^3$	$56X10^3$
26.	*Euglena elastica*	$28X10^3$	$112X10^3$	$56X10^3$
27.	*Euglena acus*	$196X10^3$	$84X10^3$	$112X10^3$
28.	*Leponcilis ovum*	$56X10^3$	$84X10^3$	$56X10^3$
	Total	$5600X10^3$	$9072X10^3$	$6748X10^3$
	Average during different periods	$200X10^3$	$324X10^3$	$241X10^3$

algae are the characteristics of tropical water and largely forms the water blooms, because these waters are usually eutrophic in nature. In the present study the second successive mixed bloom was appearing in the months of December to February. Dominated by diatom such as *Cyclotella meneghiniana, Navicula rhomboidis, Gamphonena gracile, Synedra ulna* (Table 2). The other associated forms like *Actinastrum hantzsthi, Schroederia indica* (Table 2) observed. The other members in less concentration was also observed during this period as that the forms appeared during Cyanophycean bloom.

Many workers have made an attempt to correlate various physico-chemical factors with the abundance of diatoms (West and West, 1912; Pearsall, 1923; Atkins, 1923; Lind, 1938; Hustdt, 1938; Rao, 1953; Singh, 1964; Philiphose, 1959 and Zafar, 1967). Pearsall, (1923) and Zafar, (1967) correlated high calcium contents favoured the growth of diatoms. Rutner, (1953) observed excess phosphate favoured the abundance of diatoms. Singh, (1964), West and West, (1912); Munawar, (1970) and Bharati and Hosmani, (1975) opined that low level of organic matter favoures the growth of diatoms. However in the present study it has been observed that the bloom dominated by diatoms may possibly be activated by, low organic matter phosphate and carbondioxide alongwith high nitrite and alkalinity level.

Table 3 : Annual rainfall (mm) during the period of investigation *i.e.* (94-95)

Feb.	Nil
Mar.	Nil
April	42.5
May	2.8
June	46.3
July	55.5
Aug.	61.1
Sep.	32.2
Oct.	222.5
Nov.	1.5
Dec.	Nil
Jan.	83.7
Total	548.0

Total rainfall is 548.0 mm/Annum.

Algal blooms are harmful in several ways, some algae when present in large concentration produce unpleasant taste and odour to drinking water, or their extra cellular product may liberate substances that are deterious to aquatic life. Water blooms are reported to have caused the mortality of fish and animals. Toxic bloom forming algal species like *Microcystis aerugenosa, M. flosaquae, Oscillatoria* species and *Anabaena* species etc., (Gorham, 1964) can cause these effects.

During present study two distinct type of algal blooms have been recorded. Both blooms are mixed (Table 2), first one is dominated by Cyanophycean members like *Microcystis aerugenosa, M. pulverae, var, incerta, Anabaena spiroids, Spirulina princeps, S. major,* and *Merismopedia minima, M. glauca,* and *Oscillatoria* species (Table 2). The other associated forms like some diatoms, Chlorococcales, Euglenineae members are very few in numbers.

The second bloom was dominated by diatoms like, *Cyclotella meneghiniana, Navicula rhomoboidis, Synedra ulna, and Gamphonema gracile.* The associated forms are the members of chlorococcales, cynophyeeae and Euglenineae, (Table 2). It was also observed that during the time of Cyanophycean bloom the

Table 4 : Monthly average values of physico-chemical parameters of Jagat tank during the period of study (1994-95)

Parameters Months		*pH*	*Temperature (°C)*		*Free CO_2 (ppm)*	*Alkalinity (ppm)*			*DO_2 (ppm)*
			Air	*Water*		*PA*	*MOA*	*TA*	
1	2	3	4	5	6	7	8	9	*10*
Feb.	Site A	8.95	26.5	23	13.2	16.5	206.4	222.9	4.7
	Site B	8.97	25.95	22.25	14.41	18.2	206.0	224.2	5.0
March	Site A	8.99	32.5	28.25	5.5	24.1	177.1	201.2	3.0
	Site B	9.05	31.25	27.5	8.36	21.0	168.0	189.0	3.6
April	Site A	9.0	31.5	28.5	8.0	11.0	180.0	19.0	6.0
	Site B	9.1	32.0	28.0	8.6	15.5	198.0	213.5	6.5
May	Site A	8.5	30.0	26.0	12.2	18.0	195.0	213.0	7.5
	Site B	8.6	30.5	26.5	12.4	19.0	192.0	211.0	7.8
June	Site A	9.0	30.0	28.0	10.5	14.5	220.0	234.5	8.0
	Site B	8.8	29.80	28.0	11.2	14.0	220.0	234.0	8.2
July	Site A	7.8	28.0	25.0	15.0	16.8	230.0	246.1	10.6

Contd..

Table 4 (Contd.)

1	2	3	4	5	6	7	8	9	10
	Site B	8.0	28.0	25.8	15.5	15.0	235.0	250.0	11.0
August	Site A	8.5	29.5	26.0	13.0	12.0	185.0	197.0	7.0
	Site B	8.7	28.9	26.5	13.4	12.0	193.0	205.0	6.6
Sept.	Site A	8.9	26.5	25.5	11.66	19.0	181.4	200.4	5.4
	Site B	8.9	26.25	24.5	12.29	19.6	188.0	202.6	6.0
Oct.	Site A	8.8	28.5	27.0	11.0	12.2	279.1	291.3	8.6
	Site B	8.8	28.0	27.0	11.77	14.7	267.6	282.3	9.0
Nov.	Site A	9.0	25.5	23.5	19.80	14.7	306.0	321.2	11.4
	Site B	9.1	24.75	23.5	19.58	14.7	301.8	316.4	11.4
Dec.	Site A	8.7	21.0	20.0	1.1	11.3	262.0	273.3	7.8
	Site B	8.8	20.0	20.0	AB	12.5	275.0	287.5	8.2
Jan.	Site A	8.0	22.5	21.0	AB	10.1	213.1	223.2	9.4
	Site B	8.8	21.5	21.0	AB	11.1	227.3	238.4	9.2

Contd..

Table 4 (Contd.)

Parameters Months		*Chlo-rides (ppm)*	*Dissolved organic matter (ppm)*	*Nitrites (ppm)*	*Phosphates (ppm)*	*Hardness (ppm)*		
						Per. H	*Temp.H*	*Total H*
1	*2*	*3*	*4*	*5*	*6*	*7*	*8*	*9*
Feb.	Site A	193.9	9.7	0.02	1.14	169.5	55.5	225
	Site B	193.9	9.6	0.027	1.12	166.0	57.5	223.5
Mar.	Site A	209.0	9.9	0.033	0.88	168.0	101.0	269.0
	Site B	206.15	10.0	0.038	0.83	173.5	96.5	270.0
Apr.	Site A	210.10	8.5	0.52	0.50	180.0	70.5	250.5
	Site B	211.00	8.8	0.55	0.53	182.0	70.0	252.0
May	Site A	220.00	9.2	0.110	0.70	140.0	105.0	245.0
	Site B	219.16	9.2	0.102	0.76	141.5	102.0	243.5
June	Site A	240.0	8.0	0.119	0.38	130.0	128.5	258.5
	Site B	232.0	8.2	0.123	0.44	133.0	130.5	263.5

Contd..

Table 4 (Contd.)

1	2	3	4	5	6	7	8	9
July	Site A	201.0	9.0	0.140	0.85	150.0	114.5	264.5
	Site B	215.5	9.0	0.146	0.97	153.0	120.0	273.0
Aug.	Site A	195.5	8.8	0.85	0.30	161.0	119.0	280.0
	Site B	198.0	9.1	0.92	0.27	162.5	120.5	283.0
Sept.	Site A	259.0	9.5	0.069	0.52	184.0	88.5	272.5
	Site B	255.0	9.6	0.072	0.505	180.2	100.5	280.5
Oct.	Site A	228.0	9.7	0.129	0.36	18.0	95.0	275.0
	Site B	229.05	9.7	0.135	0.39	184.5	94.5	279.0
Nov.	Site A	207.3	9.1	0.116	0.18	138.0	144.0	284.5
	Site B	207.38	8.9	0.118	0.20	165.5	124.5	290.0
Dec.	Site A	208.0	8.1	0.215	0.23	140.5	98.5	239.0
	Site B	212.6	7.9	0.225	0.245	144.0	97.0	241.0
Jan.	Site A	198.85	8.1	0.186	0.59	144.0	106.5	250.5
	Site B	195.8	8.2	0.191	0.65	151.5	105.5	257.0

Table 5 : Net primary productivity, Gross primary productivity and Community respiration ($gC^{-3}d^{-1}$) in site A and B during 94-95 season

Months		*NPP*	*GPP*	*CR*
Feb.	Site A	1.93	1.74	-0.34
	Site B	1.78	2,06	0.06
March	Site A	2.65	2.4	0.195
	Site B	3.2	2.53	0.0
April	Site A	1.01	1.60	0.50
	Site B	1.10	1.51	0.64
May	Site A	0.80	0.74	0.75
	Site B	0.95	0.95	0.62
June	Site A	1.30	2.0	0.38
	Site B	1.28	1.95	0.43
July	Site A	1.40	1.20	0.60
	Site B	1.36	1.35	0.52
August	Site A	0.70	0.85	0.40
	Site B	0.74	0.98	0.44
September	Site A	4.96	5.6	0.56
	Site B	4.26	4.55	0.395
October	Site A	8.95	9.58	0.67
	Site B	10.35	10.85	0.46
November	Site A	5.03	5.35	-0.7
	Site B	5.1	5.4	0.26
December	Site A	2.4	2.2	-0.13
	Site B	1.86	1.66	0.2
January	Site A	2.19	2.33	0.13
	Site B	1.98	2.2	0.195

Cyclotella meneghiniana occurs less frequency (Table 2). Vance, (1965) found that Microcystis species produces inhibitory substances that check the growth of other algae, such as Navicula, Chlamydomonas and others. The present study is an agreement with an observation made by Vance, (1965).

The total number of different forms occurred in different times have been shown in Table 2. The corresponding physico chemical factors during different periods are also shown in Table 1.

Conclusion

Limnological study of heavily polluted waterbody situated in Gulbarga has been made with regard to phytoplankton periodicity, productivity and physico-chemical parameters, carbondioxide, dissolved oxygen, dissolved organic matters, chlorides, hardness, pH, temperature, alkalinity, phosphates and ni-trates have been assessed regularly for a period from February 1994 to January 1995. Attempts have been made to correlate the possible factors responsible for the formation of blooms, and the eutrophication. The results and conclusions drawn are as follows.

Two major blooms have been recorded present study, one belonging to Cynophycean members and another represented by the diatoms. During the present study, when the maximum rain falls occurred, the rise in carbondioxide, dissolved oxygen, hardness and alkalinity, alongwith high phosphates dissolved organic matters favoured the growth of Cyanophycean members, while nitrite contents also increased just after this bloom may be due to blue-green algae and nitrifying bacteria.

The bloom of diatoms mainly represented by *Cyclotella meneghiniana* appeared just after the gradual disappearance of the Cyanophycean bloom. The low temperature, organic matter, phosphates, carbondioxide, slightly low pH, with higher nitrites and alkalinity may be responsible for the growth of *Cyclotella meneghiniana*, as a bloom.

From the results obtained above one may conclude that Jagat Tank is a heavily polluted waterbody, pollution being of organic origin. This tank needs to be protected from animals and human activities. In order to maintain the purity and potability of water, the inflow of sewage and dumping of city waste towards Western end has to be avoided. The macrophytic growth has to be controlled either manually or by any other means. If this work is not taken up at an early stage, the eutrophication which has reached already an alarming stage far from aesthetic values, the tank may reach an irreversible state due to further eutrophication. Immediate measures to stop this eutrophication only can bring back the tank to its original state, preventing conversion of it into land. Thus the city of Gulbarga can maintain the Jagat tank as a beautiful recreational center and fish nursery for the future.

Acknowledgement

The authors wish to express their sincere thanks to the Chairman, Department of Botany, Gulbarga University, Gulbarga for providing necessary facilities.

References

Alikunhi, K.H. 1957 : Fish culture in India. *Farm Bulletin*, New Delhi., 20, I.C.A.R. pp.1-6.

APHA, AWWA, WPCF, 1985 : Standard methods for the examination of water and waste water, *American Public Health Association Inc.*, New York, 16th Edn.,p.1268.

Atkins, W.R.J. 1923 : The phosphate content of fresh and salt waters and its relationship to growth of algal plankton. *J. Mar. Biol. Assoc.* 13:pp. 119-150.

Ayyappan, S. and Gupta, T.R.C. 1985 : Limnology of Ramsamudra Tank - Primary production. *Bull. Bot. Soc.* Sagar,32 : pp 82-88.

Bharati, S. G. and Hosmani, S.P. 1973 : Hydrobiological Studies in Ponds and Lakes of Dharwad, Part-I. *J of K. Uni. Sci.* XVIII :pp. 246-254.

Bharati, S. G. and Hosmani, S.P. 1975 : Limnological Studies in Ponds and Lakes of Dharwad, Part-III, Occurrence of Two Euglenoid blooms, *Ibid.* 20 : pp. 353-382.

Birge, E.A. and Judey,C. 1931 : A second report of the phosphorus contents of Wisconsin lake waters. *Trans Wis. Acad. Sci. Arts. lett.* 26 : pp. 353-382.

Clarke, F.W. 1924 : The data of geochemistry *Ist Edn. Bull U.S. Geol. Surv.*770, 841, pp. 229. 550, 553, 554, 555, 566, 571, 728, 756, 791 and 821.

Cole, G.A. 1975 : Text Book of Limnology, Published by *C.V. Mosby Co.* pp. 283.

David, A. Rao, N.G.S. and Roy, P 1974: Tank fishery resources of Karnataka, *Bull. Gen. Inland fish Res. Inst.*, Barrackpore, 20 : pp. 87.

Desikachary, T.V. 1959 : Cyanophyta. *Pub. I.C.A.R.* New Delhi.

Fogg, G.E. and Wolfe, W.C., 1954, The metabolism of blue-green algae (Myxophyceae) *Symp. soc. Gen. Microbiol.*, 4: pp. 99-120.

Forbes., S.T. 1887 : The lake as a microcosm. *Bull. Peoria (illinios Scientific Assocn. 1887* : pp. 77-87., Reprinted In *Bull. ill. Nat. Hist. Surv.*, 15 : pp. 537-550.

Forel, F.A. 1892 : Le. Le'man: Monographic limnologique, F. Rogue, Laussane.

Fritsch, F.E. 1907 : The subarial and freshwater algal flora of the tropics. *Ann. Bot.*, 21 : pp. 235-275.

Fritsch, F.E. and Rich, R. 1913 : A four year observation of a freshwater pond (Bartons Ponds) near Harpenden. *Ann. Biol. Lacust.*, 6 : pp. 1-183.

Gadded, S.M., Prabhakar, M.N. and Rodgi, S.S. 1983 : Ecological studies of two polluted bodies of water at Gulbarge. *Poll. Res.*, 2(2) : pp 49-52.

Ganapati, S.V. 1940 : Ecological study of garden pond containing abundant Zooplankton. *Proc. Indian Acad. Sci.* (Bombay), 17(2) : pp. 41-58.

Ganapati, S.V. 1957 : Limnological studies of the upland waters in Madras State *Arch. Hydrobiol.*, 53 : pp. 3-61.

Ganapati, S.V. 1960 : Ecology of tropical waters. *Proc Symp. Algology. I.C.A.R.*, New Delhi, pp. 204-218.

George, J.P. Venugopal, G. and Venkatashvaran. 1986 : Anthropogenic Eutrophication in a perannial tank : effect on the growth of *Cyprinus Carpio communis. Ind. J. Environ. Hlth.*, Vol. 28 : No. 4 : pp. 303-313.

Godward, M.B. 1934 : An investigation of the casual distribution of algal epiphytes. *Bein Bot. Centralbl.* 52 : pp. 506-539.

Gonzalves, E.A. and Joshi, D.B. 1946 : Freshwater algae near Bombay. *J. Bom. Nat. Hist. Soc.*, 46 : pp. 154-176.

Gorham, P.R. 1964 : Toxic algae. In algae and man. (Ed. D.F. Jackson) Plennum. Press, New York, pp. 307-306.

Gosh, A. Rao, L.H. and Banerjee, S.C. 1974 : Studies of the hydrological condition of a sewage fed pond with a note on their role in fish culture. *J. Inland Fish Soc.* India V - pp 51-61.

Hodgetts, W.J. 1922 : A study of some of the factors controlling the periodicity of freshwater algae in nature. *New Phytologist.*, 30 : pp. 221-65.

Horne, A.J. 1978 : Nitrogen fixation in Eutrophic lakes. In *Water pollution Microbiology.* Vol. 2 (Ed. R. Mitchell) John Wiley and Sons. New York pp. 1-30.

Hosmani, S.P. and Mulgand, R.K. 1978 : Diurnal studies in a pond at Dharwad. *J. Kar. Univ. Sci.* 23 : pp. 58-62.

Hosmani, S.P. and Bharati, S.G. 1980a : Algae as indicator of organic pollution. *Phykos.*, 19 : pp. 23-26.

Hosmani, S.P. and Bharati, S.G. 1980b : Limnological studies in ponds and lakes of Dharwad. *Phykos.*, 19 : pp. 27-43.

Huber-Pestalozzi, G. 1938 : Das phytoplankton des, susswasser. In Thienmann. Minnergewasser., 16(1) p. 342.

Hustedt, F. 1938 : Systematische und okologische untersuchungen uber die. Diatomeen flora von. Java, Bali and Sumatra. *Arch. Hydrobiol. Suppl.*, 16 : pp. 1-55.

Hutchinson, G.E. 1957 : A treatise on limnology, Geography, Physics and Chemistry. Vol. I. *John Wiley and Sons Inc.*, (USA).p. 1015.

Hutchinson, G.E. (1967) : A treatise on limnology. Introduction to lake biology and limnoplankton Vol. No. II. *John Wiley and Sons Inc.*, (USA) p. 1115.

Iyengar, M.O.P. 1938 : The vegetation of Madras and its environs. *In scientific survey of Madras and its environs. univ. Madras*, pp. 52.59.

Ketchum, B.H. 1969 : The development and restoration of deficiencies in phosphorus and nitrogen content of unicellular plants. *I.Cellular Comp. Physiol.*, 5 : pp. 55-74.

Kenwood, H.R. 1920 : Public Health Lab. Work. *H.K. Lewis and Co. Ltd. London*

King, D.L. 1970 : The role of carbon in Eutrophication, *J. Water Poll. Controll. Fed.* 42 : pp. 2035-2051.

Khan, J.A. and Siddiqui, A.Q. 1971 : Primary production in tropical fish pond at Aligarh. *Op. Cit.*, 37(3-4) : pp. 447-456.

Lind, E.M. 1938 : Studies on the periodicity of the lagae in the beach of ponds, Sheffield. *J.Ecol.*, 26 : pp. 257-274.

Munawar, M. 1970 : Limnological studies on freshwater, certain polluted and unpolluted environments *Hydrobiol.*, 39(1) : pp. 105-128.

Munawar, M. 1972 : Ecological studies of Eugleninea in, certain polluted and unpolluted environments water *Hydrobiol.*, 39 : pp. 105-138.

Nazneen, S. 1980 : Influence of Hydrobiological factors on seasonal abundance of phytoplankton in Kinjhar lake, Pakistan. *Int. Revue. Ges. Hydrobiol.*, 65(2) : pp. 269-282.

Olania, M.S. Saxena, K.L. and Sharma, H.C. 1976 : Pollution studies of Chambal river and its trubutaries at Kota. *Ind. J. Environ. Hith.*, (18(3) : pp. 219-226.

Pearsall, W.H. 1923 : A theory of diatom periodicity *J. Ecol.*, 9pp. : 241-253.

Pearsall, W.H. 1932 : A suggestion as to the factors influencing the distribution of Free floating vegetation. *J. Ecol.pp*, : 241-253.

Philipose, M.T. 1959 : Freshwater phytoplankton of Inland Fisheries. *Proc. Symp. Algology I.C.A.R.*, pp 272-253.

Prasad, D.Y. 1990 : Primary productivity and energy flow in upper lake, Bhopal. *Ind. J. Environ. Hlth*, Vol 32 No. 2 pp. 132-139.

Prescott, G.W. 1938 : Objectionable algae with reference to the killing of fish and other animals. *Hydrobiol.*, 1 : pp. 1-13.

Rao, C.B. 1953 : On the distribution of algae in a group of six small ponds. *J.Ecol.*41 : pp. 62-71.

Rao, C.B. 1955 : On the distribution of algae in a group of six small ponds. *J.Ecol.*43 : pp. 291-308.

Round, F.E. 1979 : The ecology of algae. Cambridge, p. 653.

Roy, J.C. 1955 : Periodicity of the plankton diatoms of the Chilka lake for the year 1950 and 1951. *J. Bomb. Nat. Hist. Soc.*, 52 : pp. 112-123.

Rutter, F. 1953 : Fundamentals of limnology. *Univ. of Toronto, Press Toronto Canada* pp. 242.

Sawyer, C.N. 1966 : Basic concepts of eutrophication. *Sewage Industr. Wastes*, pp 38, 736.

Schindler, D.W. 1971 : Light, Temperature and Oxygen regimes of selected lakes in the experimental lakes area, North Wast Ontario. *J. Fish. Res. Bd. Can.*, 28(2) : pp. 157-169.

Schindler, D.W. and Fee, E.J. 1975 : The role of nutrient cycling and radiat energy in aquatic communities. *I.B.P. Symposium on photosynthesis and production in different environment.*, Ed., Cooper, J.P. pp. 323-344.

Seenayya, G. 1971 : Ecological studies in the plankton of certain fresh water ponds of Hyderabad, India. I. Physico chemical complexes. *Hydrobiologia*, 37 : pp. 7-31.

Seenayya, G. and Subbaraju, N. 1972 : Taxonomy and biology of blue-green algae. *Univ. Madras* (Ed. T.V. Desikachary) pp. 52-57.

Sewell, R.B.S. 1934 : Fauna of the Chilka lake. Copepoda, *Mem, Ind. Mus.*, 5 : pp. 771-857.

Shapiro, J. 1973 : Blue-green algae, why they become dominant, *Science.*, 179 : pp. 382-384.

Sharma, K.D. Lal, N. and Pathak, P.D. 1981 : Water quality of sewage drain entering Yamuna at Agra. *Indian J. Environ. Hlth.*, 23 : pp. 118-122.

Singh, C.S. 1964 : Seasonal fluctuation in diatom population. *Trop Ecol.* 5: pp. 42-51.

Singh, S.R. and Swarup. 1979 : Limnological studies of Suraha lake, (Ballia) II. The periodicity of phytoplankton. *J. Indian Bot. Soc.*, 58 : pp. 319-329.

Singh V.P. 1959 : Phytoplankton Ecology of inland waters of Uttar Pradesh. *In. Proc. Symp. on Algology, I.C.A.R* New Delhi., pp. 243-271.

Thresh, J.C., Suckling, E.V. and Beal, J.F. 1944 : In the examination of water supplies. (Ed. Taylor, E.W.).

Tripathi, A.K. and Sinha, B.K. 1985 : Algal pollutants of Unnao ponds, I. Cyanophyceae. *In advances Applied Phycology Op. Cit.*, pp. 278-279.

Vance, B.D. 1965 : Composition and succession of Cyanophycean water blooms. *J. Phycol.*, 1 : pp. 81-86.

Venkateswaralu, V. 1969: An ecological study of the algae of the river Mossi, Hyderabad, (India), with special reference to water pollution. I. Physico chemical comples. *Hydrobiologia.*, 33 : pp. 117-143.

Verma, A.K., Tyagi and Dalela, R.C. 1978a, Physico chemical and biological characteristics of Kedarbad drain. *Op. Cit.*, 20 : pp. 1-13.

Verma, S.R. and Shukla, G.R. 1969, Pollution in perennial stream "Khala" by the sugar factory effluent near Laskar District Saharanpur, U.P. *Ind. J. Environ. Hlth.*, 11 : pp. 145-162.

Vijayakumar, K. and Paul, R. 1992, Comparative study of primary production in two standing Waterbodies of Dharwad, Karnataka. *Environ and Ecolo.*, 10(2): pp 262-264.

Vijayakumar, K. 1992, Limnological studies on perannial and seasonal Waterbodies of Gulbarga area. *Ph.D. thesis – 1992.*, Gulbarga University, Gulbarga.

Vijaya Raghvan, S. 1971, Studies on diurnal variations in physico chemical and biological factors in Teppalulam Tank. *Proc. Ind. Acad. Sci.*, 6 : pp. 63-73.

Watanabe, M.F. and Masayki Watanabe. 1986, The distribution and relative abundance of bloom forming *Microcystis* species in several eutrophic waters, Japan, *J. Limnol.*, 47, 1: pp. 87-93.

Welch, P.S. 1952, Limnology, McGraw Hill Book Co. New York, Toronto and London (Second Edition) : p 538.

West, W. and West, G.S. 1912, A monograph of British Desmidaceae, Vol.IV. The Roy Soci. London.,

Zafar, A.R. 1964, On the ecology of algae in certain fish ponds of Hyderabad, India I. Physico chemical complex. *Hydrobiologia.*, 23 : pp. 176-196.

Zafar, A.R. 1966, Limnology of Hussain sagar lake, Hyderabd, India. *Phycos.*, 1: pp. 115-126.

Zafar, A.R. 1967, On the ecology of algae in certain fish ponds of Hyderabad, India III.The periodicity. *Hydrobiologia.*, 30 : pp. 96-112.

5

Limnological Studies of a Village Pond in Tamil Nadu

M.B. Raghunathan and K. Revathi

Introduction

With a view to study the plankton fauna of ponds in Chingleput district, Tamil Nadu, long term investigations were carried out in twelve ponds. Three ponds were of temporary nature and water was noted only during monsoon seasons. Among the remaining nine ponds investigated, the presence of algal blooms was a continuous feature. Besides algal blooms in Kovur pond macrophytes were also noted during most of the seasons.

During the investigation period macrophytes were manually removed in May 1981. Inspite of this, macrophytes continued to spread the pond. The pond receives the sewage water from the nearby houses situated. The water was mainly used for washing clothes and cattle. Many dead fishes were floating on the surface during May 1981.

Kovur Pond

Kovur Pond is rectangular in shape with an area of 0.5 ha. and an average depth of 0.75 m. The pond has only one inlet on the northern side and no outlet. The pond mainly depends on rain water besides the flow from the inlet coming from Chambarambakkam tank. Kovur is 20 km. west of Madras city and fringed with paddy fields. During the period of investigations macrophytes such as *Pistia* and *Neptunia* were noted. There is no regular fishing in the pond and the occasional hook and line catches reveal the following species namely *Puntinus sarana*, *Puntius dorsalis*, *Channa* sp. and *Mystus* sp. No wind action was noted hence no mixing of waters.

Material and Methods

Studies were undertaken from October, 1979 to March, 1982 by collecting monthly samples. For plankton collection one 0.25 m, nylon net (0.3 mm. mesh size) was utilised. The net was thrown so that the attached rope (2.55 m.) was in the water and in this manner plankton was collected (Hebert, 1977). For each sample four throws were made. The collected samples were preserved in 5 per cent formalin and made up to 250 ml. From this a subsample of 2 ml. was examined in a Sedgewick-rafter cell for counting purposes. Total number of specimens collected were given. For taxonomic studies cladocerans were separated and identified.

Surface water samples collected were analyses for dissolved oxygen and pH. Dissolved oxygen was estimated by Winklers method with azide modification. pH was measured in the field by using wide and narrow range BDH papers. The values were again corroborated in the laboratory by using Philip's pH meter. Transparency values were noted by using a Secchi disc. Further other parameters like depth, air temperature, water temperature and weather conditions were noted.

RESULTS

Physico-chemical parameters (Table 1)

The air temperature range was between 23.0 and 32.0 °C with minimum in December 1981 and maximum in March 1980. Water temperature range was between 23.5 and 31.5 °C with minimum in December 1981 and maximum in April 1980. Depth of water was in the range of 0.60 to 1.20 m. The transparency values could be taken only upto October 1981 and subsequently the surface of the pond was covered with macrophytes. Transparency values were in the range of 0.10 to 0.50 m. with minimum in April 1980 and maximum in September 1981. The pH values were in the range of 6.8 to 9.0 with minimum in February 1982 and maximum in March 1980. The dissolved oxygen values were in the range of 0.7 to 7.5 mg/l. with minimum in October 1981 and maximum in March 1980.

Table 1: Physico-chemical data

Month		*A. Temp.* °C	*W. Temp.* °C	*Transp.* M	*pH*	*D.O.* mg/l
Oct.	1979	27.0	29.0	0.30	8.5	4.1
Nov.	1979	26.0	27.0	0.40	8.0	3.7
Dec.	1979	28.0	27.0	0.35	8.5	4.4
Jan.	1980	26.0	25.5	0.30	8.3	0.8
Feb.	1980	26.5	27.5	0.30	8.2	2.3
Mar.	1980	32.0	29.5	0.20	9.0	7.5
Apr.	1980	30.0	31.5	0.10	8.5	3.9
May	1980	28.5	31.0	0.20	8.0	1.3
June	1980	27.0	29.5	0.30	8.0	6.5
July	1980	30.0	29.8	0.20	8.0	5.7
Aug.	1980	29.0	29.6	0.19	7.8	2.35
Sep.	1980	31.0	30.5	0.30	7.9	6.2
Oct.	1980	27.0	28.0	0.20	7.7	5.1
Nov.	1980	26.0	26.0	0.25	7.3	3.7
Dec.	1980	24.0	24.7	0.25	7.7	6.9
Jan.	1981	23.5	25.0	0.30	7.2	6.5
Feb.	1981	24.0	26.5	0.35	7.5	5.2
Mar.	1981	27.0	29.0	0.25	7.3	4.6
Apr.	1981	29.0	29.0	0.35	7.4	1.5
May	1981	29.0	30.0	0.30	7.3	2.9
June	1981	30.0	29.5	0.15	7.6	2.1
July	1981	27.0	27.5	0.15	7.2	1.9
Aug.	1981	28.5	29.5	0.20	6.9	1.4
Sep.	1981	28.5	29.5	0.50	7.0	1.8
Oct.	1981	28.5	27.0	0.30	7.0	0.7
Nov.	1981	26.0	26.5	-	6.9	1.7
Dec.	1981	23.0	23.5	-	6.9	1.2
Jan.	1982	24.5	24.0	-	7.0	0.9
Feb.	1982	26.0	24.5	-	6.8	1.9
Mar.	1982	29.5	27.0	-	6.9	1.1

Explanation : Physico-chemical data. Air temperature, Water temperature, Transparency, pH and dissolved oxygen.

Biological factors (Table 2)

The following eight species of Clacocerans were encountered in the samples : *Ceriodaphnia cornuta* Sars; *Ceeiodaphnia laticaudata* P.E. Muller; *Moina micrura* Kurz; *Chydorus reticulatus* (Daday); *Diaphanosoma sarsii* Richard; *Disphnosoma escisum* Sars; *Alona quttata* Sars and *Macrothrix spinosa* King. Besides cyclpoid copepods were always available in the samples. Calanoid copepods were noted only for a short period. The two rotifers namely *Brachionus* sp. and *Testudinella* sp. were present in the collections.

Discussion

Ceriodaphnia cornuta was noted in the samples from October 1979 to March 1981 with maximum in October 1980. During this month the water was without any algal bloom and temperature (27-28 °C), pH (7.7) and D.O. (5.1) were also moderate. On the other hand *Ceriodaphnia laticudata* started appearing in the pond after the spread of macrophytes with maximum in August 1981. During this month the dissolved oxygen values were very low (1.4 mg/l). *Moina micrura* was noted in the samples from October 1979 to December 1981 with maximum in August 1980. During this month the water was without any algal bloom and dissolved oxygen values (2.35 mg/l) were also less.

The other species namely *Diaphanomous excisum, Diaphanosoma sarsii, Macrothrix spinosa, Chydorus reticulatus* and *Alona quttata* were noted only after the spread of macrophytes.

It is probable that *Moina micrura* and *Ceridaphnia laticaudata* can exist and increase in enormous numbers even at low dissolved oxygen values. Krishnamurthy (1967) also observed that dissolved oxygen values so low as 0 to 2.0 mg/l were suitable for the increase of *Moina* sp. On the other hand *Ceriodaphnia cornuta* can increase in numbers only at higher dissolved oxygen values from 4.0 mg/l (Raghunathan, 1983). From Chingleput tank more number of *C.cornuta* were recorded by him when dissolved oxygen values were between 4.0 to 8.1 mg/l.

Also this study has created an awareness among the villagers to keep the pond clean. Periodical cleaning of macrophytes is currently undertaken. A request was made by the villagers to the Fisheries department to stock culturable fishes in this pond.

Table 2 : Cladocerans

Month		*Cer.C.*	*C.lat.*	*M.m.*	*D.ex.*	*D.sar*	*Mac.S.*	*Chy.r.*	*Al.g.*	*Remark*
Oct.	1979	125	-	250	-	-	-	-	125	CLEA
Nov.	1979	5875	-	4125	-	-	-	-	-	"
Dec.	1979	250	-	500	-	-	-	-	-	"
Jan.	1980	1000	-	1125	-	-	-	-	125	"
Feb.	1980	8625	-	5875	-	-	-	-	-	Microc
Mar.	1980	500	-	-	-	-	-	-	-	"
Apr.	1980	-	-	16250	-	-	-	-	-	"
May	1980	-	-	24000	-	-	-	-	-	"
June	1980	10500	-	31750	-	-	-	-	-	"
July	1980	1125	-	15750	-	-	-	-	-	"
Aug.	1980	8250	-	102375	-	-	-	-	-	CLE
Sep.	1980	15042	-	20625	-	-	-	-	-	"
Oct.	1980	31844	-	9458	-	-	-	-	-	"
Nov.	1980	5833	-	39758	-	-	-	-	-	Microc
Dec.	1980	4875	-	25708	-	-	-	-	-	"
Jan.	1981	4375	-	6250	-	-	-	-	-	"
Feb.	1981	958	-	1000	375	-	-	-	-	Pist
Mar.	1981	560	-	875	281	344	-	-	-	"

Contd..

Table 2 (Contd.) : Cladocerans

Month		*Cer.C.*	*C.lat.*	*M.m.*	*D.ex.*	*D.sar*	*Mac.S.*	*Chy.r.*	*Al.g.*	*Remark*
Apr.	1981	-	-	335	2792	-	-	125	625	Pist.
May	1981	-	-	1906	-	-	-	-	-	Waters
June	1981	-	37187	2406	-	-	-	500	-	Microc
July	1981	-	7915	21542	125	-	-	920	-	"
Aug.	1981	-	43969	4092	1031	-	-	-	-	Pist.
Sep.	1981	-	7156	2469	2312	-	-	125	-	"
Oct.	1981	-	2844	125	406	-	-	1031	-	"
Nov.	1981	-	14125	594	200	-	-	250	-	"
Dec.	1981	-	562	250	300	-	375	-	-	"
Jan.	1982	-	594	-	-	-	532	601	-	"
Feb.	1982	-	812	-	125	-	1312	125	281	"
Mar.	1982	-	625	-	125	-	125	200	125	"

Cer. c. - *Ceriodaphnia cornuta.*, C.lat. - *Ceriodaphnia laticudata.*, M.m. - *Moina micrura.*, D.ex. - *Diaphanosoma excisum.*, D.sar - *Diaphanosoma sarsii.*, Mac.s. - *Macrothrix spinosa.*, Chy.r. - *Chudorus retuculatus.*, Al. gut. - *Alona guttata.*

Acknowledgements

The authors are grateful to the Director, Zoological Survey of India, Calcutta and to the Officer-in-Charge, Southern Regional Station, Zoological Survey of India, Madras for encouragement and facilities.

References

Herbert, P.D.N. 1977, A revision of the taxonomy of the genus *Daphnia* (Crustacea: Daphnidae) in southeatern Australia. *Aust. J. Zoo. 25* : 371-398.

Krishnamurthy, K.P. 1967, Studies on *Moina dubia* Gurney and Richard (Daphnid: Microcrustacea) from sewage oxidation ponds at Nagpur India. *Symposium on Crustacea Part II. CMERI Cochin.*

Ragunathan, M.B. 1983 : Studies on some planktonic Cladocerans (Crustacea) of Tamil Nadu, India. Ph.D. Thesis, Madras University.

6

Planctonic Cladocera (Crustacea) and Species Associations

M.B. Raghunathan and K. Revathi

Introduction

An aggregation is a group of animals which forms as a result of individuals being independently attracted to a particular environmental feature such as source of food. A swarm of waterfless gathered in some area rich in food is an example for aggregation. They are attracted to a common food source and they certainly react to one-another's presence for example by spacing themselves out so that they do not touch but they do not constitute a society. All animal groups whether aggregations or true societies confer an advantage on the members that constitutes them. The work of Allee (1938) was of great importance in revealing numerous examples of the way in which aggregation benefits the individuals which comprises it. Waterfleas cannot survive in alkaline water, but the respiratory product of a large group of them are sometimes sufficiently acid to bring down the alkalinity to accepted levels. Thus a group can survive where a few individuals could not.

Material and Methods

Species association among Cladocera (Crustacea) is analysed on the basis of studies carried out in Tamil Nadu from 1976 to 1982 (Raghunathan, 1983). A total of 470 plankton samples were collected from 47 different stations comprising all types of Waterbodies. Plankton samples were collected from the following types of habitats : A) estuaries (116 samples); B) irrigation tanks (66 samples); C) Village ponds (206 samples); D) templetanks (29 samples); E) high altitude lakes (24 samples); F) reservoirs (14 samples); G) small streams (10 samples); and H) fortmoats (5 samples).

Fourteen species of Cladocera belonging to ten genera have been recorded from Tamil Nadu. Among the fourteen species of Cladocera collected during the present investigation the number of samples in which these Cladocera recorded are given below : *Diaphanosoma excisum* (52 samples); *Diaphanosoma sarsii* (5 samples); *Penilla avirostris* (8 samples); *Daphnia lumholtzi* (11 samples); *Ceriodaphnia cornuta* (70 samples); *Ceriodaphnia pulchella* (2 samples); *Ceriodaphnia laticaudata* (35 samples); *Monia micrura* (78 samples); *Bosmina longirostris* (6 samples); *Macrothrix spinosa* (8 samples); *Macrothrix laticornis* (1 sample); *Chydorus reticulatus* (23 samples); *Alona quttata* (17 samples); and *Evadne tergestina* (15 samples). The percentage of co-occurrence (PC) is calculated by the method of Green (1971) as follows :

$$PC = \frac{C}{(a+b)} \times 100; \qquad \text{where}$$

C = Number of samples containing both species (A and B);
a = Number of samples containing species A;
b = Number of samples containing species B.

Results

The most dominant species is *Moina micrura* and this species is compared with other species collected together for percentage of co-occurrence.

Ceriodaphnia cornuta	-	87.34
Diaphanosoma excisum	-	41.30
Ceriodaphnia laticaudata	-	34.52
Chydorus reticulatus	-	20.23
Bosmina longirostris	-	9.09
Daphnia lumholtzi	-	8.50

Alona quttata	-	7.95
Macrothrix laticornis	-	1.28
Ceriodaphnia pulchella	-	1.26
Diaphanosoma sarsii	-	1.21

The second dominant species is *Ceriodaphnia cornuta* and this species is compared with all the other species collected together.

Moina micrura	-	87.34
Diaphrnosoma excisum	-	34.06
Chydorus reticulatus	-	14.81
Daphnia lumholtzi	-	10.95
Alona guttata	-	8.75
Bosmina longirostris	-	8.57
Diaphanosoma sarsii	-	1.35

The third dominant species is *Diaphanosoma excisum* and this species is compared with all the other species collected together.

Moina micrura	-	41.30
Ceriodaphnia cornuta	-	34.06
Chydorus reticulatus	-	31.57
Ceriodaphnia laticaudata	-	20.83
Alona quttata	-	6.15
Daphnia lumholtzi	-	5.00
Bosmina longirostris	-	3.57
Diaphanosoma sarsii	-	1.78

The fourth dominant species is *Ceriodaphnia laticaudata* and this species is compared with all the other species collected together.

Chydorus reticulatus	-	65.51
Moina micrura	-	34.52
Diaphanosoma excisum	-	20.83
Macrothrix spinosa	-	26.47
Alona quttata	-	6.12

The other co-occurrence values noted during the present study are the following:

Daphnia lumholtzi with *Chydorus reticulatus*	3.03
Daphnia lumholtzi with *Alona quttata*	3.70
Daphnia lumholtzi with *Bosmina longirostris*	13.33
Daphnia lumholtzi with *Ceriodaphnia pulchella*	8.33
Alona quttata with *Chydorus reticulatus*	2.56
Alona quttata with *Macrothrix spinosa*	4.16
Macrothrix spinosa with*Chydorus reticulatus*	6.89
Evadne tergestina with *Penilia avirostris*	27.77

Discussion

The percentage of Co-occurrence values indicate that *M. micrura* and *C. cornuta* have the maximum value of 87.34, *C. laticaudata* and *C. reticulatus* with 65.51 occupy the next position. *M. micrura* with *D. excisum* occupy the third position with 41.30 followed by *M. micrura* with *C. laticaudata* (34.52).

In Ennore estuary, the maximum number of Cladoceran species available in a single sample is only two (*E. targestina* and *P. avirostris*). On the other hand in Chingleput tank five species of Cladocera are seen together in the same sample in four instances (*C. cornuta, M. micrura, D. excisum, Chydorus reticulatus, A. quttata* - two samples); *C. cornuta, M. micrura, D. excisum, A. quttata, D. lumholtzi* together in one sample and *C. cornuta, M. micrura, D. excisum, C. reticulatus, D. lumholtzi* together in one sample. In two instances four species of Cladocera are seen together (*C. cornuta, M. micrura, D. excisum, C. reticulatus* in one sample and *C. cornuta, Moina micrura, C. reticulatus, Alona quttata* in another sample). From Chambarambakkam tank three species are seen together in a single sample. (*Diaphrnosoma excisum, Moina micrura, Alona quttata*) With reference Kovur pond five species of Cladocera are seen together only once (*Ceriodaphnia laticaudata, Diaphrnosoma excisum, Chydorus reticulatus, Moina micrura, Alona quttata*) and four species are seen together on may occasions.

During the earlier studies in Lake Phewa, Nepal, Swar and Fernandco (1979) it was observed that five species of Cladocera are seen to occur together (*Diaphrnosoma excisum, Daphnia longispina, Daphnia lumholtzi, Ceriodaphnia cornuta, Bosmina longirostris*) on four occasions. Further studies on Lake Begana and Rupa, Pokhravalley, Nepal (Swar and Fernando, 1980) it was observed in Lake Begana that five species of Cladocera are seen together (*Daphnia lumholtzi, Ceriodaphnia cornuta, Bosmina longirostris, Ceriodaphnia pulchella, Moina micrura*) on two occasions and from Lake Rupa, five species are seen together (*Diaphrnosoma excisum, Daphnia longispina, Daphnia lumholtzi, Ceriodaphnia cornuta, Bosmina longirostris*) is only once. Hence earlier studies also indicate that so far only a maximum of five species of Cladocera are seen together in single sample as observed during the present studies in Tamil Nadu.

Very little is known about the competitive interactions among the species of cladocera that co-existence in one waterbody and exclusion in others. According to Gannon and Stenberger (1978) the increase of planktivores and acceleration of nutrient loading can change the balance of competition between species causing changes in the community composition. But according to JanBengtson (1987) large Cladoceran species are better competitors under high food conditions and/or at lower temperatures while small species are superior when food is scarce or at higher temperatures. Further the impact of predator on zooplankton species composition is likely to be greater than the effects of competition in spite of the fact that competition is intense.

Acknowledgements

The authors are grateful to the Director, Zoological Survey of India, Calcutta and to the Joint Director, Southern Regional Station, Zoological Survey of India, Madras for encouragement and facilities. Also thanks are due to Dr. N. Murugan, Madura College, Madurai for help in literature and suggestions.

References

Allee, W.C. 1938, The social life of Animals, Norton, New York.

Gannon, J.E. and Stenberger, R.S. 1978, Zooplankton (especially Crustaceans and Rotifers) as indicators of water quality. *Trans. Amer. Micros. Soc. 97* : 16-35.

Green, J. 1971, Association of Cladocera in the zooplankton of the lake sources of the White Nile *J.Zool. Lond. 165* : 375-414.

JanBengston, 1987, Competitive dominance among Cladocera : Are single factor explanations enough. An examination of the experimental evidence. *Hydrobiologia.145* : 245-257.

Raghunathan, M.B. 1983 : Studies on some planktonic Cladocerans (Crustacea of Tamil Nadu. Ph.D. Thesis, Madras University.

Swar, D.B. and Fernando, C.H. 1979, Seasonality and fecundity of *Daphnia lumholtzi* Sars in Lake Phewa, Nepal. *Hydrobiologia*;: 64(3) : 261-268.

Swar, D.B. and Fernando, C.H. 1980, Some studies on the ecology of limnetic crustacean zooplankton in Lake Begana and Rupa, Pokhra valley, Nepal. *Hydrobiologia*. 70 : 235-245.

7

Role of Rotifers as Bioindicators in Freshwater Lentic Ecosystems

Arvind Kumar and A.K. Singh

Introduction

A bioindicator integrates its response to environment through time and reacts to all synergistic and antagonistic effects of accompanied pollutants and its usefulness is based on its sensitivity. Certain living organisms serve the purpose of monitoring the environmental pollutional conditions. These are termed as bioindicators and are capable of measuring the actual response of organisms or populations to the environmental quality. The physiological and biological diversity of species allow is wide range of indicator species for various environmental situations. In the recent time, both toxic and non-toxic pollutants have been found to distort the normal structure of biological communities. A definite clue regarding the typical organisms is given by the survey and analyses of various types of waters and by correlating their physico-chemical properties with existing fauna and flora. It has been shown by many investigators that the degree of water pollution can be estimated almost exactly by using typical organisms. Turner (1927) classified the degree of water pollution into various categories on the basis of the presence and absence of typical organisms. A variety of aquatic organisms have been widely used as biological indicators to ascertain the quality of water (Woodwiss, 1964; Torzwell, 1965; Hussainy and Abdulappa, 1967; Hakkari, 1972; Verma and Dalela, 1975; Ivanova, 1976; Kutikova, 1976; Polivanayya and Sergeyeva, 1978; Azizhussain, 1979; Saksena and Kulkarni, 1985; Saksena, 1987; Kumar, 1994a and Kumar, 1997).

The rotifers are microscopic animals mostly living in freshwaters and they are characterised by the presence of an anterior wheel-like rotating structure called "Corona". These organisms are globally known as bioindicators of water quality since long time by a number of workers such as Thunmark (1945), Berzins (1949), Lillieroth (1950), Jarnefelt (1952), Pejler (1965), Hakkari (1972), Ganon (1972), Radwan (1976), Macmets (1983). The Indian workers have also fairly used rotifers as bioindicators (Arora, 1961; Rao and Mohan, 1977; Saksena, 1987; Ali *et al.* 1990; Sharma, 1992; Kumar, 1964b and Unni, 1996). In this investigation, an attempt has been made to give an insight into the existing rotifer fauna from different waters to evaluate the role of rotifers as bioindicator, in relation to certain physico-chemical parameters of the habitat.

Study Area

The present investigations were carried out in two freshwater ponds known as Rasikpur pond (nontribal pond) and Nakati pahar pond (tribal pond). Both ponds are rainfed perennial pond and situated between 25^0 18' NL and 87^0 59' EL. The Rasikpur pond is used as pisciculture farm. Its mean depth ranged from 3.2 to 5.4 feet during different seasons. The tribal pond is unmanaged waterbody which receives domestic sewage with its depth varying from 2.8 to 6.9 feet during different months of the year. This pond is exploited by tribal people for various life activities such as bathing, washing of clothes, cleaning of utensils as well as cattle dumping activities. The water of this pond is slightly yellowish in colour.

Materials and Methods

Plankton were collected monthly (five samples in each month) from January to December, 1996 by filtering 100 litres of water through a plankton collecting net (bolting silk no. 25, mesh size-64) and were preserved in 4 per cent formalin. Qualitative and quantitative enumeration of rotifers was made according to Lackey's (1930) drop microtransect counting method, as modified by Edmondson (1974). Physico-chemical factors such as, temperature, transparency, pH, alkalinity, DO (Dissolved Oxygen), Free CO_2 and chloride were estimated according to APHA (1989) and Trivedy and Goel (1986). Correlation between total rotifers and physico-chemical parameters were also computed.

Results and Discussion

The results of the physico-chemical characteristics have been illustrated in Fig. 1 and 2. The qualitative and quantitative aspect of the zooplankton were done and the results have been expressed numerically in units per litre. Among zooplankton, rotifer have been given special emphasis showing two peaks in tribal ponds (April and November) and one peak (June) in non-tribal pond during the period of investigations. Ten species of rotifers belong to four genera, *Brachionus, Filinia, Testudinella* and *Keratella* in descending order of their abundance in tribal pond whereas in non-tribal pond, only nine species belonging to the same genera in descending order of their abundance were observed (Table 1 and 2).

The major contributors in both the peaks of tribal pond were *Brachionus calyciflorus* and *B. plicatil* is which were recorded throughout the tenure of investigation, but in non-tribal pond, the peak was shared by *Brachionus calyciflorus ,B. plicatilis, Keratella canadensis* and *Testudinella* sp. (Table 1 and 2). All the four species of genus Brachionus namely *B. calyciflorus, B. plicatilis, B. biodentata* and *B. quardridenta* and *Filinia terminalis* were always found to be present in non-tribal pond, whereas in tribal pond, only two species of genus *Brachionus i.e. B. calyciflorus* and *B. plicatilis* were recorded throughout the year of investigation. George (1961) has also reported that certain forms of rotifers were observed throughout the year, while others have noted good numerical fluctuations in the population density of rotifers. Michael (1968) observed that lericate rotifers were very high in number while non-loricates were found in winter. *Brachionus angularis* and *Keretella tropica* were found throughout the winter in tanks at Palani (Nayer, 1970). Nasar (1977) has reported that *Brachionus angularis* and *Brachionus quadridentata* were found during the winter months in a freshwater pond of Bhagalpur.

The average mean value of surface water temp. varied from 14.6 to 43.5 ^{0} C and 14.02 to 32.5 ^{0} C in non-tribal and tribal ponds respectively. The maximum Sechi's disc transparencies were noted in March, 1996 in both the ponds and minimum in the months of July (non-tribal pond) and August, 1996 (tribal pond).

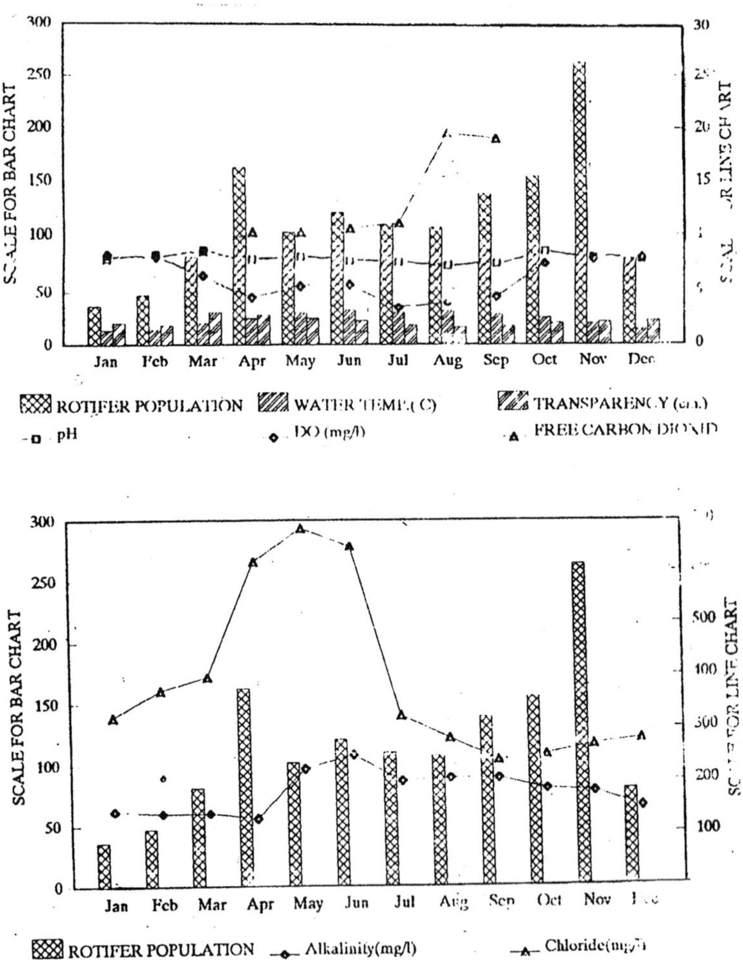

Fig. 1 : Monthly fluctuations of rotifers along with some physico-chemical characteristics of water during 1993

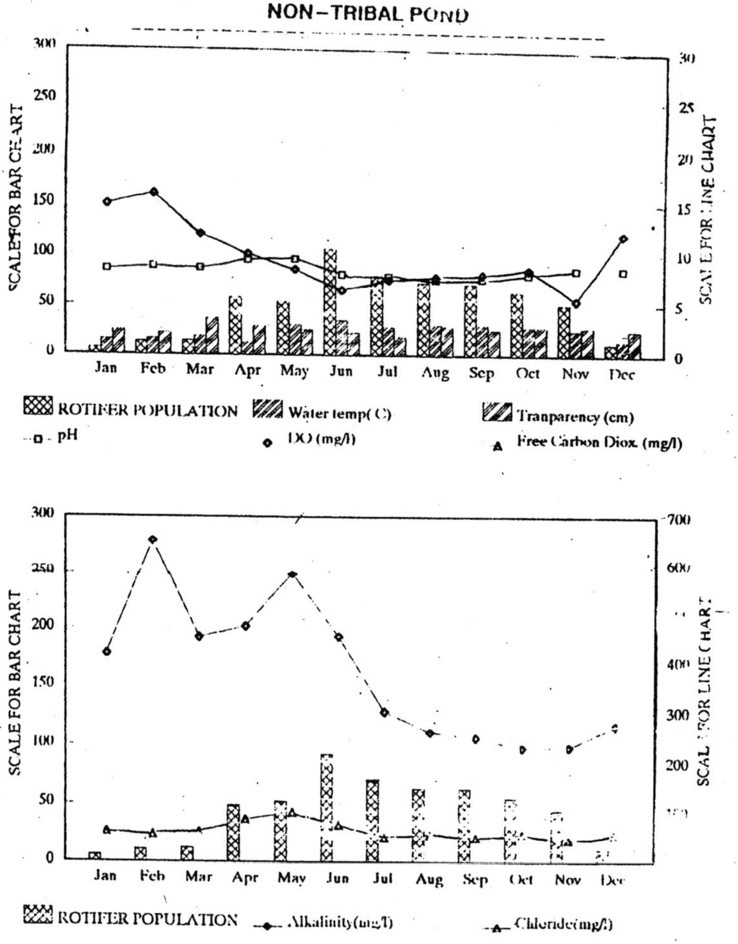

Fig. 2 : Monthly fluctuations of rotifers along with some physico-chemical characteristics of water during 1994

Table 1 : Monthly Variations in rotifer population in Tribal pond from January to December, 1996 (Nos./l)

Months *Species*	*Jan.*	*Feb.*	*Mar.*	*April*	*May*	*June*	*July*	*Aug.*	*Sept.*	*Oct.*	*Nov.*	*Dec.*
Brachionus calyciflorus	13	16	25	30	23	28	26	24	23	36	72	30
B. bidenata	-	-	-	7	15	7	8	5	12	5	8	-
B.pilicatilis	23	26	50	105	35	38	25	27	30	33	81	40
B.havanaensis	-	-	-	-	3	6	8	8	12	15	10	1
B. quadridentata	-	2	1	8	5	15	3	5	6	4	20	3
B.angularis	-	-	-	-	-	1	5	3	8	2	2	-
Keratella canadensis	-	-	-	-	-	5	8	6	13	15	20	-
Filinia terminalis	-	2	3	1	4	5	9	7	10	12	10	5
F. longseta	-	1	2	5	4	2	1	4	5	7	12	-
Testudinella sp.	-	-	-	6	12	13	16	17	19	25	28	-
Total	36	47	81	162	101	120	109	106	138	154	263	79

Table 2 : Monthly Variations in rotifer population in Non-Tribal pond from January to December, 1996 (Nos./l)

Months *Species*	*Jan.*	*Feb.*	*Mar.*	*April*	*May*	*June*	*July*	*Aug.*	*Sept.*	*Oct.*	*Nov.*	*Dec.*
Brachionus calyciflorus	2	4	5	15	18	14	14	13	11	11	8	1
B. bidenata	1	1	1	2	3	5	10	8	3	8	5	1
B.pilicatilis	-	2	3	15	5	14	13	18	20	12	13	8
B. quadridentata	1	2	2	1	2	1	1	1	3	2	2	1
Keratella canadensis	-	-	-	5	8	15	2	1	4	2	1	-
K. cochlearis	-	-	-	-	3	8	2	1	4	-	-	-
K. valga	-	-	-	1	2	5	1	2	3	5	8	-
Filinia terminalis	2	4	3	12	10	8	20	18	4	3	10	1
Testudinella sp.	-	-	-	6	2	35	15	10	19	21	4	-
Total	6	13	14	57	53	105	78	72	71	64	51	12

The pH varied from 7.3 to 8.6 in tribal pond and 7.5 to 9.4 in non-tribal pond. The minimum values of pH coincided with monsoon months of this Santhal Parganas. The dissolved oxygen content of the pond was observed to vary from 5.5 to 16.0 mg/l (non-tribal pond) and 3.50 to 8.20 mg/l (tribal pond). Higher concentrations of dissolved oxygen were recorded during January, February and December, 1996 in both the ponds. Free Carbondioxide was noted during March to September, 1996 in tribal pond only.

According to Yousuf (1989), the rotifers are opportunistic in their abundance and their population seemed to be influenced by factors other than the temperature of water. But, in the present investigation, the ponds showed a substantial decrease in water temperature during winter which coincided with the lowest population densities of rotifers in both the ponds. It is obvious that the decline in the population density was mainly due to disappearance of *B. angularis, Keretella canadensis, K. valga* and *Testudinella* sp,. Reappearance of the aforesaid species from March onwards was mainly due to getting favourable environmental conditions for their reproduction and development (Fig. 1 and 2).

The total rotifer population density in both the ponds showed an inverse relation with transparency ($r = -0.53$, $p < 0.01$) and total alkalinity ($r = -0.46$, $p < 0.01$). These parameters were higher in non-tribal pond in comparison to tribal pond. The tribal pond gets domestic sewage and other organic waste materials which reduce its transparency as well as alkalinity.

The population of the rotifer fauna of both the ponds were found insensitive to pH. Das and Srivastava (1956) and Moitra and Bhowmick (1968) have shown that the most rotifer species are sensitive to pH. But Hutchinson (1967) considered that it is not possible to draw conclusion from pH, because in natural conditions, it never changes alone.

Dissolved Oxygen values were quite high throughout the year in both the ponds showing no direct effect on the rotifer population. Much higher concentration in non-tribal pond as compared to tribal pond was mainly due to less comtamination and luxuriant growth of macrophytes near the shore line.

The chloride ion was much more obvious in tribal pond due to the discharge of domestic sewage, bathing, washing of clothes, cleaning of utensils and cattle dumping activities. According to Thresh, *et al.* (1994), the higher chloride content of water is an index of pollution of animal origin. Thus, tribal pond is more contaminated than the non-tribal pond and showed pronounced effect on the rotifer population in the ponds. However, *Brachionus calyciflorus* and *Brachionus plicatilis* population is high in tribal pond which showed much resistance to adverse environmental conditions.

Results and Discussion

While studying a number of Swedish lakes of different trophic levels, Pejler (1957) observed that *Brachionus* sp. *Keratella* sp. and Filinia sp. are inhabitants of mesotrophic to eutrophic waters. Hutchinson (1967) stated that the genus *Brachionus* is characteristic of hard water.

Rotaria rotatoria was collected in abundance from polluted waters in Nagpur (Arora, 1961). A complete absence of *R. rotatoria* from pure and clean waters indicates its habitat exclusively in polluted water. He further suggests that dissolved oxygen alone influences the survival of rotifer fauna. In another study, Arora (1966) has pointed out that *Brachionus angularis, B. calyciflorus, Filinia longiseta* and *F. terminunalis* are common forms predominantly occurring in eutrophic waters. *Keratella tropica* in a temperature region corresponds to K. quadrata in a temperate region, and *Polyarthra multiappendiculata* occurring in moderately polluted waters may be regarded to correspond with *Polyarthra euryptero,* which is designated as an indicator of eutrophy in cold climate. Michael found that the development of *Brachionus, Keratella* and *Anuraeopsis fissa* is greater where alkalinity and temperature are higher. Davis (1968, 1969) reported that the abundance of Brachionus is conclusive of the eutrophic nature of lake Eris.

Euchlanis, Monostyla, Distyla and *Pterodina* occur predominantly in the Daha river (Bihar, India), which receives effluents from nearby sugar distilleries in industries, yet none of these can be solely regarded as a pollution indicator (David and Ray, 1966). Most abundant of rotifers in the East Central lake Mechigan were *Polyarthra vulgaris* and *Keratella cochlearis, Kellicottia longispina, Notholca squamula, Keratella earlinae, K. quadrata* and *Synchaeta stylata* were some other important sps.

present, while the eutrophication indicator *Brachionus* was rarely present (Duffy and Listen, 1978). *Brachionus plicatilis* was found to occur in highly polluted industrial effluents around Vishakhapatnam and harbour back waters on the East Coast of India (Rao and Mohan, 1977), *B. plicatilis* was found to occur in a proportion of 7.4 per cent to 17.8 per cent. Rotifer forms such as *Keratella quadrata, K. cochlearis, K. earlinae, Pompholyx sulcata* and *Brachionus anqularis* were found predominantly in some shallow eutrophic lakes in Western Canada, Baker.

Philodina and *Rotaria* are of most common occurrence in eutrophic lakes of Nainital in North India (Das, 1978) while Singh and Sahai (1978) concluded that a numerical superiority of rotifers among zooplankton and a high concentration of nitrate and phosphate is a indication of eutrophy.

On the basis of a survey of rotifers, Sampath *et al.* (1979) have characterised the Cauvery river (India) water saprobity. Polysaprobic zones at sewage and distillery outfalls were inhabited by a pronounced *R. rotatoria* population. Saksena and Sharma (1981) have found that eutrophic impondments harbour *Brachionus* and *Filinia* predominantly.

In Canadian lakes situated in the Mount Tremblant Provancial Park, Nogrady (1980) found that the association of species of rotifers clearly indicate their trophic level. The total absence of the genera *Brachionus, Pompholyx, Filinia* and of *Polyarthra euryntera* is in keeping with neutral of law pH low Ca^{2+}-ion concentration and law trophy. On the other hand, numerous acidiphlic sps. such as *Euchlanis menete, Keratella serrulata, K.taurocephala, Monommata phoxa, Notommata peridia, Polyarthra minor* and *Tetrasiphon hydrocora* are observed. His study on a chain of reclaimed sewage fed lakes revealed that a rich rotifer fauna dominated by *Brachionids, Keratella* and *Trichocerca* was supported in these lakes (Nagrady 1982). However, slightly alkaline lakes in Eastern rift valley in Kenya showed a qualitatively rich typical warm water rotifer association dominated by *Brachionids* and *Filinia* while high saline soda lakes were dominated by several populations of *Brachionus dimidiatus* (Nagrady, 1983). A mesotrophic lake, Lake Glubekeye in USSR, had a rotifer community consisting of *Trichocerca, Synchaeta, Polyarthra, Keratella, Conochilus, Filinia, Asplanchoa* and *Euchlanis* (Mateeva 1983), Saksena and Kulkarni (1985) found a qualitative abundance of *Brachionus, Lecane* and *Monostyla* in two sewage fed channels. *Trichocerca* cylindrica was a common species in both the channels at Gwalior. In a study made by Bil grami and Dutta Munshi (1985), it was reported that *Keratella cochlearis, Brachionus plicatilis, Filinia* spec. are some of the common rotifer indicaters of a eutrophic environment.

In conclusion, it is quite clear that the tribal pond is more contaminated than nontribal pond. We can, thus, conclude that the rotifers are apparently the good indicators of water quality. Their numerical strength can be used as a bioindicator for assessment of the water quality of freshwater ecosystems.

References

Ali, S., Choudhary, A. and Ray, A.R. (1980), Ecology and seasonal abundance of zooplankton in a pond in Tongi, Dacca. Bangladesh J. Zool. 8 (1) : 41-49.

A.P.H.A.(1989). Standard Methods for the Examination of Water and Wastewater. 17th, ed. American Public Health Association, Washington D.C.

Arora, H.C. (1961), Rotifers as indicators of pollution, CPHERI Bull, 3 and 4 : 24.

Arora, H.C. (1966), Rotifers as indicators of trophic nature of environment. Hydrobiologia IXVII, (1-2) : 146-159.

Azizhussain, M. (1979), The role of protozoa in polltion ecology - A review. Proc. Symp. Environ. Biol. : 315-325.

Baker, R.L. (1979), Birth rate of plankton rotifer in relation to feed concentration in a shallow eutrophic lake in Western Canada, Can. J. Zool. : 57 : 1206-1214.

Berzins, B. (1949), Zur Limnologie der Seen Siidostlettlands Schweiz. Z. Hydrol. 11 : 583-607.

Bilgrami, K.S. and Datta Munshi, J.S. (1985), A comparative evaluation of some freshwater ecosystems functioning under stress environment. Proc. 73rd Ind. Sc. Cong. Part IV : 58-59.

Das, S.M and Srivastava, V.K. (1956), Quantitative studies of freshwater plankton in a fish tank in Lucknow, India. Proc. Nat. Acad. Sci., India. 2 (6), 85-92.

Das, S.M. (1978), High pollution in lake Nainital, Uttar Pradesh as evidenced by biological indicators. Sci. and Cult. 44 :236-237.

David, A. and Ray, P. (1966), Studies on the pollution of the river Daha, N. Bihar by sugar and distilery wastes. Environ. Hlth. 8 : 6-35.

Davis, C.C. (1968), The July, 1967 Zooplankton of lake Erie. Proc. 11th conf. Great Lakes Res. Internet. Assoc. Great Lakes Res, 61-75.

Davis, C.C. (1969), Seasonal distribution Constitution and abundance of zooplankton of Lake Erie. J. Fish. Res. Bd. Can. 26 : 2459-2476.

Duffy, W.G. and Listen, C.R. (1978), Seasonal abundance of planktonic rotifers in Near Shore area of Central lake Mechigan. J. Great. Lakes Res. 4 (1), 46-49.

Edmondson. W.T. (1974), A manual on method for measuring primary production in Aquatic invertebrates as indicators of stream pollution. Public Hlth, Rep. 67 : 57-64.

Ganon, J.E. (1972), A contribution to the ecology of Zooplankton Crustacea of Lake Michigan and Green Bay. Ph.D. desertation University of Wisconsin Madison : 257.

George, M.G. (1961), Observation on the rotifers from shallow ponds in Delhi. Curr. Sci. 30 : 268-269.

Hakkari, L. (1972), Zooplankton studies as indicators of environment. Aqua. Fenn: 46-54.

Hussainy, S.V. and Abdulappa, M.K. (1967), Expression of biological data in water pollution research. Environ. Hlth. 9 : 210-219.

Hutchinson, G.E. (1967), A treatise on Limnology Vol. 1 John Willey and Sons, New York. U.S.A.

Ivonova, M.B. (1976), The effect of pollution on plankton crustacea and the possibility of using them for determining the pollution in a river. In : Metady biol. analiza presnykh Ved. Leningrad : 68-80.

Jarnefelt, H. (1952), Plankton als Indicator der Trophiegruppen der Seen. Ann. Acad. Sci. fenn. Ser. A. 4(18), 1-29.

Kumar, A. (1994a), Role of species diversity of aquatic insects in the assessment of pollution in wetlands of Santhal Parganas (Bihar). Jr. Environ. and Poll. 1 (3and4), 117-120.

Kumar, A. (1994b), Periodicity and Abundance of rotifers to certain physico-chemical characteristics of two ecologically different ponds of Santhal Parganas (Bihar). Indian J. Ecol. 21 (1), 54-59.

Kumar, A. (1997), Biomonitoring of pollution by aquatic insect community and Micro-organisms in freshwater ecosystems. Ecotechnology of Pollution Control and Environmental Management (Eds. R.K. Trivedy and A. Kumar), Enviromedia, Karad. (In Press).

Kutidova, L.A.(1976), Rotifers of river plankton as a water quality indicator. In : Metody biol. analiza Presnykh Ved. Leningrad : 80-90.

Ladkey, J.B. (1930), The Manipulation and Counting of river plankton and changes in some organisms due to formalin preservation. Pub. Hlth. Dept. 53 : 2080.

Lillieroth, S. (1950), Uber die Folgen kulterbedingter Wasserstandssenkungen fur Madrophyten und Panktoggeneinschaften in Seichten Seen des sudschwedischen Oliogotrophiegebiotes. Acta limnol. Lund. 3 : 188.

Macmets, A. (1983), Rotifers as indicators of lake types in Estonia. Hydrobiologia 104 : 357-361.

Mateeva, L.K.(1983), Community structure of planktonic rotifers in mesotrophic lake. Hydrobiologia. 104 : 353-356.

Michael, R.G.(1968), Studies on the zooplankton of a tropical fish pond, India. Hydrobiologia. 32 : 47-68.

Moitra, S.K. and Bhowmick, M.L. (1968), Seasonal cycle of rotifers in a freshwater fish pond and their role in fish culture, Hydrobiologia. 2531 : 54-60.

Nasar, S.A.K. (1977), Investigation on the seasonal periodicity of zooplankton in a freshwater pond in Bhagalpur, India. Acta. hydrochem et hydrobiol 5 (6): 577-584.

Nayar, C.K.G. (1970), Studies on the rotifer population of two ponds at Palani, Rajasthan. J. Zool. Soc. India. 22 (1and2) : 21-34.

Nagrady, T. (1980), Canadian Rotifers II. Parc. Mont Tremblant, Quebec. Hydrobiologia. 71 : 35-46.

Nagrady, T. (1982), Correlation of rotifer associations in a chain of lakes fed by reclaimed sewage. Hydrobiologia. 89 : 277-284.

Nagrady, T. (1983), Succession of planktonic rotifer populations in some lakes of the eastern Rift Valley. Hydrobiologia. 98 : 45-54.

Pejler, B. (1957), Taxonomical and ecological studies an planktonic Rotatoria from central northern Swedish Lapland. K. Svenska Vetensk Akad. Handl. Ser. 4.6 (7), 52-120.

Pejler, B. (1965), Regional-Ecological studies of Swedish freshwater Zooplankton. Zool. Bidrag. Uppsala. 36 : 407-515.

Polivanayya, M.F. and Sergeyava, O.A. (1978), Zooplankters and bio-indicators of water quality. Hydrobiol. Journ. 14 (3) : 39-43.

Radwan, S. (1976), Planktonic rotifers as indicators of lake tropy. Ann. Univ. Mariae Curie Sklodowska, Sectio C. 31 : 227-235.

Rao, K.R. and Chandramohan, P. (1977), Rotifers as indicators of pollution. Curr. Sci. 46 (6), 120.

Saksena, D.N. and Sharma, S.P. (1981), Zooplanktonic fauna of some lentic Waterbodies of Gwalior I. Govind Sagar, Chhatri tank, Savarkar Sarovar and Matsya Sarovar. Environ. India. 4 : 13-17.

Saksena, D.N. and Kulkarni, N. (1985), On the rotifer fauna of two sewage channels in Gwalior; Limnologia 11 (1), 65-72.

Saksena, D.N.(1987), Rotifers as Indicators of water Quality. Acta hydrochim. hydrobiol. 15 (5), 481-485.

Sampath, V. Sreenavasan, A. and Ananthanarayanan, R. (1979), Rotifers as biological indicators of water quality in Cauvery river. Proc. Symp. Environ. Biol : 441-462.

Sharma, S.P. (1992), Systematics, Distribution and Ecology of freshwater Rotifers in West Bengal, pp 231-273. In S.R. Mishra and D.N. Saxena (ed) Aquatic ecology, Ashish Publ. Delhi.

Singh, S.B. and Sahai, R. (1978), Seasonal fluctuations of zooplankton population in relation to certain physico-chemical factors of a pond. Geobics. 5(5) : 228-230.

Tarzwell, C.M. (1965), Biological problems in water pollution. Third Seminar U.S. Deptt. of Health Education and Welfare, Cincinnati, Ohio.

Thresh, J.C. Suckling, E.V. and Beale, J.F. (1994), The examination of waters and water supplies. Ed. Taylor, E.W. (1949).

Thunmark, S. (1945), Die Abwasserfrage der Vaxjo-seen in hydrobiologischer Beleuchtung. Grundziige in der regionalen planktologie von Siidschweden. Medd. Lunds. Univ. Limnol. Inst. 4 : 239.

Trivedy, R.K. and Goel, P.K. (1986), Chemical and biological methods for water pollution studies, Environ. Publ. Karad. pp 247.

Turner, G.L. (1927) Biological survey of Fax, Wisconsin and Flambeau River, Wiscensin, with special reference to pollution. In : Stream Pollution in Wisconsin. Splel. Rept. Wisconsin State Bd. of Health : 2420.

Unni, K.S. (1996), Ecology of River Narmada, A.P.H. Publishing Corporation, New Delhi.

Verma, S.R. and Dalela, R.C. (1975), Studies on the pollution of the Kalinadi by industrial wastes near Masurpur. Part II : Biological index of Pollution and the biological characteristics of the River. Acta Hydrochem. Hydrobiol. 3 : 274-295.

Woodwiss, F.S. (1964), The biological system of stream classification used by the Trent river Board. Chem. Ind. 443-447.

Yousuf, A.R. (1989), Zooplankton studies in India with special references to North India : a critical review. In Management of Aquatic Ecosystem (Eds. Agrawal, V.P. Desai, B.N. and Abidi, S.A.H.) Society of Biosciences, Muzaffarnagar, India. 309-324.

8

Correlation of Rotifers with Physico-chemical Factors at Harike Reservoir (Punjab-India)

Kaur, H., Dhillon, S.S., Bath, K.S. and Mander, G.

Introduction

Rotifers range from 40 microns to 2.5 mm in size and occur in a variety of aquatic habitats. In most freshwater ecosystems, rotifers dominate the zooplankton populations in terms of density and diversity due to their remarkable adaptations to varied physico-chemical environments. For studying the relationship of rotifers with their abiotic parameters, the calculation of correlation coefficient has been extensively employed (Patil *et al.*, 1985; Saha and Pandit, 1988). In the present studies also, the correlation coefficient ('r') was calculated statistically to depict the relationship of rotifers with nine basic physico-chemical parameters (Temp, DO, pH, T.A., T.H., Chl., Kl., Nit. and Phos.). These studies are a part of detailed limnological studies undertaken at Harike reservoir (A RAMSAR wetland) of Punjab, India, during the period extending from January 1990 to December, 1993.

Materials and Methods

For physico-chemical studies, samples were collected in 500 ml plastic bottles, once in a month from six collection sites of the reservoir. The physico-chemical parameters were analysed as per procedures of APHA (1989).

For the collection of biological samples, a 250 ml polytrene bottle was tied to a planktonic net and about 50 litres of water was allowed to pass through the net to collect one sample. Samples were preserved in 5 per cent formalin and were brought to the laboratory for microscopic examination.

Correlation coefficient (r) was calculated using the formula:

$$r = \frac{\Sigma(x-\bar{x})(y-\bar{y})}{\sqrt{\Sigma(x-\bar{x})^2(y-\bar{y})^2}}$$

A total of 44 (n=44) observations were taken at each of the six sites during the four years (n=6 x 44). Test of significance (t) was calculated using the formula :

$$t = r\frac{n-2}{\sqrt{n-r^2}}$$

The value of calculated 't' was compared with the critical value of 't' obtained upto n-2 degrees of freedom at 5 per cent level while referring to student's 't' distribution table. For the present studies, the value of 't' came to be 1.96, which becomes an indication of significance for the purpose of calculating relationships.

Results and Discussion

Rotifers belonging to 43 genera were identified at Harike reservoir from January, 1990 to December, 1993. During the year 1990, 25 genera were recorded and in 1991, 19 genera were identified. In the year 1992, 15 genera were enlisted while in 1993, only 12 genera were recorded. The genera which were dominant and recorded in good numbers included : *Brachionus, Keratella, Lecane, Monostylla, Filinia, Platyias, Notholca, Cephaledella, Asplanchna, Pleosoma* and *Philodina.*

The physico-chemical parameters which included water temperatures, pH, dissolved oxygen, total alkalinity, total hardness, chlorides, flourides, nitrates and phosphates were also studied alongside. The ranges of fluctuations in various physico-chemical parameters during the four years of study are depicted in Tables 1 and 2.

Table 1 : Range of fluctuations in physico-chemical parameters at Harike from 1990 to 1993

S.No.	*Parameter*	*1990*	*1991*	*1992*	*1993*
1.	Water temperature (^{0}C)	17-38	12-37	14-34	15-35
2.	pH	7.20-8.60	7.16-8.90	7.40-9.40	7.90-8.22
3.	Total alkalinity (mg/l)	56-105	30-118	30-130	40-130
4.	Total hardness (mg/l)	64-112	68-140	80-230	105-250
5.	Dissolved Oxygen (mg/l)	7.00-8.00	6.80-8.80	3.10-8.95	3.00-8.20
6.	Chloride (mg/l)	16-46	5-70	Traces-75	Traces-.72
7.	Flouride (mg/l)	0.20-1.08	0.40-1.20	0.50-1.60	0.56-1.50
8.	Nitrate (mg/l)	0.02-3.20	0.04-3.10	0.02-2.00	0.00-1.90
9.	Phosphate (mg/l)	0.00-1.00	0.00-0.58	0.01-2.40	0.05-1.00

Water temperature followed usual seasonal pattern being higher in summers and lower in winters. During the full study period, water temperature ranged between 12 and 38^{0}C. It is revealed that species diversity of rotifers was more in winters, whereas, population density was higher during in summer. The low rate of multiplication during colder month might be due to the slow metabolic rates during that season (Kaur *et al.*, 1996). From the statistical calculation of coefficient of correlation, it was revealed that a very weak and insignificant positive correlation existed between water temperature and rotifers density. A large number of workers including Elliot (1976). Sampath *et al.*, (1978), Yousuf and Qadri (1981), Patil *et al.*,(1985), Chauhan (1988) and Saha and Pandit (1988) observed an increase in rotifer populations with an increase in temperature.

pH value at Harike ranged between 7.16 and 9.40, being higher during summer and winter months and lower during the monsoon period. Rotifers were found to be independent of changes in pH values within the observed range. King (1981), Vasisht (1988) and Vasisht and Monga (1981) noted a positive relationship between pH and rotifers.

Table 2 : Results of correlation of rotifers with physico-chemical parameters

S.No.	*Physico-chemical Parameter*	*Value of Coefficient of Correlation (r)*	*Value of test of Significance (t)*
1.	Water temperature (^{0}C)	+0.1425*	0.946
2.	pH	+0.0974*	0.641
3.	Dissolved Oxygen (mg/l)	+0.0343*	0.759
4.	Chloride (mg/l)	+0.1298*	0.955
5.	Flouride (mg/l)	+0.0886*	0.644
6.	Total alkalinity (mg/l)	+0.1119*	0.581
7.	Total hardness (mg/l)	+0.0799*	0.605
8.	Nitrate (mg/l)	+0.3076	2.095
9.	Phosphate (mg/l)	+0.3878	2.727

* denotes insignificant correlation
'r' is significant at t > 1.96.

Dissolved oxygen values (3.00 - 8.95 mg/l) were found to be higher during winters and lower during summers. Rotifers exhibited insignificant positive relationship with dissolved oxygen. This suggests that population density of rotifers is not affected by variations in dissolved oxygen values within the ranges observed at Harike. Elliot (1976) noted that rotifers can withstand wide range of fluctuations in dissolved oxygen values.

Chlorides ranging between traces and 75 mg/l at Harike, did not show any definite seasonal trend. The statistical analysis also showed that population density of rotifers was not significantly affected by fluctuations in chloride contents. Mechael (1964), Arora (1966) and Sampth *et al.* (1978) have correlated rotifers and chlorides positively.

Flourides showed a range of 0.20 to 1.60 mg/l at Harike. The values were seen to increase after the monsoon months with large scale drainage of water into the reservoir. A significant correlation of flourides and rotifers was not established at Harike.

Total hardness and total alkalinity ranged from 64 to 250 mg/l and 30 to 130 mg/l, respectively. The values of both these parameters were found to be high during summers and low during the monsoons. However, a significant correlation of these parameters with rotifers was not established. Michael (1964), Prabhavati and Sreenivasan (1977) and Sampath *et al.* (1978) noted that rotifers increased in number with the increase in hardness and alkalinity of water. Hutchinson (1967) and Bayly and Williams (1975) noted a positive correlation of rotifers with alkalinity.

Nitrate values varied from traces to 3.20 mg/l at Harike while the phosphate values ranged between traces and 2.40 mg/l. Nitrates values did not show any marked seasonal trend while phosphates were observed to increase after the monsoon period. Both nitrates and phosphates were found to be positively and significantly related to rotifer populations at harike. Both of these nutrients are essential for the growth of plants and at higher concentrations, cause eutrophication of water. Therefore, it can be deduced that increase in nutrient values and eutrophicated conditions are favourable for the growth of rotifers. Chauhan (1988) and Saha and Pandit (1988) also observed an increase in the number of rotifers with an increase in the values of nutrients.

From these studies, it can be concluded that rotifers can withstand a large range of fluctuations in the basic physico-chemical parameters. The observed ranges of water temperature, pH, alkalinity, hardness, chloride and flouride do not significantly affect the growth and survival of rotifer populations. These studies also point out that increasing values of phosphates and nitrates arefavourable for the rotifers. In addition, some other factors such as season, current, water volume, vegetation, food and diurnal migrations may also be responsible in determining the population density and species diversity of rotifers.

Acknowledgements

The authors are thankful to CSIR and PSCST for financial assistance.

References

A.P.H.A.(1989), American Public Health Association, Standard Methods for the Examination of Water and Wastewater. 17th, ed. A.P.H.A Washington, U.S.A.

Arora, H.C. 1966, Responses of rotifera to variations in some ecological factors. *Proc. Ind. Acad. Sci.* 58: 57-63.

Bayly, I.A.E. and Williams, W.D. 1975, Inland waters and their ecology, Logmans, Australia, 1-316.

Chauhan, R. 1988, Seasonal abundance of zooplankton in Rewalsar lake (H.P.) *Geobios New Reports*, 7 : 117-121.

Elliot, I.J. 1976, Seasonal changes in the abundance and distribution of planktonic rotifers in Grasmere (English lake District). *Fresh Water Biol.*, 8 : 144-166.

Hutchinson, G.E., 1967, A treatise on Limnology - II. Introduction to lake biology and the limnoplankton. John Wiley and sons Inc. New York.

Kaur, H.I. Dhillon, S.S., Bath, K.S. and Mander, G. 1996, Seasonal distribution and population dynamics of rotifers in the Harike reservoir (Punjab-India). Paper communicated.

King, J.M. 1981, The distribution of invertebrate communities in a small South African river. *Hydrobiologia*, 83 : 4365.

Michael, R.G. 1964, Limnological investigations on pond plankton, macrofauna and chemical constituents of water and their bearing on fish population. Ph.D. Thesis, Calcutta Univ., Calcutta.

Patil, S.G., Harshey, D.K. and Singh, D.F. 1987, Limnological studies of a tropical freshwater fish tank at Jabalpur, *M.P. Geobios New Reports*, 4(2) : 143-148.

Prabhavati, G. and Sreenivasan, A. 1977, Ecology of warm freshwater zooplankton of Tamil Nadu. Proc. of Sym. on warm water Zooplankton, N.I.O., Goa, Spl. Publication : 319-399.

Saha, L.C. and Pandit, B. 1988, Density of zooplankton between lentic and lotic ecosystem. *Geobios New Reports*, 7(1) : 31-35.

Sampath, V. Sreenivasan, A. and Ananthanarayanan, R. 1979, Rotifers as biological indicators of water quality in Cauvery River. *Proc. Symp. on Environ. Biol.*, Muzafarnagar.

Vasisht, H.S. 1968, Limnological studies of Sukhna Lake, Chandigarh, *Proc. Symp. Rec. Adv. Tropical Ecology*, 9 : 316-325.

Vasisht, H.S. and Monga, K.S. 1981, The hydrobiology of Bhakhra Main Line Canal (Narwana Branch). *Ind. J. Ecol.*, 8(2) : 313-314.

Yousuf, A.R. and Qadri, M.Y. 1981, *Inland Fish Soc. India*, 13(1) : 36-45.

9

Cyclomorphosis, the Morphological Responses in Cladocera (Waterfleas) to Certain Environmental Factors: A Review

Asif A. Khan and Aftab Alam

The term cyclomorphosis has been applied to the cyclic morphological changes that occur fairly strict seasonality in certain freshwater plankton. Seasonal changes in the morphology within populations involve the alternation of different morphs in time. Forms with external protuberances like helmets, crests, peaks of the head, jumps of spikes of the carapace, spines, elongation of the antennules usually alternate with forms lacking these exuberances.

The term cyclomorphosis was coined by Leauterborn (1904), but Daday (1885) was probably the first to note and appreciate the variations in plankton. Later, Zacharias (1894) and Wesenberg-Lund (1908) published their reports on cyclomorphosis. Although Ostwald (1904), Woltereck (1909), 1913), Wagler (1936), Coker and Addlestone (1938), Coker (1939), Brooks (1946, 1947), Lieder (1951), Hrbacek (1959) and Jacobs (1961, 1966) have published their observations on cyclomorphosis, but the detailed resume of Hutchinson (1967) provides a comprehensive account on this subject.

After that, in the last few decades lot of new findings and valuable investigations on the causes and biological significance of seasonal morphological changes have been presented by Hazelwood (1966), Zaret (1972 a, b), Dodson (1974), Kerfoot (1977), O'Brien and Vinyard (1978), O'Brien an Schmidt (1979), Black (1980), Grant and Bayly (1981), Krueger and Dodson (1981), Havel and Dodson (1984), Lee (1984), Havel (1985), Herbert and Green

(1985), Venkatraman and Krishnaswamy (1986), Mort (1989), Pijanowska and Tundisi (1990), Benzei (1991), Black (1993) and Manea and Tognota (1993).

Seasonal changes in morphology occur in warm season mostly in temperate waters, they disappear in cold season. It was believed that in subtropical waters cyclomorphosis is weak or does not occur, except where there is large temperature change between winter and summer. The same has been reported for tropical Waterbodies (Zogo, 1976; Mitchell, 1978), but there are many cases that have been studied in tropical and subtropical regions, like *Daphnia gibba* and *D. barbata* in Africa (Wagler, 1936) and *D. carinata* in India (O'Brien and Vinyard, 1978) which are highly cyclomorphic species.

Cyclomorphic Species of Cladocera

Cyclomorphosis is most imbringing and puzzling of all cladoceran problems. Seasonal polymorphism occurs in cladocera perhaps more conspicuously than in any other group. Cladocerans have been found to favourable material and easily cultured among all the plankton. Therefore, much of the work has dealt with the freshwater cladoceran females because males occur rarely and reproduction takes place usually by parthenogenesis.

Several genera among cladocera have been studied for cyclomorphosis like *Bosmina, Ceriodaphnia, Scapholebris, Chydorus* and of course *Duphnia*. Cyclomorphosis in these genera other than *Daphnia* is much less distinct.

In *Bosmina longirostris,* changes often occur by means of the formation of transparent dorsal hump with no increase in Length, reduction in antennule length and number of segments. Length of mucro (caudal spine) and number of sutures may also be reduced as shown in Fig. 1b (Kerfoot, 1977; O'Brien and Schmidt, 1979; Black, 1980). Cyclomorphosis has also been studied on another species of this genus, *B. coregoni* by Lieder (1951) as shown in Fig. 1F, I.

In *Ceriodaphnia reticulata* mark reduction in size during summer with a striking increase in the relative height of the head and of the diameter of the eye has been reported. These small, large eyed high headed forms are referred as *forma kurzii* (Hartmann, 1915) shown in Fig. 1 A,B. Zaret (1972 a, b) studied C. *cornuta* and reported horned and unhorned forms (Fig. 1 J, K).

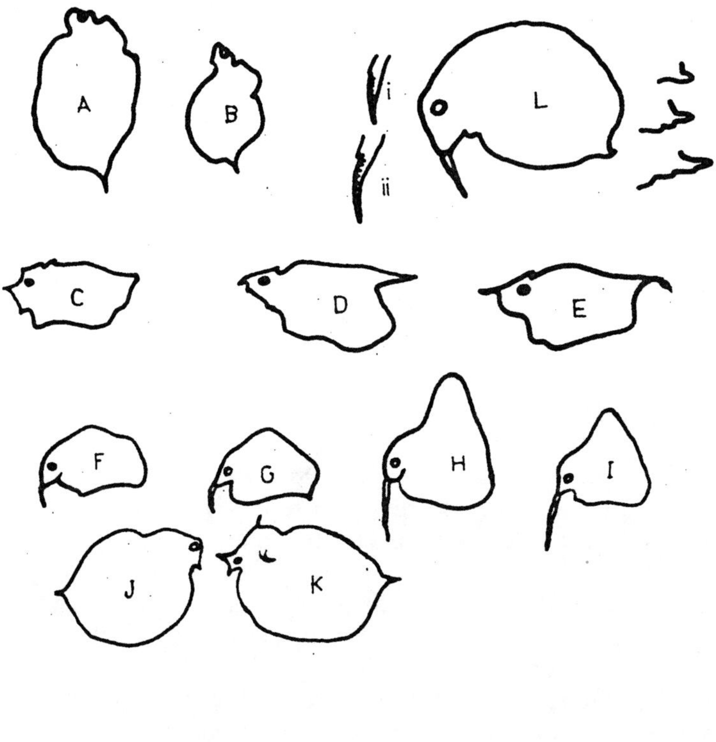

Fig. 1 : Cyclomorphosis in different cladocerans (A-B: *Ceriodaphnia reticulata* and its small large eyed cyclomorphic form Kurzii; C: *Scapholebris mucronata*. fronto-laervi form without anterior little spine; D: Typical form of *S. mucronata* with little spine; E: Form cornuta of *S. mucronata* with marked spine (Redrawn from Hutchinson, 1967); F-I: *Bosmina coregoni* showing development of dorsal hump and increase in antennule length (Redrawn from Welch, 1952); Typical and spined morphs of *Ceriodaphnia cornuta* (Redrawn from Zaret, 1972); L: Mature female of *Bosmania longirostris* in the middle with on the left (i) antennule with nine stgments, (ii) cyclomorphic antennule with twelve segments, and on the right is mucro showing varying sutures (Redrawn from Black, 1980).

Table 1 : Rearing data of *Ceriodaphnia cornuta* (Reconstructed from Zaret (1972) (after slight modifications)

Rearing data	*Unhorned*	*Horned*
General viability	12 of 12	9 of 17
Percentage survival	100 per cent	52.9 per cent
Days of maturity	4.9	6.4
Range	4-9	6-8
Sample size	12	9
Average life span	18.1	11.2
Range	13-37	7-28
Sample size	9	8
Fecundity (Eggs/life span)	28.1	11.1
Range	1-86	1-45
Sample size	10	9

Chydorus spaericus showed marked variation in size. These plankton reach a length in winter some 30 per cent greater than that achieved in summer (Hartmann, 1915). Posterior spine and anterior horn also manifest cyclomorphic changes (Green, 1967). Hartmann (1915) has also reported that *Alonella nana, Alonella exijera* and *Alana rectangula* reach smaller size in summer than in late autumn. A similar variation in size was noted in *Acroperus haspae* by Hartmann (1915), where variations in the height of the head and the distance of ocellus from the vertex and the apex of the rostrum occur.

The neuston feeding small cladoceran, *Scapholebris mucronata* has two forms. The form with a well marked anterior spine pointing forward from the head is known as form *cornuta.*, while the form without spine is known as form *fronto-laevi.* Although typical form has only a small spine (Lilljeborg, 1900) as shown in Fig. 1 C-E.

Most extensively studied cases of cyclomorphosis are among the species of genus *Daphnia.* From the dying years of last century until the present day this genus has attracted so many limnologists for various types of studies including cyclomorphosis. Most of the *exuberances* in this genus are manifested by development of helmet or crest on the head. The species extensively studied for phenotypic variations are *D. cucullata* (Ostwald, 1904; Lampert and

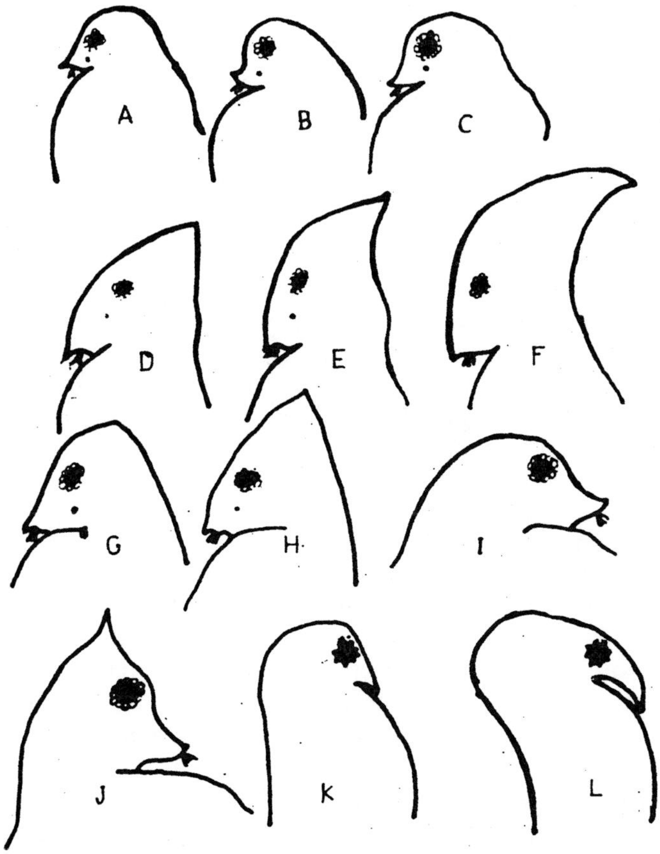

Fig 2 : Cyclomorphic helmets or crests in some species of the genus Daphnia (A-C : *D. pulex*; D F: *D.retrocurva*; G - H: *D. galeta mandotae*; (I-J: *D. ambigua* (Redrawn from Pennak, 1978); K-L: *D.carinata* (Redrawn from O'Brien and Vinyard, 1978).

Wolf, 1986), *D. galeata mandotae* (Jacobs, 1961; Mort, 1989), *D. longispina* (Coker and Addlestone, 1938), *D. retrocurva* (Brooks, 1947), *D. ambigua* (Hebert and Green, 1985), *D. carinata* (Grant and Bayly, 1981; Hebert, 1978a), *D. pulex* (Krueger and Dodson, 1981; Black 1993), *D. gibba* and *D. hyalina* (Manea and Tognota, 1993). All these species show remarkable variation in the crest development. Lampert and Wolf (1986) in *D. cucullata* and Mort (1989) in *D. Galeata mandotae* have reported changes in the tail spine length. Similar changes in small features like anal spines and antenal setae have also been reported by Egloff (1968). Fig. 2 A-L shows different crested and uncrested morphs of various *Daphnia* species.

The variable size and shape of the cyclomorphic strudtures like helmets. crest, humps etc. could be due to an increase either in number or size or both of the epidermal cells. This is also important to know that the change of form appears in the passage of one generation into one or more succeeding generations and not by growth changes within a single generation.

Cyclomorphosis in Indian waters

Morphological changes in phenotypes have also been studied in Indian waters, where remarkable fluctuations occur in temperature. O'Brien and Vinyard (1981) studied typical form of *Daphnia carinata* (form typica) and exuberant form of *D. carinata var. cephalata*. They pointed out the differential predation by notonectids as an important factor for the distribution of two varieties of *D.carinata*. Further, in the life table experiments, they claimed higher survival rate of forma typica than that of variety cephalata. In another study in Indian waters, Venkataraman and Krishnaswamy (1986) conducted experiments on *Anisops bouvieri*, as predator, to investigate the advantages of cyclomorphic helmeted form of *D.cephalata* and results were compared with non-helmeted species, *D.similis*. These workers concluded that the helmeted *D.cephalata* avoids *A. bouvieri* predation better than the non-helmeted *D.similis*. The predator selectively preys upon smaller sized *D.similis* than the larger *D. cephalata*.

Causes of Cyclomorphosis

Investigations on cyclomorphosis especially concerned with the elucidation of its causes and adaptive significance to the plankton.

The proximal causes of cyclomorphosis are not well understood. Since the beginning of this century, temperature was known to be the first factor contributing in the phenotypic variations among the population of the same species. Ostwald (1904) on *D.cucullata* and Coker and Addlestone (1938) on *D. longispina* suggested that temperature has a role in cyclomorphosis. Jacobs (1961) claimed a clear positive correlation between temperature and the degree of relative head length both in neonatal and postnatal stages. He also demonstrated that temperature exposure during the second half of empryonic development influences relative head length of neonates. But Brooks (1946) found no direct correlation between water, temperature and the rate of head growth of *D. rectrocurva*. He denied any direct influence of temperature on postnatal growth. Grant and Bayly (1981) thought that temperature did have a secondary effect on crest size, but whether this influence was direct or acted by reducing food supply remains unsolved. Coker (1939) and Brooks (1946) pointed out the presence of unknown factor absent in cultures. The proposed a joint function of temperature and unknown factor which might depend on the temperature threshold.

Woltereck (1913) claimed the degree of nutrition to be the main determinant of cyclomorphosis in cladocera. He thought that embryonic helmet growth is influenced by a food dependent maternal substance secreted in the brood pouch. Hebert (1978 b) correlated food availability with that of crest size. However, according to Jacobs (1961), nutrition is only a secondary determinant, but Grant and Bayly (1981) observed no development of crest in well fed population with large clutch sizes. Thus, there is not clear evidence for any role of nutrition in cyclomorphosis.

Turbulence, as one of the causes of cyclomorphosis, was presented by Brooks (1947) and Hrbacek (1959). Jacobs (1961) also reported the role of turbulence in seasonal morphological changes, but Herbert (1978 a) obtained negative correlation between turbulance and relative head size in *D.carinata var. cephalata*.

Above mentioned causes were not found satisfactory because even in warm turbulent waters with excess food, laboratory cultured cladocerans usually have smaller proturbances than wild population at similar temperature. Therefore, limnologists have had to evoke an unknown field factor to account for the difference in cyclomorphic structures between the cultured organisms and those in the field population.

Gilbert (1966) reported that the rate predation by invertebrates is influenced by morphological changes in certain zooplankton. In some rotifers, cyclomorphosis was found to be a response to size-selective predation. Beauchamp (1952) and Gilbert (1966) have shown that predaceous rotifer, *Asplancha* release a protien in the water to which prey rotifer, *Brachionus* responds by growing long anterior and posterior spines. Zaret (1972 a, b) has also suggested relationship between cyclomorphic changes and predation. Dodson (1974 a) presented a hypothesis that cyclomorphosis may an adaptation to escape invertebrate predation. He claimed that enlargement of the head, tail spine and other areas of cyclomorphic cladocerans may make capture difficult by invertebrate predators. He suggested that the morphology of the prey and capturing appendages of invertebrate predators are generally fit together, so that the exuberant forms hinder the ability of predators to grasp or capture the individuals of cyclomorphic variety.

Experiments of Krueger and Dodson (1981) have suggested a water soluble factor released in the environment by the predator, *Chaoborus* larva (Diptera), causing embryo of the *D.pulex* to develop into cyclomorphic form, *D. minnehaha*. Grant and Bayly (1981) have also claimed predation as the most potent influence inducing crest development in *D.carinata.*

In addition, so many studies have confirmed the hypothesis that size-selective predation is the main factor, and predator secreted substances in the water medium bring morphological changes in prey phenotype. Esperiments conducted by Kerfoot (1977) on *Bosmina*, O'Brien and Vinyard (1978) on *D. carinata*, Black (1980) on *Bosmina longirostris*, Krueger and Dodson (1981) on *D. pulex*, Mort (1989) on *D. galeata mandotae*, Hanazoto (1991) on *D. ambiguae*, Black (1993) on *D. pulex*and manea nad Tognota (1993) on *D.hyalina* have successfully demonstrated that invertebrate predation help in inducing cyclomorphosis in cladocerans. Most of the workers have reported *Anisops* (Notonectid), *Chaoborus* larva and calanoid copepod, *Heterocope septentrionalis* as the common predator causing cyclomorphic changes among cladocerans.

Genetic Component of Cyclomorphosis

Cyclomorphosis in cladocera has been ascribed only to phenotypic plasticity. It is very important to know that some cyclomorphic species are in reality composed of many genetically differentiated species (Herbert, 1978 a, b; Black, 1980 a; Kerfoot, 1977). A number of these species often occur in the same community and seasonal changes in their relative frequencies can cause polymorphic cycles that mimic true cyclomorphosis. Black (1980) stated that generic changes within a population is indeed an important component of cyclomorphosis.

Kerfoot (1977) has described three possible mechanisms for cyclomorphosis, first phenotypic plasticity with a genotype, second clonal replacement and third succession of sibling species. But, recent generic studies of Mort and Wolf (1985) and Lampert and Wolf (1986) have supported the idea that changes occur via phenotypic plasticity only. There is no clue of significant changes in population genetics. Thus, clonal replacement and species succession have been ruled out as possible mechanisms for cladoceran cyclomorphosis.

Adaptive Significance

Wesenberg-Lund (1908) proposed buoyancy theory that the crest development functions to lessen sinking rate in less dense or less viscous water during summer. Thus, phenotypic variations are simply floating devises. But, according to Jacobs (1966), exuberant forms of *Daphnia* sink faster than the typical form. Sinking speed has been found to be dependent on its density and size rather than shape (Brooks and Hutchinson, 1950). Further, the species which typically live on or near the bottom among the aquatic weeds also show cyclomorphic changes. Therefore, the buoyancy hypothesis has been rejected unanimously.

Woltereck's (1909, 1913) theories of balancer, stabilizer and rudder functions were also controversial and unable to satisfy the limnologists. Now all the hypothesis regarding the adaptive significance of cyclomorphosis have given way to hypothesis of predator avoidance. The new hypothesis put forwarded by Dodson (1974) indicates that morphological changes in animals act as an antilock device against grasping appendages of size-selective invertebrate predators. This hypothesis has been widely accepted and confirmed by a number of limnologists through field as well as laboratory studies. In favour of this theory, O'Brien and Vinyard (1978) have found that crested individuals of *D. carinata* were better able to avoid predation by *Anisop spp*. Than the uncrested ones. Kerfoot (1977) and O'Brien and Schmidt (1979) have shown that cyclomorphic structures of *Bosmina* reduce predation by copepods. Herbert and Loaring (1980) reported two morphs of *D. middendorffiana* in arctic Canada. Morph 'A' with long tail spine and dorsal hump is less succeptible to predation by calanoid, *Heterocope septentrionalis,* than morph 'B' which has short tail spine. Grant and Bayly (1981) also confirmed the view that crested for of *D.carinata* were better able to avoid predation than the typical uncrested form. Earlier Zaret (1972 a, b) showed that exuberant horned morphs of *Ceriodaphnia cornuta* were preyed upon less often by predatory fishes. Havel (1985) demonstrated the enlarged helmets of *D. retrocurva* reduce clearance rate of juveniles by *Anisops vernalis,* while increased body size reduce predation by *Leptodera kindti*. Helmeted *D. galeata mandotae* is less susceptible to *Chaoborus* predation than unhelmeted form (Mort, 1985). According to Venkataraman and Krishnaswamy (1986), *D. carinata var. cephalata* avoided *Anisops bouveri* predation better than the non-helmeted *D. similis* (Fig. 3).

Herbert (1978 b) after studying *Daphnia* in Australian waters concluded that large crests serve two functions by allowing the development of large antennal muscles. They confer better swimming ability in warm, less viscous water and they provide an increased surface area for oxygen transfer. He rejected the predator avoidance theory, although he later reported less susceptibility of crested morphs of *D. middendorffiana* to copepod predation (Herbert and Loaring, 1980).

Beside advantages, crest growth has certain disadvantages too. The energy involved in the crest development is predicted

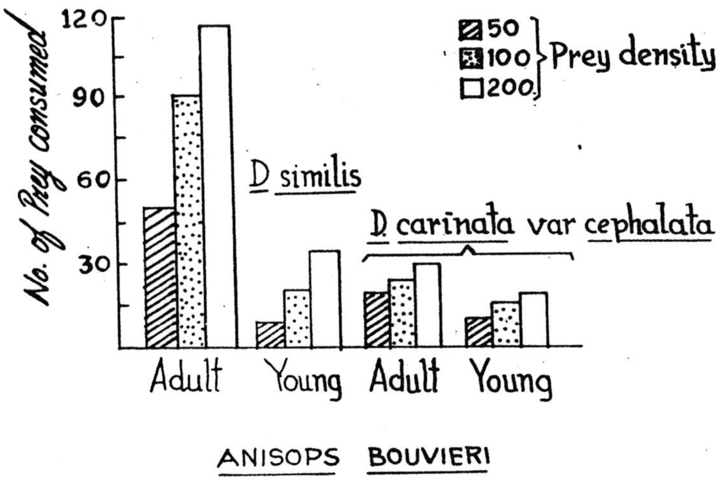

Fig. 3 : The relationship between the number of prey consumed by *Anisops bouveri* (Adult and young) exposed to equal numbers of *D.similis* and *D. cephalata* (Redrawn from Venkataraman and Krishnaswamy, 1986).

as a disadvantage by Dodson (1974). According to Grant and Bayly (1981), adults with growing crests produced fewer egges, suggesting either that crest development requires more energy or that crest development inhibits reproduction. O'Brien and Vinyard (1978) also found a lower rate of population growth in crested morphs than in uncreated ones. Venkataraman and Krishnaswamy (1986) have also reported the same findings in *D.carinata*. Zaret (1972) presented rearing data of *Ceriodaphnia cornuta* which showed inferiority of horned morphs (Fig. 3).

Conclusion

After extensive research work in cyclomorphosis in cladocera one can find it as an interesting phenomenon among certain plankton organisms. Cyclomorphosis is brought about by predator secreted substance in water and codetermined by temperature, turbulence, light and to some extent by degree of food availability, but all these codeterminents work indirectly. Exuberant forms are better able to avoid predation specially be invertebrate predators, although these forms are least productive in comparison of the typical morphs. It can also be said that phenotypic plasticity in cyclomorphosis may be caused by certain biotic factors aided by abiotic factors and this adaptive phenomenon leads the plankton to lessen size-selective predation.

Reference

*Beauchamp, P.D. (1952), In facteur de la variabiliate chez les rotifers de genere *Brachionus*. *C.R. Hebd. Seanes Acad. Sci.*, 234 : 573-575.

Benzie, J.A.H. (1991), Other environmental effects on the size of crests development by *D. carinata* (Cladocera - Daphnidae) in response to notonectid predators. *Freshwater Biology*, 26(1) : 69-76.

Black, A.R. (1993), Predator induced phenotypic plasticity in *Daphnia pulex* : Life history and morphological responses to *Notonecta* and *Chaoborus*. *Limnol. Oceanogr.* 38(5) : 986-996.

Black, A.W. (1980), Cyclomorphosis in *Bosmina longirostris*. *Ecology*, *61* : 1122-1132.

*Brooks, J.L. (1946), Cyclomorphosis in Daphnia : An analysis of *D. retrocurva* and *D.galeata*. *Ecol. Monogr.*, 16: 409-417.

*Brooks, J.L. (1947), Turbulance as an environmental; determinant of relative growth in *Daphnia*. *Proc. Nat. Acad. Sci.*, *33* : 141-148

*Brooks, J.L. and Hutchinson, G.E. (1950), On the rate of passive sinking of *Daphnia*. *Proc. Nat. Acad. Sci.*, *36* : 272-277.

Coker, R.E. (1939), The problem of cyclomorphosis in *Daphnia. Quart. Rev. Biol., 14* : 137-148.

Coker, R.E. and Addlestone, H.H. (1938), Influence of temperature on cyclomorphosis in *Daphnia longispina. J. Elisha Mitchell Sc. Soc., 54* : 45-75.

Daday, E. (1885), Beirtrage zur kenntniss der plattensee - Faune Math naturw. *Ber. Ung., 3* : 179-184.

Dodson, S.I. (1974 a), Adaptive change in plankton morphology in response to size-selective predation : A new hypothesis. *Limnol. Oceanogr. , 19* : 721-729.

Dodson, S.I. (1974 b), Zooplankton competition and predation: A experimental rest of the size efficiency hypothesis. *Ecology, 55* : 605-613.

Dodson, S.I. (1984), Predation of *Heterocope septentrionalis* on two species of *Daphnia* : Morphological defences and their cost. *Ecology, 53* : 234-257.

Egloff, D.A. (1968), The relative growth and seasonal variation of several cyclomorphic structures of *Daphnia catawba* in natural populations. *Arch Hydrobiol. , 65* : 325 - 359.

Gilbert, J.J. (1966), Rotifer ecology and empryological induction, *Science, 151* : 1234-1237.

Grant, J.W.G. and Bayly, J.A.E. (1981), Predator induction of crests in morphs of the *Daphnia carinata* King complex. *Limnol. Oceangr., 26 (2)*, 201-218.

Green, J. (1967), The distribution and variation of *Daphnia lumholtzii* (Crustacea - cladocera) in relation to fish predation in lake Albert, East Africa, *J. Zool., 151* : 181-197.

Hanazato, J. (1991), Influence of food density on the effects of a Chaoborus released chamical on *Daphnia ambigua. Freshwater Biol.*, 25(3) : 477-483.

Hanazato, J. (1992), Population dynamics and cyclomorphosis in *Bosmina longirostris* in lake Yunoko. *Japan. J. Limnol., 53* : 13-25.

Hartmann, D. (1915), Studies uber die cyclomorphose bei cladoceran. *Arch. Hydrobiol., 10* : 436-519.

Havel, J.E. (1985), Cyclomorphosis of *D, pulex* spined morphs. *Limnol. Oceangr., 30*(4), 853-861.

Havel, J.E. and Dodson, S.I. (1984), Chaoborus predation in typical and spined morphs of *Daphnia pulex* : Behavioural observations. *Limnol. Oceangr., 29*(3), 487-494.

Hazelwood, D.H. (1966), Illumination and turbulance effects on relative growth in *Daphnia. Limnol. Oceanogr., 11* : 212-216.

Hebert, P.D. (1978 a), Cyclomorphosis in natual populations of *Daphnia cephalata* King. *Freshwater Biol., 8* : 79-90.

Hebert, P.D. (1978 b), The adaptive significance of cyclomorphosis in *Daphnia*. More possibilities. *Freshwater Biol.* 8: 313-320.

Hebert, P.D. and Green, J. (1985 a), *Chaoborus* included shift in the morphology of *Daphnia ambigua. Limnol. Oceangr., 30*: 1291-1297.

Hebert, P.D. and Loaring, J.M. (1980), Selective predation and the species composition of arctic ponds. *Canad, J. Zool., 58* : 422-426.

*Hrbacek, J. (1959), Circulation of water as a main factor influencing the development of helmets in *Daphnia Cucullata* Sars. Hydrobiologia, *13* : 170-185.

Hutchinson, G.E. (1967), *A Treatise on Limnology*, Vol.II John Wiley and Sons Inc., New York, London, 1115 pp.

Jacobs, J. (1961), Cyclomorphosis in *Daphnia galeata mandotae* Birge, a case of environmentally controlled allometry. *Arch. fur Hydrobiol., 58* : 7-71.

Jacobs, J. (1966), Predation and rate of evolution in cyclomorphic *Daphnia*. *Int. Ver. Theor. Angrew. Limnol. Verh.*, *16* : 1645-1652.

Kerfoot, W.C. (1977), Implication of copepod predation. *Limnol. Oceanogr.*, *22* : 316-325.

Krueger, D.A. and Dodson, S.I. (1981), Embryological induction and predation ecology in *Daphnia pulex*. *Limnol. Oceanogr.*, *26* : 219-223.

Lampert, W. and Wolf, H.G. (1986), Cyclomorphosis in *Daphnia* cucullata : Morphometric and population genetics analysis. *J. plankton Res.*, *8* : 289-303.

Leauterbarn, R. (1904), Teil Die cyklische order temporale variation von Anuraea cochlearis. *Verh. natusch-med. Hiedelb, n.f. 7* : 529-621).

Lee, L.W. (1984), Environmentally controlled phenotypic plasticity of morphology and polypeptide esperiments in two population of *Daphnia pulex*. *Oceanloga*, 63 (2), 207-214.

*Lieder, U. (1951), Der stand der zyklomorphosse forschung. *Nature*, *38* : 39-44.

*Lilljeborg, W. (1900), *Nova Acta R. Soc. Scient. Upsal.* (Ser. 3), *19* : 1-701.

Manea, M. and Tognota, G. (1993), Seasonal changes in morphology and size of *Daphnia Hyalina*. Hydrobiologia, *264*(3), 154-167.

Mitchell, B.D. (1978), Cyclomorphosis in *Daphnia carinata* King from two adjacent sewage lagoons in South Australia. *Aus. J. Mar. and Freshw. Res.*, *29* : 265-576.

Mort, M.A. (1989), Cyclomorphosis in *Daphnia galeta mandotae*: Variation and stability in phenotypic cycles. *Hydrobiologia*, *171*(2), 159-170.

Mort, M.A. and Wolf, H.G. (1985), Enzyme variability in large lake *Daphnia* populations. *Heredity*, *55* : 27-36.

O'Brien, W.J. and Schmidt, S. (1979), Arctic *Bosmina* morphology and copepod predations. *Limnol. Oceanogr.*, *24* : 264-568.

O'Brien, W.J. and Vinyard, G.L. (1981), Polumorphism and predation: The effect of invertebrate predation on the distribution of two varieties of *Daphnia carinata* in South Indian ponds. *Limnol. Oceanogr.*, *23*(3), 452-460.

Ostwald, W. (1904), Experimentelle iiber den sai son polymorphismus Daphniden. *Arch. Entw.mech.*, *18* : 415-451.

Pennak, R.W. (1978), *Freshwater in invertebrates of United States*. 2nd ed. John Wiley and Sons, New York. 366 pp.

Pijanowska, J. and Tundisi, J.G.(1990), Cyclomorphosis in *Daphnia* : An adaptation to avoid invertebrate predation. In Dunmont, H.J. and Roche, K. (Eds) *Intrazooplankton Predation. 41-50 pp.*

Venkataraman, K. and Krishnaswamy, S. (1986), *Anisops bouveri* predation and advantages of cephalic expansion in *Daphnia carinata* under laboratory conditions. *Proc. Ind. Acad. Sci.* animal Aci.) *95*(5), 509-513.

*Wagler, E. (1936), Die systemetic und geographsche verbreitung des genus *Daphnia* mit besonderer Berucksichtigung der Sudafrika nischen Arten.*Arch. Hydrobiol.*, *30* : 505-556.

Wesenberg-Lund, C. (1908) Plankton investigations of Danish lakes. General part. *Dan.Freshw. Biol. Lab.*, No.5, Nordisk Forlag, Copenhagen.

Woltereck, R. (1909), Wietere experimentlle Untersuchungen uber Artveranderung speziell uber dar wesen quantitative. Artunterschiede bei Daphniden. *Verh.D.Zool. Ges. 19* : 110-173.

*Woltereck, R. (1913), Uber function herkunft und Entstehungsursache der sogenanten Schwebefot saeze pelagischer cladocera. *Zoologica*, *67* : 474-550.

*Zacharias, M. (1894), Beobachtungen um plankton des Grossen ploner Sees. *Forschber Biol. Stn. Plon*, 2 : 91-137.

Zago, M.S.A. (1976) A preliminary investigation on the cyclomorphosis of *Daphnia gessneri* in a Brazilian reservoir. *Balm. Zoll. Univ., S. Paulo*, : 147-160.

Zaret, T.M. (1972 a), Predator, invisible prey and the nature of Polymorphism in the cladocera (Crustacea). *Limnol. Oceanogr.*, 17 : 171-184.

Zaret, T.M. (1972 b), Predator-prey interaction in a tropical lacustrine ecosystem. *Ecology*, 53 : 234-257.

* Cross references.

10

Phytoplanktonic Dynamics in Bhadrakali Lake, Warangal, Andhra Pradesh

Y. Kalyani and M.A. Singara Charya

Introduction

Phytoplanktons are ecologically significant, as they trap the radiant energy of sunlight to convert chemical energy, *i.e.*, organic materials. Many herbivores graze upon phytoplanktonic food organisms and thus these organisms will become the basis of food in the aquatic ecosystems. The role of phytoplankton in managing bio-energetics of lake, their role as bio-indicators, and their secretions of enzymes in purification of polluted aquatic habitats has known for a long time (Thunmark, 1945; Anderson, 1967; Munawar *et al.*, 1978; Marriaze *et al.*, 1991; Shastree, 1992). Zafar (1964), Singh and Ahmed (1990), Satyamohan (1991) and Suvarna and Charya (1995) explained the spatial and temporal variations of phytoplankton in different lakes and ponds and related their frequency with seasonality of production. In the present study the ecology of phytoplanktons was undertaken in Bhadrakali Lake, a drinking water resource, to study its role in maintaining the quality of lake waters. Various diversity indices were calculated to establish the phytoplankton as efficient biological pollution indicators.

Material and Methods

Bhadrakali lake, situated in Warangal city between the altitude of 17^0 19' and 18^0 36' N and longitude of 78^0 48' and 80^0 43' E. The water samples were collected in four corners of lake and designated as site I, II, III and IV. In the enumeration of phytoplanktonic organisms one litre of water was filtered and filtrate washed into a beaker with 100 ml distilled water and fixed in Lugol solution (KI +I). From this one ml of sample was drawn quickly with a wide mouthed pippette and plankton numbers were counted and the average of three samples was taken into consideration. Among algal indices, the Kothe's species deficit index and odum's algal index was calculated as suggested by Odum (1971). The Nygaard's algal indices as methods adopted by Nygaard (1949) and Palmer's algal pollution indices as suggested by Palmer (1969). The sequential comparison index (Cairns, 1968) and Shannon's species diversity index (Shannon and Weaver, 1949) were also analysed.

Results and Discussion

The phytoplankton abundance with monthly fluctuations were studied and reported in Table 1. The mean monthly abundance of phytoplankton varied from 28 to 37 x 10^3 during October 1994 to September 1995. The maximum abundance was recorded in site IV (88 x 10^3) followed by Site III (55 x 10^3). The site I and II could record moderate phytoplanktonic populations during this study period. These populations showed two peaks (bimodel) *i.e.*, summer peak occurring in April and Winter peak in December. Similar to the present observations January (1973), Sheath and Munawar (1974), Shastree (1992) also showed great individual variations in phytoplankton abundance in the lakes studied by them.

Besides, the phytoplankton total enumeration, the individual groups, such as, chlorophyceae, cyanophyceae, bacillariophyceae and euglenophyceae members were individually counted and reported. All the other forms, which could be identified were kept under the category of 'Others". The cyanophycean members (98 per cent) were maximum in site I during 1994-95 followed by chlorophycean members (69 per cent). Bacillariophyceae dominated in one or two months and absent in 6-7 months. Euglenophyceae could detect only in months of June and September with 17-28.6 per cent. In site II, the variations in percentages of different groups of algal species showed with dominance of chlorophycean members followed by cyanaphycean organisms. In site III, the high incident of algal species were in chlorophyceae with their maximum numbers in the months of March (summer peak) and August. The site IV was rich with cyanophycean species, with frequent algal blooms. These were maximum throughout the year except in June (80 per cent). Among these members, the *Microcystis* blooms were quite prevalent in the entire year. Gopal *et al.*, (1981), Dash *et al.*, (1993) also enumerated a diversified algal general in various Indian lakes and reservoirs. Joshi and Adoni (1993) in order to review the water quality of central Indian lakes, signified the algal members, in the evaluation of the trophic status of the lakes.

Table 1 : Phytoplankton Abundance (1×10^3) in Four Sites of Bhadrakali Lake during 1994-95

Month	*Site I*	*Site II*	*Site III*	*Site IV*
1994				
October	22.0	27.5	55.0	27.5
November	27.5	27.5	27.5	11.5
December	19,4	30.3	25.7	38.5
1995				
January	22.0	29.3	24.8	25.7
February	18.3	44.0	11.0	19.3
March	18.3	27.5	16.5	23.8
April	27.5	16.5	33.0	88.0
May	16.5	11.0	22.0	27.5
June	84.3	33.0	41.2	39.3
July	24.8	41.8	24.8	79.2
August	28.6	30.8	22.0	30.8
September	30.8	33.0	35.8	33.0
M.V.	28.3	29.4	28.3	37.0
S.D.	18.2	9.1	11.7	23.1
S.E.	5.3	2.6	3.4	6.7

Diversity Indices

Species diversity is a statistical abstraction with two components, reflecting the numbers of species and distribution of individual of all species at a particular site (Read, 1993).

(i) **Kothe's Index :** It records the levels of pollution in various sites of an aquatic ecosystem. It varied from -321 to -650 during October 1994 to September 1995 (Table 2). The site IV could record moderate index (-1700) values, in the month of July, followed by -1150 in June. Site II and III could record minimum Kothe's index in the months March and February respectively. The maximum index value -2550 was recorded in month of December in site I indicated the growing levels of pollution in the lake.

Table 2 : Kothe's Index (Algal) in Four Sites of Bhadrakali Lake during 1994-95

Month	*Site I*	*Site II*	*Site III*	*Site IV*
1994				
October	-100	-150	-800	-300
November	-300	-150	-150	-950
December	-2550	-2650	-250	-250
1995				
January	-150	-300	-350	-250
February	-150	-300	-100	-250
March	-150	-150	-50	-550
April	-150	-50	-200	-850
May	-1500	-200	-200	-150
June	-1050	-350	-650	-1150
July	-350	-850	-350	-1700
August	-550	-650	-200	-600
September	-600	-650	-550	-800
M.V.	-513	-533	-321	-650
S.D.	702	708	232	464
S.E.	202	204	67	133

(ii) **Odum's Algal Index :** Odum (1971) determined the algal index per thousand individuals and established certain principles to determine the quality criteria of waters in lentic habitats. In the present study, diversified values were recorded in four sites. The maximum index (800) was recorded for algal genera, *Microcystis*, *Spirulina* and *Ulothrix* (Table 3). The moderate range (400-650) was seen in certain algal genera: *Cladophora*, *Gomphonema*, *Achanthus*, *Spirogyra*, *Tabellaria*, *Hyalotheca*, *Chlorogonium* etc. The latest incidence (<100) was observed with algal species, such as *Mougeotia*, *Gomphonema*,

Table 3 : ODUM'S Algal Index per 1000 Individuals in Four Sites of Bhadrakali Lake During 1994-95

Month	*Site I*		*Site II*		*Site III*		*Site IV*	
Oct.**94**	*Gomphonema* (B)	500	*Microcystis* (Cy)	600	*Micocystis* (Cy)	880	*Spirogyra* (Ch)	200
	Achnanthes (B)	500	*Spirogyra* (Ch)	400	*Spirulina* (Cy)	200	*Microcystis* (Cy)	800
Nov.	*Cladophora(Ch)*	400	*Cladophora (Ch)*	600	*Spirulina(Cy)*	800	*Microcystis (Cy)*	857
	Ulothrix (Ch)	600	*Spirogyra(Ch)*	400	*Ulothrix(Ch)*	200	*Eugluna(E)*	200
Dec.	*Micocystis* (Cy)	925	*Micocystis* (Cy)	982	*Tabellaria* (B)	286	*Microcystis* (Cy)	857
	Synedra (B)	18.9	*Zygnema* (Ch)	18	*Spirogyra* (Ch)	143	*Spirogyra (Ch)*	143
	Pleurocapsa (Cy)	56.6			*Microcystis* (Cy)	571		
Jan.**95**	*Dichotomosiphonch* (Ch)	200	*Microcystis (Cy)*	375	*Chaemaesiphon*(C)	111	*Microcystis (Cy)*	439
	Fragilaria (B)	600	*Gomphonema* (B)	250	*Fragilaria* (B)	222	*Mougetoia*	143
	Spirulina (Cy)	200	*Ankistrodesmus* (Ch)	375	*Scizothrix* (Cy)	333	*Ankistrodesmus* (ch)	429
Feb.	*Tabellaria* (B)	400	*Ankistrodesmus* (Ch)	250	*Ankistrodesmus (Ch)*	250	*Spirogyra (Ch)*	428
	Aphanizomenon (Cy)	200	*Cladophora* (Ch)	250	*Cladophora (Ch)*	250	*Aphanizomenon (Ch)*	286
	Hyalotheca (Ch)	400	*Aphanizomenon* (Ch)	250	*Aphanizpmenon (Cy)*	250	*Bulbochatae* (Ch)	143
			Nannadrium (Chara)	125	*Spirogyra (Ch)*	250	*Ulothrix (Ch)*	143
			Spirogyra (Ch)	125				

Contd..

Table 3 (Contd.) : ODUM'S Algal Index per 1000 Individuals in Four Sites of Bhadrakali Lake During 1994-95

Month	*Site I*		*Site II*		*Site III*		*Site IV*	
Mar.	*Aphnizomenon* (Cy)	400	*Spirulina* (Cy)	200	*Spirogyra* (Ch)	333	*Aphanizomenon*	231
	Spirogyra	200	*Ulothrix*	800	*Chlorogonium*	666	*Tabellaria*	154
	Spirulina (Cy)	400					*Spirogyra*	231
							Gomphonema (B)	77
							Chlorogonium (Ch)	154
Apr.	*Microcystis* (Cy)	800	*Spirogyra* (Ch)	333	*Microcystis* (Cy)	500	*Microcystis*	700
	Staurastrum (Ch)	200	*Cladophora* (Ch)	666	*Ulothrix*	500	*Botryococcus* (B)	250
May	*Ulothrix* (Ch)	666	*Ulothrix* (Ch)	166	*Microcystis* (Cy)	500	*Microcystis(Cy)*	800
	Microcystis (Cy)	333	*Melosira* (B)	166	*Aphanizomenon* (Cy)	166	*Cladophora* (Cy)	200
			Cladophora (Ch)	333	*Fragilaria* (B)	333		
			Tabellaria (B)	166				
			Microcystis (Cy)	166				
Jun.	*Microcystis* (Cy)	696	*Fragillaria* (B)	444	*Mougeotia (Ch)*	133	*Spirogyra* (Ch)	80
	Spirogyra (Ch)	130	*Synedra (B)*	333	*Pedistrum* (Ch)	400	*Melosira* (B)	320
	Euglena (E)	174	*Ulothrix (Ch)*	222	*Ankistrodesmus* (Ch)	200	*Nevicula* (B)	120
					Synedra (B)	266	*Clostridium* (Ch)	200
							Pediastrum (Ch)	120
							Zygnema (Ch)	40
							Ankistrodesmus (Ch)	120

Contd..

Table 3 (Contd.) : ODUM'S Algal Index per 1000 Individuals in Four Sites of Bhadrakali Lake During 1994-95

Month	*Site I*		*Site II*		*Site III*		*Site IV*	
Jul.	*Nitzschia* (B)	222	*Ulothrix* (Ch)	210	*Closteridium* (Ch)	33	*Microcystis* (Cy)	777
	Mougeotia (Ch)	222	*Ankitestrod-esmus* (Ch)	263	*Aphanizomenon*	322	*Gomphonema* (B)	56
	Zygnema (Ch)	333	*Spirogyra* (Ch)	105	*Microcystis* (Cy)	222	*Mougeotia*	56
	Synedra (B)	222	*Closteridium* (Ch)	315	*Melosira* (Ch)	222	*Aphanizomenon* (Ch)	56
			Aphanizomenon(Cy)	105			*Achnanthes* (Cy)	56
Aug.	*Spirogyra* (Ch)	308	*Zygnema* (Ch)	286	*Scenedesmus* (Ch)	333	*Microcystis*	214
	Microcystis (Cy)	154	*Volvox* (Ch)	143	*Botrycoccus* (Ch)	166	*Euglena* (E)	143
	Ankistrodesmus (Cy)	231	*Euglena*	143	*Spirogyra* (Ch)	500	*Spirulina* (Cy)	286
	Cladophora (Ch)	154	*Nodularia* (Cy)	143			*Chlamydomonas*	143
	Tolypothrix (Cy)	154	*Navicula* (B)	286			*Spirogyra* (Ch)	214
Sept.	*Spirulina* (Cy)	286	*Ankistrodesmus* (Ch)	133	*Synedra* (B)	385	*Closteridium* (Ch)	222
	Euglena (E)	286	*Cladophora* (Ch)	267	*Chlorogonium* (Ch)	153	*Scendesmus* (Ch)	111
	Tetradron (Ch)	71	*Clostridium* (Ch)	267	*Spirogyra* (Ch)	154	*Euglena* (E)	111
	Ankistrod-esmus (Ch)	143	*Lemenea* (Ch)	133	*Zugnema* (Ch)	308	*Ankistrodesmus* (Ch)	278
			Spirogyra (Ch)	200			*Microcystis* (Cy)	167
							Navicula (B)	111
M.V.		325		292		324.2		264
S.D.		213		197		175		236
S.E.		34.9		30.8		29		35

Cy = Cyanophaceae; B = Bacillariophyceae; E = Euglenophyceae; Ch = Chlorophyceae.

Tetradron etc. The mean Odum's algal index ranged between 292 to 325. Uhlaman *et al.* (1976) and Salman (1993) predicted the phytoplankton development and water quality in eutrophicated waters and analysed the constituent diversity of algal species and monitored water quality with the help of Odum's algal index.

(iii) Nygaard's Algal Indices : Nygaard (1949) has proposed five indices for biological monitoring of aquatic habitats, based on the individual populations of myxophyceae, chlorophyceae, bacillariophyceae, euglenophyceae and the mixture of all these populations together. The mean range of variation of myxophyceae index in four sites of the lake was from 0.06 to 1.00 (Table 4). In majority of the months this index was not recorded, due to absence of required number of populations. The mean chlorophycean index varied between 0.5 to 0.85 with the incidence of their members only in 3-4 months. The average range of euglenophycean index was in between 0.02 to 0.04 and the compound index of all populations put together was 0.36 to 3.23. The statistical analysis of these different indices clearly indicated the slow progress of this lake towards eutrophic status. Many researchers extensively used Nygaard's indices to understand the quality criteria of the water (Seth, 1975; Biswas, 1981; Ragothaman and Jaiswal, 1995).

(iv) Palmer's Algal Index : Palmer (1969) reviewed a composite rating of algae, tolerating organic pollution and developed an index to establish the status of the aquatic body. This pollution index varied with sites and seasons. The site IV considered to be more diversified compared to other sites and gradually reaching to a polluted habitat (pollution index 16), with mean monthly average of this index between 1.5 to 4.0 (Table 5). Palmer's algal index at site I was between 1-7 and at site II 1-8 and at site III 1-4. Sinha *et al.*(1993) calculated the Palmer's index in certain polluted lentic waters. Veronique (1983) and George *et al.*(1986) studied the anthropogenic eutrophication in some perennial tanks and successfully related the increased organic pollution status with that of the genus included in Palmer's algal pollution index.

Table 4 : Nygard's Algal Index in Four Sites of Bhadrakali Lake During 1994-95

Month	*Site I*				*Site II*				*Site III*				*Site IV*			
	A	*B*	*C*	*D*	*A*	*B*	*C*	*D*	*A*	*B*	*C*	*D*	*A*	*B*	*C*	*D*
Oct.94	0	-	-	-	-	-	-	-	-	-	-	-	-	-	-	-
Nov.	-	-	-	-	-	-	-	-	-	-	-	-	-	-	0.16	-
Dec.	-	-	-	-	-	-	-	-	-	-	-	-	-	-	-	-
Jan.95	0.3	0.3	-	0.6	1.5	1.5	-	3.0	3.0	0.5	-	4.5	-	-	-	-
Feb.	0.5	1.0	-	1.5	-	-	-	-	-	-	-	-	-	-	-	-
March	-	-	-	-	-	-	-	-	-	-	-	-	0.6	0.8	-	6.5
April	-	-	-	-	-	-	-	-	-	-	-	-	-	-	-	3.0
May	-	-	-	-	0.5	1.5	-	2.0	2.0	-	-	3.0	-	-	-	-
June	-	-	0.2	-	-	0.3	-	1.3	-	2.8	-	3.8	-	-	-	2.3
July	-	1.3	-	1.3	-	-	-	-	2.0	1.5	-	4.5	-	-	-	18.0
Aug.	-	-	-	-	-	2.0	0.3	3.5	-	-	-	-	-	-	-	-
Sep.	-	-	-	-	-	-	-	-	0.4	1.6	-	2.6	-	-	-	9.0
M.V	0.06	0.9	0.03	0.4	0.4	0.4	0.02	0.8	0.6	0.5	-	1.2	1.6	0.6	0.04	3.2
S.D	0.2	0.5	0.14	0.6	2.9	0.8	0.07	1.3	1.1	0.9	-	1.7	4.5	1.3	0.07	5.5
S.E	0.04	0.1	0.04	0.2	0.8	0.2	0.02	0.4	0.3	0.3	-	0.5	1.3	0.4	0.03	1.6

A = Myxophycean Index; B = Chlorophycean Index;
C = Euglenophycean Index; D = Compound Index

Table 5 : Palmer's Index in Bhadrakali Lake during 1994-95

Month	*Site I*	*Site II*	*Site III*	*Site IV*
1994				
October	1	1	1	1
November	-	-	-	6
December	3	1	1	1
1995				
January	-	4	1	3
February	-	2	1	-
March	-	-	-	1
April	1	-	1	1
May	1	2	1	-
June	6	2	4	7
July	5	3	3	2
August	3	8	4	10
September	7	3	2	16
Mean	2.2	2.1	1.5	4.0
S.D.	2.5	2.2	1.4	4.9
S.E.	0.7	0.6	0.4	1.4

(v) Sequential Comparison Index (SCI) : It was proposed by Cairns (1968) where the organism are collected from healthy and polluted areas and their diversity is used to determine the index. The comparison of values of different sites is made to assess the effect of pollution. The mean variation of the index among the sites fluctuated from 2.4 to 3.8 with maximum values at site IV (Table 6). In majority of the months though the algal species were detected they could not have a position in SCI. Generally, the months June to September recorded high pollution index over other months.

(vi) Shannon's Species Diversity Index : Shannon and Weaver (1949) observed a correlation with algal species and pollution status of ecosystem based on their information theory. The mean variation of Shannon's index varied from 0.487 to 0.672 in entire lake ecosystem (Table 7). Site IV recorded high diversified algal species over other sites, where its fluctuation was from 0.409 to 1.19. The minimum and maximum Shannon's index at site I was 0.093 and 1.079 and at site II 0.091 and 1.08, and at site III it was 0.50 and 1.42. Wetzel (1981), David *et al.* (1990), Sinha *et al.* (1993) estimated the cultural entrophication of lake ecosystems and related the species diversity indices in general and Shannon's index in particular with that of the organic pollution levels in the lakes.

Table 6 : Sequential Comparison Index in Four Sites of Bhadrakali Lake during 1994-95.

Month	*Site I*	*Site II*	*Site III*	*Site IV*
1994				
October	1	0	0	0
November	0	0	0	0
December	0	3	0	0
1995				
January	0	4	2	3
February	0	3	3	0
March	0	0	0	1
April	0	0	0	0
May	0	2	0	0
June	1	0	6	8
July	4	3	2	1
August	2	1	4	1
September	4	3	3	9

Table 7 : Shannon's Index in Four Sites of Bhadrakali Lake during 1994-95.

Month	*Site I*	*Site II*	*Site III*	*Site IV*
1994				
October	0.0	0.672	0.0	0.5
November	0.0	0.0	0.5	0.409
December	0.093	0.091	0.953	0.409
1995				
January	0.948	1.08	0.848	0.682
February	1.054	0.563	0.562	0.682
March	0.256	0.5	0.00	1.06
April	0.256	0.00	0.692	0.562
May	0.636	1.01	0.636	0.5
June	0.821	0.529	0.579	0.685
July	0.686	0.335	1.42	0.425
August	0.616	0.670	0.0	0.991
September	1.079	0.392	0.665	1.19
Mean	0.537	0.487	0.568	0.672
S.D.	0.402	0.352	0.420	0.271
S.E.	0.116	0.102	0.121	0.078

(vii) Margalaf's Community Diversity Index : This diversity index is based on information theory of the functional aspect of (bioenergetics) the biotic community (Margalef, 1965; Odum, 1971). At site 1 this index fluctuated from 0.47 to 0.54 among various algal genera (Table 8). At site II this variation was 0.65 to 4.65 and at site III it was 1.78 to 2.97. The site IV which was designated as polluted area in the lake showed its index value between 0.26 to 5.95. Rai and Kumar (1980), Venkateshwarly (1986) and Shastree (1992) established the dynamics of phytoplanktonic fluctuations in lentic Waterbodies using Mergalef's community diversity index.

Table 8 : Margalef's Community Diversity Index in Four Sites of Bhadrakali Lake, During 1994-95

Month	*Sites*	*Chloro-phyceae*	*Cyano-phyceae*	*Bacill-ariphy-ceae*	*Euglen-ophyceae*	*Other*
1	*2*	*3*	*4*	*5*	*6*	*7*
1994						
October	I	-	-	4.98	-	-
	II	1.44	2.86	-	-	-
	III	-	9.00	-	-	-
	IV	-	4.29	-	-	-
November	I	5.72	-	-	-	-
	II	5.72	-	-	-	-
	III	-	4.29	-	-	-
	IV	-	12.86	-	1.51	-
December	I	-	29.58	-	-	-
	II	-	30.45	-	-	-
	III	-	3.55	1.18	-	-
	IV	-	5.92	-	-	-
1995						
January	I	-	-	2.86	-	-
	II	2.21	2.21	2.21	-	-
	III	-	5.24	1.05	-	-
	IV	3.55	2.37	-	-	-
February	I	1.43	-	1.43	-	-
	II	5.54	1.11	-	-	-
	III	5.54	1.10	-	-	-
	IV	3.55	2.37	-	-	-
March	I	-	4.29	-	-	-
	II	4.29	-	-	-	-

Contd..

Table 8 Contd...

1	2	3	4	5	6	7
	III	4.19	-	-	-	-
	IV	3.59	1.79	4.49	-	-
April	I	-	4.29	-	-	-
	II	4.19	-	-	-	-
	III	2.57	2.57	-	-	-
	IV	-	9.14	2.49	-	-
May	I	2.095	-	-	-	-
	II	2.57	-	1.29	-	-
	III	-	3.86	1.28	-	-
	IV	-	4.29	-	-	-
June	I	1.47	11.0	-	2.20	-
	II	1.05	-	6.29	-	-
	III	8.50	-	2.55	-	-
	IV	9.29	19.9	0.64	-	-
July	I	4.19	-	4.29	-	-
	II	12.5	0.78	-	-	-
	III	2.09	3.14	1.05	-	-
	IV	0.64	19.9	0.64	-	-
August	I	7.18	2.69	-	-	-
	II	6.11	-	2.62	0.872	-
	III	6.43	-	-	-	-
	IV	3.49	5.24	-	0.87	-
Sept.	I	4.36	2.62	3.49	-	-
	II	10.20	-	-	-	0.085
	III	6.28	-	-	-	-
	IV	6.37	3.19	0.79	0.79	-

(viii) Menhinick's Index : This index was based on the ratios between the number of species and total number of organisms present in a unit area. The mean range of this index at site I was from 0.48 to 1.48 (Table 9). The index values were generally high during months July to September. The range of variation at site II was 0.42 to 1.77 and at site III 0.49 to 1.23. The range at site IV was in between 0.72 to 2.06. Harris (1986), Elser and Carpenter (1988) and Padhi (1995) in their studies on developments and operations of large limnotic enclosures and their contribution to phytoplankton ecology indicated the importance of implementing this biomonitoring, using algae as bioassay model.

Table 9 : Menhick's Index in Four Sites of Bhadrakali Lake, During 1994-95

Month	*Sites*	*Chloro-phyceae*	*Cyano-phyceae*	*Bacill-ariphy-ceae*	*Euglen-ophyceae*	*Other*
1	*2*	*3*	*4*	*5*	*6*	*7*
1994						
October	I	0.0	0.0	2.0	-	-
	II	0.89	1.34	-	-	-
	III	-	3.16	-	-	-
	IV	0.45	1.79	-	-	-
November	I	2.24	-	-	-	-
	II	2.24	-	-	-	-
	III	0.45	1.79	-	-	-
	IV	-	3.92	-	0.65	-
December	I	-	7.14	0.14	-	-
	II	0.13	7.28	-	-	-
	III	0.38	1.51	0.76	-	-
	IV	0.38	2.27	-	-	-
1995						
January	I	0.45	0.45	1.34	-	-
	II	1.06	1.06	0.70	-	-
	III	0.33	2.00	0.66	-	-
	IV	1.51	1.13	-	-	-
February	I	0.89	0.45	0.89	-	-
	II	2.12	0.71	-	-	-
	III	2.12	0.71	-	-	-
	IV	1.51	1.13	-	-	-
March	I	0.45	1.79	-	-	-
	II	1.79	0.45	-	-	-
	III	1.73	-	-	-	-
	IV	1.39	0.83	1.66	-	-
April	I	0.45	1.79	-	-	-
	II	1.73	-	-	-	-
	III	1.22	1.22	-	-	-
	IV	-	3.00	1.00	-	-
May	I	1.55	0.58	-	-	-
	II	1.22	0.40	0.82	-	-
	III	-	1.63	0.82	-	-
	IV	0.45	1.79	-	-	-
June	I	0.63	3.34	-	0.83	-
	II	0.66	-	2.33	-	-
	III	2.84	-	1.03	-	-
	IV	2.80	-	2.20	-	-

Contd...

Table 9 Contd...

1	2	3	4	5	6	7
July	I	1.67	-	1.33	-	-
	II	3.90	0.46	-	-	-
	III	1.00	1.33	1.33	-	-
	IV	0.33	5.30	0.33	-	-
August	I	2.50	1.11	-	-	-
	II	2.14	-	1.07	0.53	-
	III	2.45	-	-	-	-
	IV	1.34	1.87	-	0.53	-
September	I	1.60	1.07	-	1.07	-
	II	3.36	-	-	-	0.52
	III	2.22	-	1.39	-	-
	IV	2.12	1.18	0.47	0.47	-

(ix) **Index of Similarity and Dissimilarity :** Shastree (1992) developed a quotient of similarity in phytoplankton between two habitats and the index of dissimilarity with the help of coefficient of similarity (1-5). The similarity index ranged between 0.12 to 0.2 among all the sites of the lake. It wad noticed that majority of algal species observed in one site, invariably found in other three sites, which resulted for the little value for this dissimilarity a index (0.8 to 0.9).

In this was the enumeration and diversities of algal genera in Bhadrakali lake explained their role as bio-indicators and also strengthened the principles on conservation of water resources.

Summary

Phytoplanktonic dynamics were studied in Bhadrakali lake, Warangal during 1994-95. The planktonic abundance with monthly fluctuations showed in between 28 to 37 x 10^3 during October 1994 to September 1995. Among four sites of the lake the maximum pollution loads and algal densities (88 x 10^3) were recorded in site IV. Generally, the cyanophycean members were maximum followed by chlorophycean populations. *Microcystis* blooms were quite prevalent in site IV in the entire year. Various diversity indices were calculated to assess the nature of the lake and to establish the algal role in bio-indication. The maximum Kothe's index (-2550) was recorded in the month of December. The least Odum's algal index (<100) was noted in *Mougeotia, Gomphonema* and *Tetradron*. Nygaara's algal indices were not recorded in majority of months due to absence of required number of populations. Palmer's algal index varied substantially with changes in sites and seasons. Similarly sequential comparison index, Shannon's species diversity index, Margalef's community diversity index, Menhinick's index, and index of similarity and dissimilarity also substantiated their role of algal species in understanding the status of this lake.

Acknowledgements

Authors are thankful to Prof. V. Thirupathaiah, Head, Department of Botany, Kakatiya University for providing laboratory facilities. One of us YK is grateful to Principal and management of Chaitanya Degree College, for their encouragement and help.

References

Anderson, G. 1967, in. *Soil Biochemistry*, A. Donglas McLaren and G.H. Peterson (Eds), Edward Arnold, London, p. 67.

Biswas, AK. 1981, Water for the third world. In C.K. Varshney (Ed.) *Water Pollution and Management Review*.

Cairns, J. 1968, The sequential comparison under a simplified method for non-biologists to estimate relative differences in biological diversity in stream pollution studies. *J. Water Poll. Control. Fed. 9* : 1607-1613.

Dash, J., Wilson E.J. and Miller, C.T. 1993, Modelling of trace organic pollutants in ground water. *Ground Water. 31* : 439-443.

David, R. Bayne Wendy, Seesock, C. Clifford, E. Webber and John, A McGuire 1990, Cultural eutrophication of west point lake - a 10 year study. *Hydrobiologia*. 199 : 143 - 156.

Elser, J.J. and Carpenter, S.K. 1988, Predation driven dynamics of zooplankton and phytoplankton communities in a whole lake experiment. *Oecologia. 76* : 148-154.

George, J.P., Venugopal, G. and Venkateswaran, K. 1986, Anthropogenic eutrophication in a perennial tank. Effect of growth of *Cyprinus carnio-communis. Indian J. Enviorn. Health 28* : 303-313.

Gopal, B, Sharma, K.P. Goel, P.K. and Trivedy, R.K. 1981, *Limnological* studies of a freshwater water reservoir, Jamwa Ramgarh, *Hydrobiologia* 62 : 187-190.

Harris, G.P. 1986, *Phytoplankton ecology*. Structure, function and fluctuation. Chapman and Hall, New York, pp. 384.

Jana, B.B. 1973, Aquatic environment. *Hydrobiologia* 5 : 127-143.

Joshi, G. and Adoni, A.D, 1993, Studies on some water quality parameters of central Indian lakes and evaluation of their trophic status. *Ecology and Pollution of Indian Lakes and Reservoirs 10* : 225-236.

Margalef, R. 1965, Ecological correlations and the relationship between primary productivity and community structure. *Mem.Ist. ital. Hydrobiol. 18* : 355-364.

Mariazzi Aldo, Victor Conzonno, Ricardo Echenique and Hector Labollits 1991, Physical and chemical characters, phytoplankton and primary production of Ezequiel Ramos Mxia Reservoir (Argentina), *Hydrobiologia* 209 : 107-116.

Munawar, M. Munawar, I.F., Culp, L.R. and Dupuis, G. 1978, Relative importance of nanoplankton in lake superior phytoplankton biomass and community metabolism. In : M.Munawar (ed.). Limnology of lake superior, *J. Great Lakes. Res. 4* : 462-480.

Nygaard, G. 1949, Hydrobiological studies on some Danish ponds and lakes. II. The quotient hypothesis and some new or little known phytoplankton organisms. *Dat. Kurgl. Danske Vid. Sel. Biol. Skr.* 7 : 1-293.

Odum, E.P. 1971, 'Fundamentals Ecology' W.B. Saunders Co., Philadelphia and Toppan Co. Ltd., Tokyo, pp. 574.

Padhi, 1995, Algal environment of polluted and unpolluted freshwater ponds. Kargupta, An and Siddiqui,E.N. (Eds.) *Algal ecology*. An overview International book Distributor, Dehradun, India.

Palmer, C.M. 1969, A composite rating of algae tolerating organic pollution, *J. Phycol. 5* : 78.

Ragothamam, S. and Jaiswal 1995, Studies on the hydrobiology of Tapi river from Jalgaon Region (Maharashtra) with reference to phytoplankton. *Poll.Res. 14* : 181-194

Rai, L.C. and Kumar,A. 1980, Effect of certain environmental factors on the toxicity of Zinc to *Chlorella vulgaris*. *Microbios. Letters. 13* : 79-84.

Read, W.U. 1993, The biodiversity prospecting, using genetic resouces for sustainable development. *World Resources Institute*, Washington.

Salman, S. 1993, Deterministic model of a shallow tropical lake undergoing eutrophication. *Ecology and Poll. of Indian Lakes and reservoirs. 6* : 99-112.

Satyamohan, K. 1991, Limnology of a mid-lake station of lake Mir Alam, Hyderabad. *A perspective current trends in Limnology* - I : 121-132.

Seth, G.K. 1975, *Water pollution control in India Science Reporter. 12* : 474-477.

Shannon, C. and Weaver, W. 1949, The mathematical theory of communication. Univ. of Illiois Press, Chicago.

Shastree, Nalin, K. 1992, Dynamics of phytoplanktonic fluctuations in a lentic waterbody. *Aquatic environment. 6* : 59-85.

Sheath, R. and Munawar, M. 1974, Phytoplanktonic compositon of a small sub-arctic lake in the North-West Territories, Canada, *Phycologia. 13* : 149-161.

Sinha, M.P. Sinha Rathna, Mehrotra, P.N. 1993, Species diversity indices and related habitat characteristics. *Pollu. Res. 12* (4) :" 207-212.

Singh, A.K. and Ahmed, S.H. 1990, A comparative study of the phytoplankton of the river Ganga and Pond of Patna (Bihar) India. *J.Indian.Bot. Soc., 69* : 153-158.

Suvarna, B. and Singara Charya, M.A. 1995, Biological control or water pollution I. Biodegradation of three aquatic macrophytes. *J. Ecobiol.*. 7 : 49-52.

Thunmark, S. 1945, Zur Soziologie des Subwasser Planktons. Eine Methodologisch Okologischa Studie. *Fol. Limnol. Scand. 3* : 1-66.

Uhlaman, D. Benndrof, J. and Gnauck, A. 1976, Development of models for predicting phytoplankton and water quality in eutrophicated compound water. *Proc. Internat. symp. Entro. Sym. 76* : III GDR, 387-407.

Venkateshwarlu, V., 1986, Ecological studies on rivers of Andhra Pradesh with special reference to water quality pollution. *Proc. Indian Acad. Sci.* (Plant Sci.). 96 : 495-508.

Vernoique, M., 1983, Facterus hydrologique et. phytoplankton en Baie de Morloaise (Manche Ocidentale) *Hydrobiologia 102* : 131-143.

Wetzel, R.G., 1981, Longterm dissolved and particulate allealine phosphate activity in a harwater lake in relation to lake stability and phosphorus environments. *Verhandlungen der Internationalen Vereinigung fur theoretische und angwandt Limnologie*, 21 : 337-349.

Zafar, A.R. 1964, On the ecology of algae in certain fish ponds of Hyderabad, India. Distribution of unicellular and colonial forms. *Hydrobiologia.* 24 : 556-566.

11

Pollution Status of a Leachate Reservoir Receiving Effluents from a Thermal Power Plant with Reference to Plankton Population

Rajeev Kumar Gaur, Asif A. Khan, M. Afaque Alam and Aftab Alam

Introduction

Most Waterbodies around industrial and urban centres receive effluents which affect water quality. Society's basic interest in pollution is owing to it's effects on living organisms, hence the assessment of pollution is basically a biological problem. In the aquatic environment living organisms provide convenient full time monitoring of pollution as one can determine effects over time and discover the presence of pollutants by analysing the population structure of organisms (Gunale, 1991). Plankton, specially phytoplankton are potential indicators of pollution because of their quick response to the toxicants and other chemicals (Walash and Merril, 1984) and, therefore, they are widely used as indicator of water pollution nowadays.

In the present study, the pollution status of a leachate reservoir having water-filled area of 13.5 ha and receiving effluents from Harduaganj Thermal Power Plant (HTPP), was assessed on the basis of plankton studies. This power plant (HTPP), located about 15 km. from A.M.U. campus and with a total capacity of power generation of 700 MW, uses sulpher rich bituminous coal as fuel at the rate of 11,65,069 tonnes/annum. The pollutants, being released in a man-made leachate reservoir through candensation process using water, include certain gases, fly ash and traces of heavy metals. Khan (1985) have reported that the pollutants from this power plant adversely affected the growth, primary productivity and seed production of agricultural crops. Hence, it is obvious that this may also exert tremendous impact on the aquatic ecosystem. Plankton, specially algae, being potential indicators of pollution, may clearly indicate the pollution status of the leachate reservoir.

Materials and Methods

Five stations A, B, C, D, and E were selected based on the inflow of effluents from HTPP into the leachate reservoir and outflow of effluents from leachate reservoir into the near-by drain (Fig.1). The samples for phytoplankton were fixed in Lugol's iodine solution and left overnight and concentrated. The samples for zooplankton were fixed in 5 per cent formaldehyde solution. The samples were later identified and counted in the laboratory following the works of Ward and Whipple (1959) and Needham and Needham (1962).

The results were computed on the basis of diversity index and scales of pollution. Although there are a number of diversity indices but Shannon-Weaver's diversity index (1949) is the best index of pollution (Sarkar and Krishnamoorthi, 1985). For the present study this index has been followed and determined as,

$$\overline{D} = - \Sigma \, pi \log_e pi$$

where, $\overline{D}$ = Diversity Index

pi = ni/N (ni = Number of individuals in species i

N = Total number of individuals in sample)

In addition to this index, the scales of pollution given by Wilhem and Dorris (1966) and Staub *et al.* (1970) were also considered while computing the results.

Results and Discussion

Results are presented in Table 1 and 2 and illustrated in Figs. 2 and 3.

Phytoplankton : Thirty-seven genera of phytoplankton belonging to 4 major groups, namely Myxophyceae, Chlorophyceae,

Desmidiaceae and Bacillariophyceae were observed. Fig. 2 reveals that Myxophyceae was dominant group at station D and least at A. Presence of myxophyceae in large numbers indicates that station D is approaching towards the phase of eutrophication. Chlorophyceae was dominant at station B and least at D whereas Bacillariophyceae was dominant at C and least at D. Desmidiaceae was found to be most dominant at station A and C while least at D whereas Bacillariophyceae was dominant at C and least at D. It may be attributed to the fact that some of the desmid species withstand certain degree of eutrophication but are absent in polluted waters (Munawar, 1970).

Table 1 reveals that among genera, *Polycystis* was most abundant and *Staurastrum* and *Oedogonium* were least. Among different groups, Myxophyceae showed abundance of Polycystis at all the sampling stations while *Phormidium* was least and found only at station A, B and E. Among Chlorophyceae, *Botryococcus* was most abundant and *Oedogonium* was least and was present only at station A and B only. Among desmidiaceae, *Closterium* was most abundant at all stations while *Staurastrum* was least, occurring only at station B and C. Among Bacillariophyceae, *Nitzschia* was most abundant and occurred at all the sampling stations whereas, *Melosira* was recorded in least numbers and recorded at stations B and C only. The pollution indicating genera recorded during the present investigation in order of decreasing abundance are *Closterium, Nitzschia, Botryococcus, Ankistrodesmus, Coelosphaerium, Oscillatoria, Navicula, Microspora, Coelastrum, Scenedesmus, Melosira* and *Phormidium*.

Zooplankton : Nine genera of zooplankton representing three groups namely, Copepoda, Cladocera and Rotifera were observed. Eggs and nauplii were also noted in appreciable numbers at various study stations. Among the major groups, Copepoda was the most dominant while the rotifera was least (Fig. 3). This may be attributed to the probable deterioration of the environment which suppressed further population growth by decreasing egg production and increasing mortality (Hoffman, 1977) or the scarcity of the food affected the fertility of females (Patalas, 1972) thereby, affecting the reproductive process of the females.

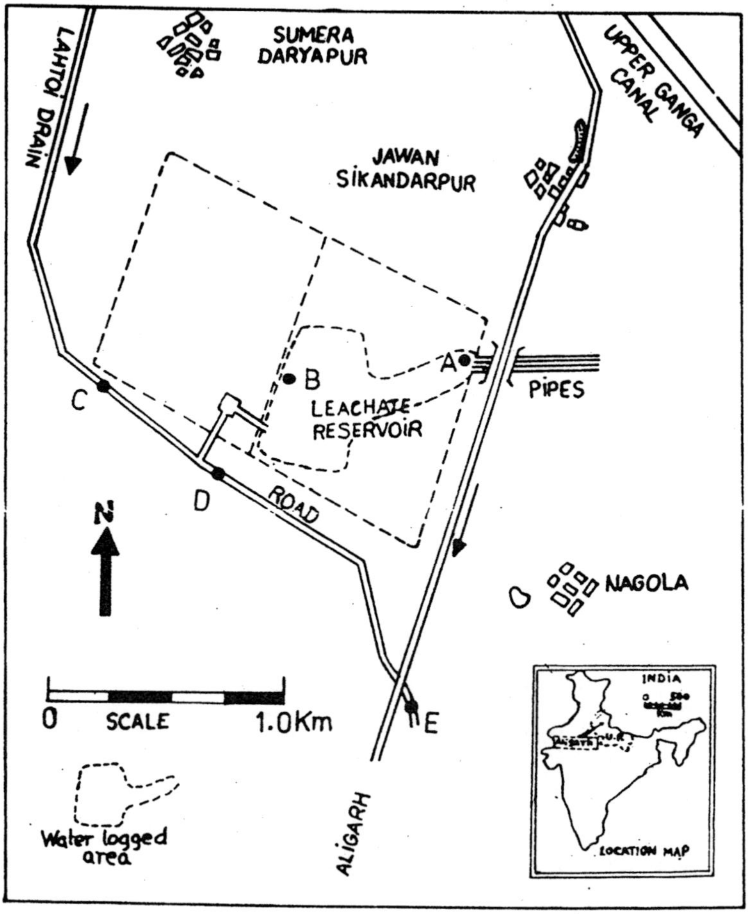
LAHTOI DRAIN
SUMERA
DARYAPUR
UPPER GANGA
CANAL
JAWAN
SIKANDARPUR
A
B
LEACHATE
RESERVOIR
PIPES
C
D
ROAD
N
NAGOLA
0 SCALE 1.0Km
E
Water logged
area
ALIGARH
INDIA
LOCATION MAP

Fig. 1

Table 1 : Abundance and Diversity Indices in the phytoplankton genera at various stations

Phytoplankton	*STATIONS*									
Genera	*A*		*B*		*C*		*D*		*E*	
	No./ml	$\overline{D}$	*No./ml*	$\overline{D}$	*No./ml*	$\overline{D}$	*No./ml*	$\overline{D}$	*No./ml*	$\overline{D}$
Myxophyceae										
Anabaena	-	-	36	0.1708	9	0.0847	12	0.1704	10	0.1158
Coelosph-aerium	15	0.1571	18	0.1065	30	0.1930	12	0.1704	19	0.1782
Agmenellum	3	0.0487	9	0.0638	3	0.0364	-	-	10	0.1158
Nostoc	6	0.0825	45	0.1964	-	-	6	0.1063	-	-
Oscillatoria	6	0.0825	24	0.1304	3	0.0364	6	0.1063	9	0.1075
Phormidium	3	0.0487	6	0.0467	-	-	-	-	5	0.0698
Polycystis	24	0.2110	48	0.2043	42	0.2353	51	0.3498	24	0.2058
Tetrapedia	-	-	12	0.0793	-	-	-	-	5	0.0698
Chlorophyceae										
Actinastratum	-	-	36	0.1708	6	0.0625	3	0.0637	-	-
Ankistrod-esmus	-	-	30	0.1516	21	0.1536	6	0.1063	14	0.1450
Botryococcus	15	0.1571	34	0.1646	24	0.1677	15	0.1960	24	0.2058

Contd..

Table 1 (Contd.)

Coelastrum	-	-	13	0.0268	3	0.0364	2	0.0465	15	0.1528
Cladophora	3	0.0487	-	-	-	-	-	-	5	0.0698
Crucigenia	-	-	21	0.1188	4	0.0456	-	-	19	0.1782
Dictrosph-aerium	3	0.0487	6	0.0467	3	0.0364	-	-	5	0.0698
Oedogonium	3	0.0487	3	0.0268	-	-	-	-	-	-
Microspora	6	0.0825	18	0.1065	9	0.0847	-	-	19	0.01782
Pediastrum	3	0.0487	6	0.0467	6	0.0625	-	-	-	-
Protococcus	21	0.1946	15	0.0934	6	0.0625	3	0.0637	7	0.0896
Scenedesmus	-	-	9	0.0638	3	0.0364	-	-	-	-
Selenastrum	3	0.0487	3	0.0268	3	0.0364	1	0.0268	10	0.1158
Spirogyra	12	0.1353	21	0.1188	9	0.0847	-	-	-	-
Tetraspora	3	0.0487	-	-	3	0.0364	-	-	1	0.0194
Ulothrix	3	0.0487	27	0.1413	15	0.1220	6	0.1063	3	0.0471
Zygnema	9	0.1107	18	0.1065	-	-	-	-	-	
Desmidiaceae										
Closterium	36	0.2642	36	0.1708	42	0.2353	6	0.1063	29	0.2298
Cosmarium	-	-	-	-	6	0.0625	-	-	-	-
Desmidium	-	-	3	0.0268	1	0.0148	2	0.0465	3	0.0471
Gonatozygon	12	0.1353	9	0.0638	3	0.0364	-	-	-	-
Staurastrum	-	-	3	0.0268	3	0.0364	-	-	-	-

Contd..

Table 1 (Contd.) :

Bacillario-phyceae										
Amphora	33	0.2524	30	0.1516	30	0.1930	12	01704	5	0.0698
Cyclotella	3	0.0487	6	0.0467	12	0.1044	-	-	10	0.1158
Diatoma	3	0.0487	15	0.0934	27	0.1808	12	0.1704	12	0.1314
Melosira	-	-	6	0.0467	9	0.0847	-	-	-	-
Navicula	12	0.1353	9	0.0638	21	0.1536	6	0.1063	9	0.1075
Nitzschia	30	0.2398	18	0.1065	36	0.2154	30	0.2865	19	0.1782
Synedra	9	0.1107	6	0.0467	12	0.1044	6	0.1063	-	-
Total	279	2.8766	589	3.2517	404	3.0355	197	2.5052	291	3.0148

Note : $\overline{D}$ = Diversity Index (Shannon-Weaver's, 1949).

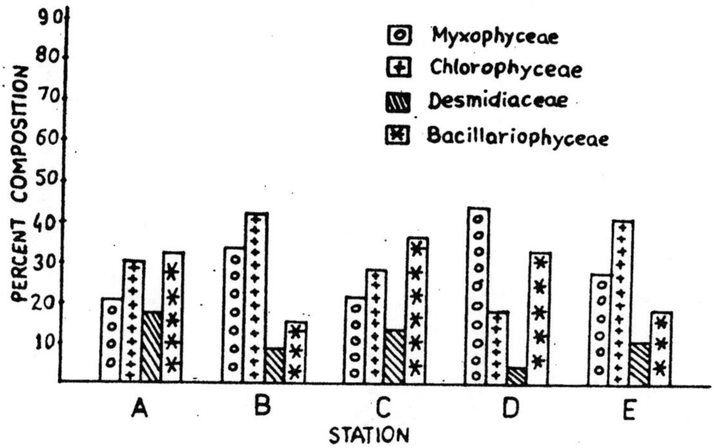

Fig. 2 : Per cent composition of different groups of phytoplankton of various stations

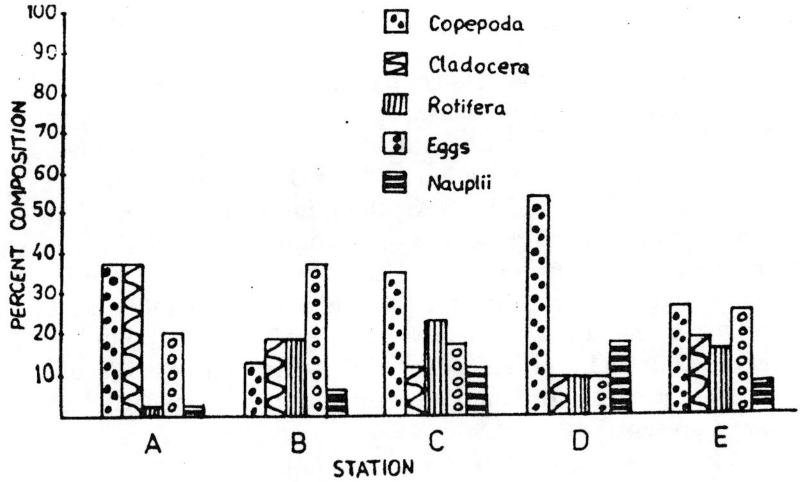

Fig. 3 : Per cent composition of different groups of Zooplankton of various stations

Table 2 : Distribution and abundance of Zooplankton at various stations (No./ L)

Zooplankton Genera	Stations				
	A	B	C	D	E
COPEPODA					
Cyclops sp.	5	1	2	2	7
Diatomus sp.	2	-	3	3	5
Mesocyclops sp.	6	1	1	1	5
Total	13	2	6	6	17
CLADOCERA					
Bosmina sp.	1	-	-	-	3
Ceriodephnia sp.	3	-	-	-	3
Daphnia sp.	2	1	-	-	-
Diaphanosoma sp.	7	2	2	1	6
Total	13	3	2	1	12
ROTIFERA					
Brachionus sp.	-	1	2	1	6
Keretella sp.	1	2	2	-	4
Total	1	3	4	1	10
EGGS	7	6	3	1	16
NAUPLII	1	1	2	2	5
GRAND TOTAL	35	16	17	11	61

Table 2 reveals that the zooplankton genera, recorded as indicator species, include *Brachionus* sp., *Mescyclops* sp. and *Daphnia* sp. only. *Brachionus* sp. was absent at Station A and least at B and D, *Mesocyclops* sp. was least at station B, C and D and *Daphnia* sp. was found absent at C, D, and E. From these findings it is clear that station D is approaching towards pollution as these indicator organisms were either absent or in least abundance at station D. Rotifers are well known indicators of pollution and are either absent or present in small numbers in areas having any kind of pollution (Ali *et al.*, 1990). In the present study, distribution of rotifers (Table 2) indicate the extent of pollution at station A and D where they were just present.

Diversity Index and Scales of Pollution

Ratios between the number of species and importance values (number, biomass etc.) of individuals are called species diversity

(Odum, 1971). Changes in species diversity in population of living organisms is a reliable and useful means of assessing biological affects of pollution (Patrick, 1972).

After making impact studies based on Shannon-Weaver's diversity index (1949) (Table 1) it appears that water of the leachate reservoir is not polluted as the diversity index is >2.5 for each sampling station but on comparison of different stations, it is clear that stations A and D are having lowest values of diversity index and hence are approaching towards pollution. Based on Wilhem and Dorris (1966) scale of pollution stations A and D are moderately polluted as the species diversity index for these stations lie in between 1 and 3. Similarly, according to Staub *et al.* (1970) scale of pollution stations A and D are the sites of light pollution as the species diversity index at these stations lie between 2.0 and 3.0.

Hence, from this study it can be concluded that station A and D are showing the positive signs of pollution.

Acknowledgements

Authors are thankful to Chairman, Department of Zoology, A.M.U., Aligarh for providing necessary laboratory facilities. Thanks are also due to ICAR, New Delhi for providing SRF to one of the author (R.K.G.).

Reference

Ali, M., Khan, A.A. and Haque, N. (1990), Population dynamics of rotifer fauna from two tropical ponds of Aligarh, India. *Proc. Nat. Acad. Sci., India, 60B,* 1 : 13-19.

Gunale, V.R. (1991), Algal communities as indicators of pollution. *J.Environ. Biol., 12* (special number) : 223-232.

Hoffman, W. (1977), The influence of abiotic environmental factors on population dynamics in planktonic rotifers. *Arch. Hydrobiol., 8* : 77-83.

Khan, F.A. (1985), *Studies on primary productivity in relation to air pollution emitted by a coal fired power plant.* Ph.D. thesis, A.M.U., Aligarh.

Munawar, M. (1970), Limnological studies on freshwater ponds of Hyderabad, India II - Biocenose. *Hydrobiologia,36(1)* : 105-128.

Needham, J.G. and Needham, P.R. (1962), *A guide to the study of freshwater Biology.* Holden-Day, Inc. San Francisco, 108 p.

Odum, R.P. (1971), *Fundamental of ecology,* Third edition. B.W.B. Saunders Co., Philadelphia, 178 p.

Patrick, R. (1972), *Aquatic communities as indices of pollution. In Indicators of Environmental Quality* (W.A. Thomas, ed.) Plenum Press, New York.

Sarkar, R. and Krishnamoorthi, K.P. (1985), The Current Concepts of Biological Monitoring of Inland Water Studies at NEERI. *IAWPC Annual Convention*, 1985, Nagpur.

Shannon, C.E. and Weaver, W. (1949), *Mathematical theory of communication* (Urbans III), University of Illinois Press.

Staub, R. Applying, J.W., Hosteiler, A.M. and Hass, I.J. (1970), The effects of industrial wastes of Memphis and Shelby Country on primary plankton producers. *Bioscience*, 20 : 905-912.

Walash, G.E. and Merril, R.G. (1984), Algal bioassays of industrial energy process effluents. pp. 329-360. In *Algae as ecological indicators* (L.E. Shubert, ed.), Academic Press, London.

Ward, H.B. and Whipple, G. C. (1959), *Freshwater Biology*. II Edition (W.T. Edmondson, ed.) John Wiley and Sons, Inc. New York.

Wilhem, J.L. and Dorris, T.C. (1966), Species diversity in a stream receiving domestic and oil refinery effluents. *Am. Medi. Nat. 76* : 427-449.

12

Water Quality and Periphytic Algal Component of Anchar Lake, Kashmir

S.G. Sarwar

Introduction

The lakes of Kashmir are socio-economically and bioaesthetically important aquatic ecosystem harbouring a rich and diverse genepool. In recent years some of these lakes have shown definite signs of cultural eutrophication due to the increase in the human activities in their catchment area (Zutshi *et al.*, 1980). The lake water, which is a basic component of our natural environment, is being polluted at an alarmingly fast rate. The erosion of lake environment has greatly accelerated during the last two decades causing serious economic loss to the entire society. Enhanced population explosion, rapid rate of encroachments and increased utilization of lake waters for tourist traffic and disposal and dilution of sewage, etc. have not only deteriorated the water quality of these lakes but have also affected the biotic flora. Most of the limnological investigations in this region have centered around Dal lake (Mukherjee, 1921, 1926; Kaul and Zutshi 1966; Kant and Kachroo 1974; Trisal, 1977; Zutshi & Vass 1978); Vass & Zutshi 1979; Trisal & Kaul 1983; Sarwar & Zutshi, 1988, 1989), while all other waterbodies, although equally important, have not received sufficient attention from limnologists. In view of this detailed limnological investigations of Anchar lake were undertaken for one year from June, 1982 to May, 1983 to provide baseline data on water quality and biotic components.

Materials and Methods

Anchar lake (alt. 1583 m. a.s.l.) is situated at a distance of 14 km. to the north-west of Srinagar within the geographical co-ordinates of lat. 34^0 20' - 34^0 26' N; long. 74^0 82' - 74^0 85' E. The morphometric features of the lake are presented in Table 1.

Table 1 : Morphometric features of Anchar lake

Maximum length	3.55	km.
Maximum breadth	0.65	km.
Mean breadth	0.60	km.
Maximum depth	3.5	m.
Mean depth	1.1	m.
Surface area	680	ha.

The lake is single basined being connected on its eastern side with Dal lake through an inflow channel "Nalla Amir Khan". A network of channels from the cold water river Sind enter the lake on its western shore forming delta. The water to the lake is also supplied by the springs present in the basin and along the periphery. Spring precipitation also contributes considerable quantity of water. A number of small channels from agricultural fields and human settlements also drain into the lake. Although there is no permanent outflow channel, excess quantities of water leave the lake through an exit in the north-east direction.

The lake has typical rural environment. The catchment comprises long stretches of elevated land on the north-west side which is used for raising different types of vegetables. Agricultural fields surround the lake on the western side. The elevated plateau is partly under paddy cultivation and partly under willow plantations.

The lake supports rich growth of *Myriophyllum spicatum, Ceratophyllum demersum, Potamogeton natans, P. crispus, P. lucens, Trapa bispinosa, Salvinia natans, Nymphoides peltatum, Hydrocharis dubia, Hydrilla verticellata, Nelumbo nucifera, Phragmites australis, Numphaea alba, Lemna gibba, L. minor* and *L. trisulca*. According to Zutshi (1975) Anchar lake has greatest diversity of plant communities with twenty associations. Kak (1981) reported 42 macrophytic species for this lake.

Sampling was done at the monthly intervals. The water samples were collected in 1 litre capacity polyethylene bottles and analysed for various parameters adopting the methods of Mackereth (1963), APHA (1979) and Golterman *et al.* (1978). pH meter Grip type No. 121, Systronics conductivity meter type 303 and systronics Double Cell Colorimeter type 102 were used for determining pH, specific conductance and other chemical characteristics of water samples respectively. Trace metal estimation was carried out on Atomic Absorption Spectrophotometer model 2900SP (Pye Unicam Ltd.).

Method of Gough and Woelkerling (1976) was adopted for the removal and quantification of periphytic algal flora from *Myriophyllum, Ceratophyllum, P. natans, Nymphoides, Typha* and *Phragmites.*

Results and Discussion

The averages and the ranges for various physico-chemical parameters are depicted in Table 2.

The lake is small sized waterbody with depth ranging from 0.02 to 3.5 m. According to Rawson (1955) lakes with a mean depth of 10 m or less are usually considered as eutrophic. Since the investigated lake has the maximum depth of only 3.5m it can be considered eutrophic. The light penetration is reduced due to the silt-loaded inflow from the river Sind (Zutshi *et al.* 1980). According to Yoshimura (1933) waters with reduced light penetration are eutrophic.

The minimum water temperature of the lake never fell below 7.5^0 C and therefore can be classified as warm monomictic or sub-tropical lake (Hutchinson, 1957; Reid and Wood, 1976). The lake water remained well mixed without depicting either summer or winter stratification. Since the maximum depth does not exceed 3.5 m and as such there is no question of stratification as according to Crumrine and Beeton (1975) summer stratification develops only when the depth exceeds 8 m.

Table 2 : Physico-chemical features of Anchar lake.

Parameter	*Unit*	*Range*			$\bar{X}$
Water temperature	°C	7.5	-	26.0	17.2
Depth	m	0.02	-	3.5	0.75
Transparency	m	0.02	-	2.0	0.45
Extinction Co-efficient	K	1.04	-	28.0	8.22
pH	-	7.2	-	9.0	7.9
Specific Conductivity at 25 °C	µS	122	-	526	316
Dissolved Oxygen	mg/l	3.2	-	13.6	9.8
Oxygen saturation	per cent	39.9	-	229.7	123.7
Free Carbon dioxide	mg/l	0.05	-	26.7	6.9
Magnesium	mg/l	1.9	-	21.8	8.8
Calcium	"	23.2	-	58.0	38.2
Sodium	"	1.3	-	14.4	5.8
Potassium	"	1.0	-	10.6	4.1
Total alkalinity $CaCO_3$	"	58.0	-	275.0	149
Chloride	"	11.0	-	38.0	24.0
Sulphate	"	0.4	-	2.6	0.9
Silicate	"	0.4	-	6.3	1.9
Iron "	"	0.06	-	0.4	0.2
Ammonical nitrogen	"	0.01	-	0.04	0.02
Nitrate-nitrogen	"	0.02	-	0.36	0.09
Orthophosphorus	"	0.006	-	0.06	0.002
Total phosphorus	"	0.10	-	0.25	0.18

The pH values are indicative of the alkaline nature of the waterbody. The seasonal fluctuations in the H-ion concentration were not significant as the waters are well buffered. Zutshi *et al.* (1980) attribute the alkaline nature of the Kashmir lakes to the presence of calcium rich rocks in the catchment areas. Higher silicate content in the rocky beds and sediments also contribute to higher pH due to the production of silicic acid by hydrolysis of the silicates.

The higher conductivity values of Anchar lake indicate high ionic concentration. According to Crumrine and Beeton (1975) lakes of highest conductivity also have greater inputs of ground water. The lower conductance values recorded during spring and summer are attributed to the intake of ions by plants during the growing season while higher values during winter are related to the abundance of nutrients which are released due to the death and decay of macrophytes.

In the investigated lake the range of dissolved oxygen was 3.2 - 13.6 mg/l ($\bar{x}$: 9.8 mg/l). Maximum DO Content was observed during winter months as (i) cold water are known to have more dissolved oxygen than warm waters (Rao, 1955); (ii) the rate of decomposition of the plant matter was retarded as the water cooled making the micro-organisms less active to respire with the result an appreciable quantity of DO content still remained in the water column, and (iii) rich periphytic algal growth and their photosynthetic activity helped to maintain appreciable amount of dissolved oxygen in water.

The ionic composition of the lake, computed on an average basis, reads as :

$$HCO_3^{1-} > Ca^{2+} > Cl^{1-} > Mg^{2+} > Na^{+} > K^{+} > SO_4^{2-};$$

which reveals the predominance of calcium and bicarbonate over other ions and in this respect it does not differ much from a "mean" global freshwater composition (Rodhe, 1949). There is slight departure from his generalised scheme ($Ca^{2+} > Mg^{2+} > Na^{+} > K^{+} > HCO_3^{-} > SO_4^{2-} > Cl^{-}$) as Cl^{-} is dominant to SO_4^{2-} in the investigated lake.

In the investigated lake total alkalinity was of bicarbonate type. According to the classification of Sawyer and McCarty (1980) Anchar lake water is moderately hard. Following the classification Phillipose (1959) and Spence (1964) it can be placed in the higher category of nutrient types.

The chloride concentration of the lake may be related to the presence of large amounts of organic matter and sewage contamination. Blum (1957) also attributes high chloride content to sewage contamination while Thresh *et al.* (1944) related it to organic pollution of animal origin. The sulphate content was quite low and is not supposed to play any key role in structuring and shaping of the biotic communities.

Divalent cations - calcium and magnesium exhibited a ratio of 4.3 : 1 while the monovalent cations - sodium and potassium appeared in the ratio of 1.4 : 1. Ohle (1934) has laid stress on the importance of calcium in classifying nutrient types of water. He considers lakes with less than 10 mg/l of calcium as poor, with 10 -25 mg/l as medium and with more than 25 mg/l as rich. Applying his classification, the investigated lake falls in the 3rd category and as such is polluted. These findings are in agreement with Zutshi (1968) and Vass (1973) who have also categorised lake waters of Kashmir as "calcium rich". Forsyth and McColl (1975) attributed high calcium values to the use of fertilizers which might also be contributing towards total hardness in the present lake as fertilizers are used in its catchment.

Magnesium content is low in comparison to calcium but it is not thought to be a limiting factor for the growth of phytoplankton, periphyton or macrophytes. The low magnesium content is possibly due to its uptake by the plants in the formation of chlorophyll - magnesium - porphyrin metal complex and in enzymatic transformations.

Table 3 : Mean ratios of salts.

Salts	*Anchar lake*	*Clarke's (1924) ratios for World freshwaters*
Cl' / HCO'	0.16	0.16
Mg^{2+}/ Ca^{2+}	0.23	0.17
Ca^{2+} / Cl'	1.58	3.60
Mg^{2+}/ Cl'	0.36	0.60
Ca^{2+} / Mg^{2+}	0.31	0.68
Alkalinity		

The mean ratios of various salts are presented in table 3. It has been observed that Cl' / HCO_3' ratio was similar to that of Clarke (1924) while other ratios, except for Mg^{2+} / Ca^{2+}, were lower than Clarke's ratios for world freshwaters.

The silicate content is low as the Kashmir lakes abound in diatoms (Sarwar 1986, 87, 88, 89) which utilize this mineral. The average iron content of the lake (0.02 mg/l) is in conformity with the values reported by Wetzel (1972) for some hard water lakes of U.S.A. Higher iron values may be associated with increase in turbidity and suspended matter as demonstrated by Kothandaraman and Evans (1978).

The relatively low content of nitrate-nitrogen in the investigated lake may be attributed to luxuriant macrophytic growth which utilise it during photosynthesis. The low orthophosphate content is due to the formation of an insoluble calcium - phosphate complex. According to Wetzel and Otsuki (1974) such a phenomenon functions as scanvenger of some inorganic nutrients and also acts as a removal agent of dissolved matter by absorption. On the basis of total phosphorus values the lake can be placed in the moderate eutrophic category (Strm 1930). Since average TP values exceed 0.18 mg/l hence there is such a profuse macrophytic growth in the lake.

Trace-elements like mangenese, cobalt, nickle, copper, cadmium and mercury were present below the detection levels. The concentration of lead varied from non-detectable level to 0.31 mg/l ($\bar{x}$: 0.14 mg/l) and that of zinc from undetectable level to 0.51 mg/l ($\bar{x}$: 0.44 mg/l). The low concentration of lead and zinc reflect natural background levels. These are within the permissible limits showing thereby that there is no heavy metal pollution.

One hundred fifty four texa, representing six algal classes were recorded from the six natural substrates. Bacillariophyceae and Chlorophyceae had the highest representation with 70 and 65 texa respectively; while Cyanophyceae (15), Dinophyceae (2), Cryptophyceae (1) and Chrysophyceae (1) were the least represented classes. Fifty texa were common to the investigated substrates.

Classwise distribution of various taxa on different substrates was as under :-

	Typha	*Cerato-phyllum*	*Phrag-mites*	*Nympho-ides*	*Myrio-phyllum*	*Potam-ogeton*
Bacillariophyceae	47	46	46	55	50	52
Chlorophyceae	31	28	36	43	31	28
Cynaphyceae	10	11	09	11	11	11
Dinophyceae	02	02	01	02	02	02
Chrysophyceae	-	-	01	-	-	01
Cryptophyceae	01	01	-	01	-	01
Total	91	88	93	112	94	101

Myriophyllum supported the maximum number of taxa while the minimum number was recorded on *Phragmites.*

Some of the dominant taxa recorded are : *Achnanthes minitissima, Amphora spp., Cocconeis placentula, Cymbella lanceolata, C. ventricosa, Fragilaria capucina, F. crotonensis, Synedra* sp., *Navicula* spp. and *Oedogonium* spp.

Rare taxa encountered on natural substrates in the investigated lake are : *Caloneis amphisbaena, Cymbella sinuata, Melosira* sp. and *Neidium dubium* (Bacillariophyceae), *Closterium moniliforme, Pediastrum ovatum, Selanastrum* sp., *Staurastrum furcigerum* (Chlorophyceae).

On the basis of degree of attachment and proximity to the substratum various algal life forms in the periphyton can be divided into three groups :-

1. 'Tightly attached algae' include forms that grow appressed to the substratum (*e.g, Cocconeis, Eunotia*);
2. 'Filamentous and long stalked algae' are also attached to the substratum but extend outward to form a mat (*e.g. Cymbella*) and;
3. Motile-planktonic and filamentous long stalked algae.

Community co-efficient of periphytic algal flora on investigated substrate of the lake ranged from 6.21 to 10.04 per cent ($\bar{x}$: 7.94 per cent) with the maximum recorded in November, 1982 and the minimum in May, 1983.

Diatoms dominated the periphytic algal component of the lake on all the substrate. Jorgensen (1957), Stockner and Armstrong (1971) and Ennis (1975) also observed the dominance of diatoms in periphyton assemblages. Numerical superiority of diatoms, both in terms of number of taxa and population density, is indicative of the eutrophic nature of the waterbody. The dominance of diatoms with sub-dominance of blue-green algae is regarded as a characteristic feature of eutrophic waters (Rawson, 1956; Lund, 1962 and Reynolds & Allen, 1968).

On all the substrates the percent contribution of diatoms was more than 80 per cent (Table 4) while the share of other classes was quite low.

The overall periphytic algal population varied from 3 x 10^4 μ/10 mg d.wt. in August, 82 to 23.2 x 10^4 μ/10 mg d.wt in April, 83 with an average value of 0.78 x 10^4 μ/10 mg d.wt. In the present

Table 4 : Average percent contribution of major algal classes

Substrate	*Bacillario-phyceae*	*Chloro-phyceae*	*Cyano-phyceae*
Myriophyllum	91.2	2.8	5.4
Ceratophyllum	85.4	6.8	7.4
Nymphoides	81.4	4.9	11.9
Potamogeton	85.3	3.9	10.6
Phragmites	83.6	5.9	8.3
Typha	95.4	1.1	3.4
On all investigated substrates	91.5	5.0	3.1

study total periphytic population depicted bimodal behaviour with major peak (18.6 x 10^4 $\mu/10$ mg d.wt) recorded in spring and another peak (13.4 x 10^4 $\mu/10$ mg d.wt) in winter. The peaks are mainly contributed by the diatoms. The abundance of diatom population in cold months is probably due to the fact that they are able to grow under conditions of weak light and low temperature which are less suitable for other algae (Lund, 1965; Zafar, 1967).

The summer-autumn decline in the total periphyton population density can be attributed to the grazing effect of zooplankters which abound in Kashmir lakes during this period.

The silicate concentration of the investigated waterbody was generally low particularly in the months following peak development of diatoms, which indicates uptake of silicates by diatoms. The low silicate content and the abundance of diatoms in the ecosystem is suggestive of the dependence of diatoms upon the availability of silica. Munawar (1974 a, b) and Wetzel (1975) have observed the depletion of silica in ponds and lakes due to diatoms. Thus it can be suggested that the diatoms, which formed the bulk of periphyton biomass, consumed the silicate content of water ($r = -0.51$).

However, it appears that the seasonal fluctuations in periphytic population is not governed by a single factor but by the cumulative effect of several physico-chemical factors. Similar observations have also been made by Reid (1961).

Variance in the population density on the investigated natural substrates was quite significant with F value recorded as 4.72 which is higher than the critical value of 2.36 at 5 per cent level of significance. Bacillariophyceae too showed quite significant variation. (F = 4.44).

In terms of average population density it has been observed that the highest average values were obtained on *Myriophyllum* followed by *Ceratophyllum* and lowest on *Phragmites* (Table 5).

Myriophyllum and *Ceratophyllum* supported large population density as they offered greater surface area due to the dissected nature of their leaves. Harrod and Hall (1962) and Cattaneo and Kalf (1979) also recorded similar observations.

Table 5 : Average population density values (μ/10 mg d. wt.) on natural substrates

Myriophyllum	40,469
Ceratophyllum	28,603
Typha	21,901
Potamogeton	13,991
Nymphoides	13,515
Phragmites	3,310

Periphyton population showed positive correlation with pH (r = 0.162), chloride (r = 0.109), sodium (r = 0.057), Sulphate (r = 0.162), orthophosphorus (r = 0.347) and total phosphorus (r = 0.246). It depicted weak negative correlation with water temperature (r = -0.371), nitrate-nitrogen (r = -0.260), ammonical-nitrogen (r = -0.430), specific conductivity (r = -0.049), iron (r = -0.008), potassium (r = -0.085), magnesium (r = 0.-215) and calcium (r = -0.41); and a significant inverse correlation with silicate (r = -0.521) and total phosphorus (r = -0.538).

In conclusion, it may be said that the investigated lake shows signs of accelerated rate of cultural eutrophication due to the stresses and strains of the biotic interferences. This important waterbody is shrinking and getting deteriorated at an alarmingly high rate due to human greed (accelerated land use along the catchment, dumping of domestic refuse, garbage and sewage, and illegal encroachments, etc.) and it is losing its character and identity as it is rapidly becoming silted up. An integrated and multidisciplinary approach is needed to conserve this lake from further degradation. The various coservation measures to be adopted should include :

(a) Reduction of the nutrients entering the lake by catchment improvement and control by diversion of sewage and inflows from the agricultural fields and orchards, etc.;
(b) Harvestation and removal of the weeds to improve the nutrient balance of the lake;
(c) Dredging of the lake;
(d) Afforestation and reforestation of the catchment area;
(e) Ban on illegal encroachments;
(f) Regulation and curtailment of the silt nutrient load flowing into the lake; and
(g) Control of the water-level fluctuations of the lake.

Summary

The paper highlights the water quality and periphytic algal composition of a Kashmir Himalayan lake - Anchar lake. The recorded observations and the conclusions drawn thereoff are based on a year long sampling and analysis of the water samples of the waterbody. Various physico-chemical and biological characteristics reveal the degraded and deteriorated nature of the waterbody. The present status of the lake is attributed to tremendous biotic pressures, negligence, carelessness and inefficiency of the concerned quarters. Curtural eutrophication of the lake is evident from its deteriorated water quality and luxuriant macro and micro-floral growth. The paper suggests certain conservation measures for the retrieval and restoration of the lake.

References

American Public Health Association (1979), *Standard methods for the examination of water, sewage and industrial wastes.* APHA, New york.

Blum, J.L. : An ecological study of the algae of the saline river, Michigan, *Hydrobiologia*, 9(4) : 361 - 408.

Cattaneo, A. and Kalff, J. (1979), Primary production of algae growing on natural and artificial aquatic plants. : A study of interactions between epiphytes and their substrate. *Limnol Oceangr.*, 24 : 1031-1037.

Clarke, F.W. (1924), The data of Geochemistry. 5th ed. *Bull, U.S. Geol. Surv.*, 770; pp. 841.

Crumrine, J.P. and Beeton, A.M. (1975), Limnology of lakes of the Sylvania Recreation Area, Ottawa National forest Special Report No. 24. Centre for Great Lake Studies, the University of Wisconsin, *Milwakee*, U.S.A., 34.

Ennis, G.L. (1975), Distribution an abundance of benthic algae along phosphorus gradients in Kooteny Lake, British Columbia. *Verh. Int. Ver. Limnol., 19* : 562-570.

Forsyth, D.J. and and McColl, R.H.S. (1975), Limnology of lake Nagahema, North Island, New Zealand, N.Z.J. Mar. *Freshwat. Res., 9*(3) : 311-322.

Golterman, H.L., Clymo, R.S. and Ohnstad, M.A.M. (1978), Methods for physical and Chemical Analysis of Freshwaters, IBP Handbook No. 8.

Gough, S.B. and Woelkerling, W.J. (1976), On the removal and quantification of algal aufwuchs from macrophyte hosts. *Hydrobiologia, 48* : 203-207.

Harrod, J.J. and Hall, R.E. (1962), A method for determining the surface area of various aquatic plants, *Hydrobiologia, 20* : 173-178.

Hutchinson, G.E. (1975), A Treatise on Limnology 1. Geography, Physics and Chemistry. New York, John Wiley and Sons, Inc.

Jorgensen, E.G. (1957), Diatom periodicity and silicon assimilation Dansk *Bot. Arkiv., 18*(1), pp. 54.

Kak, A.M. (1981), Floristic composition, structure and ecology of some high altitude lakes of Kashmir Himalayas. Proc. Abs. Silver Jubilee Symposium of Tropical Ecology "Ecology and Resource Management in the Tropics". Ed. R.S. Ambasht and H.N. Pandey, Varanasi, 106-108.

Kant, S. and Kachroo, P. (1974), Limnological studies in kashmir lakes. IV. Seasonal dynamics of phytoplankton in the Dal and Nagin. *Proc. Indian Nat. Sci. Acad., India, 39*(B) 5 : 632-645.

Kaul, V. and Zutshi, D.P. (1966), Some ecological considerations of floating islands in Srinagar lakes. *Proc. Nat. Acad. Sci. India, 36* : 273-281.

Kothandraman, V. and Evans, R.L. (1978), Physial and Chemical water quality characteristics of the Fox chain of lakes. Trains. III. *Stat. Acad. Sci., 71*(3): 239-259. Rep. Series No. 462 III Stat. Wat. Surv.

Lund, J.W.G. (1962), Phytoplankton from some lakes in Northern Sarkatchewan and from Great Slave Lake. Can. J. Bot., 40 : 1499-1514.

Lund, J.W.G. (1965), The ecology of freshwater phytoplankton. *Biol. Rev., 40*, 231-293.

Mackereth, F.J.H. (1963), Water Analysis for Limnologists. Freshwater *Biol. Assoc.*, 21 : 1-70.

Mukherjee, S.K. (1921), The Dal lake (Kashmir) - A study in biotic succession. *Proc. 8th Indian Sci. congr.* pp. 185 (Abstract).

Mukherjee, S.K. (1926), Aquatic and marsh vegetation of the Dal lake of Kashmir. *Proc. Linn. Soc.*, London, pp. 55 (Abstract).

Munawar, M. (1974 a), Limnological studeis on the freshwater ponds of Hyderabad, India. III. The Biosenose. Seasonal abundance of unicellular and colonial phytoplankton in polluted and unpolluted environments. *Hydrobiologia*, 44(1), 13-27.

Munawar, M. (1974 b), Limnological studies on the freshwater ponds of Hyderabad, India. IV. The Biocenose. Seasonal abundance of unicellular and colonial phytoplankton in polluted and unpolluted environments. *Hydrobiologia*, 45(1) : 1-32.

Ohle, W. (1934), Chemische and Physikalische Untersuchungen norddeutscher Seen. *Arch Hydrobiol., 26* : 386-464 and 584-568.

Philipose, M.T. (1959), Freshwater phytoplankton of inland fisheries. *Proc. Symp. Algol., ICAR,* New Delhi, pp. 271-291.

Rao, C.B. (1955), On the distribution of algae in a group of six small ponds. II. Algal periodicity. *J. Ecol.,* 43 : 291-308.

Rawson, D.S. (1955), Morphometry as dominant factor in the productivity of large lakes. *Verh. Int. Limnol., 12* : 164-175.

Rawson, D.S. (1956), Algal indicators of trophic lake types.*Limnol. Oceangr., 1* : 18-25.

Reid, G.K. (1961), Ecology of inland waters and estuaries. D.Van. Reinold, N.Y., pp. 375.

Reid, G.K. and Wood, R.D. (1976), Ecology of inland waters and esturies. D.Van. Nostrand Co., New York, pp. 485.

Reynolds, C.S. and Allen, S.E. (1968), Changes in the phytoplankton of Oak Mere, following the introduction of Base-rich water. *Br. Phycol. Bull., 3*(3) : 451-462.

Rodhe, W. (1949), The ionic composition of lake waters. *Verh. Int. Ver. Limnol., 10* : 377-386.

Sarwar, S.G. (1986), Species composition and seasonal variations of periphyton on *Ceratophyllum demersum* Linn. in Waskur lake, Kashmir. *Geobios New Reports, 6* : 114-118.

Sarwar, S.G. and Zutshi, D.P. (1987), Studies on periphyton population of Himalayan. lakes. 1. Species composition and community structure on natural and artificial substrates. *Proc. Indian Nat. Sci. Acad., 53* : 239-243.

Sarwar, S.G. and Zutshi, D.P. (1988), Species distribution and community structure of periphytic algae on artificial substrate. *Trop. Ecol.,29* (2), 116-120.

Sarwar, S.G. and Zutshi, D.P. (1989), Species composition and population density of periphytic algal flora of two macrophytes of Kashmir. Indian *J. Ecol.,16* (1), 68-71.

Sawyer, C.N. and McCarty, P.L. (1980), Chemistry for sanitary and alkali soils. Agri. Handbook 60, U.S. Deptt. of Agric. Washington D.C., pp. 100. In : David Keith Todd (ed.), Groundwater Hydrology. John Wiley & Sons, Inc.

Spence, D.H.N. (1964), The macrophytic vegetation of lochs, swamps and associated fens. pp. 306-425. In J.H.Burnett (ed.), The vegetation of Scotland, Oliver and Boyd, Edinburgh.

Strm, K.M. (1930), Limnological observations on Norwegian lakes. *Arch. Hydrbiol., 21* : 97-124.

Stockner, J.G. and Armstrong, F.A.J. (1971), Periphyton of the Experimental Lakes Area (ELA), northwestern Ontario, *J. Fish. Res. Board Can. V. 28* : 215-229.

Thresh, J.C., Suckling, E.V. and Beale, J.E. (1944), The examination of water and water supplies. 6th Edition (E.W.Taylor, ed.).

Trisal, C.L. and Kaul, S. (1983), Sediment composition, mud water inter-changes and the role of macrophytes in Dal lake, Kashmir. *Int. Revue ges. Hydrobiol. V. 68* : 671-682.

Vass, K.K. (1973). Studies on hydrobiology and primary productivity of some Kashmir lakes. Ph.D. Thesis, Kashmir University (Unpublished).

Vass, K.K. and Zutshi, D.P. (1979). Limnological studies on Dal lake. 1. Morphometry and physical features. *J. Inld. Fish. Soc. India,* 2 (1), 12-21.

Wetzel, R.G. (1972). The role of carbon in hard-water marl lakes. In : G.E. Likens, ed. Nutrients and Eutrophication : The Limiting-Nutrient Controversy. *Sp. Sym. Amer. Soc. Limnol. Oceanogr.*, *1* : 84-91.

Wetzel, R.G. (1975), Limnology. W.R. Saunders Co., Philadelphia, pp. 743.

Wetzel, R.G. and Otsuki, A. (1974). Allochthonous organic carbon of a marl lake. *Arch. Hydrobiol.*, *73* : 31-56.

Yoshimura, S. (1933). Rapid eutrophication within recent years of lake Haruna, Gunmen, Japan, *J. Geol. Geog.*, *2* : 31-41.

Zafar, A.R. (1967). On the ecology of certain fish ponds of Hyderabad, India. III. The periodicity. *Hydrobiologia*, *24* (4), 556-566.

Zutshi, D.P. (1968). Ecology of some Kashmir lakes. Ph.D. thesis, Jammu & Kashmir University (Unpublished).

Zutshi, D.P. (1975). Association of macrophytic vegetation in Kashmir lakes. *Vegetatio*, *30* : 61-65.

Zutshi, D.P. and Vass, K.K. (1978). Limnological studies on Dal lake. II. Chemical features. *Indian J. Ecol.*, *5* (1), 90-97.

Zutshi, D.P., Subla, B.A., Khan, M.A. and Wanganeo, A. (1980). Comparative limnology of nine lakes of Jammu and Kashmir Himalayas. *Hydrobiologia*, *72* (1-2), 101-112.

13

Population Biology of Trematode Species in Freshwater Fishes at Indore (M.P.)

Q. Ismail

Introduction

Agrawal S.M. (1974, 1984, 1987, 1991) has thrown light on the bio-chemical, immunological and Experimental studies on helminthic parasite from freshwater fishes at Raipur. Helminth parasites from freshwater-fishes have been reported by Bayer (1959) and *Wedeneyer* (1970). Freshwater fish diseases from the State of U.P., West Bengal and Maharashtra have been described in detail by Day and Kumar (1986), Chakravarty and Chaudhry (1980) and Siddique 1964. Despite the above work no report of fish parasites were made from Indore even though it contains lakes, ponds, and rivers. Trematode fauna of fishes at Indore has remained practically unknown till 1995 of the current century. Author initiated a systematic survey of Trematode parasite of fishes of Indore, which has been sadly neglected by the Zoologist at Indore.

The author has studied the dynamics of infection and the seasonal occurrence of ten species of Trematodes spread over 5 genera. A host parasite record species wise of all these are included in Table 1. Data on seasonal occurrence have been listed in Table 2. Species occurring in multiple infection have been listed in Table 3.

Work Plan and Methodology

Fishes are collected from various ponds, lakes, rivers and reservoirs. Fish market survey is conducted. They are sexed, weighted and scanned for Trematode infection. Efforts were made to examine at least ten fishes of each species for their parasitic infection.

Table 1 : Host-Parasite record of Species recovered

No.	Parasite Species	Host Species
1.	*Isoparorchis hypselobagri*	*Wallago attu*
2.	*Isoparorchis sinensis*	*Wallago attu*
3.	*Gyrodactylus gussevi*	*Heteropneustes fossilis*
4.	*Dactylogyrus brevitignus*	*Puntius sophore*
5.	*Dactylogyrus longicirrus*	*Puntius sophore*
6.	*Neodactylogyurs callaius*	*Catla catla. Labeo rohita*
7.	*Neodactylogyurs dorsalis*	*Labeo rohita*
8.	*Neodactylogyurs mirgali*	*Labeo rohita*
9.	*Phyllodistomum mansoni*	*Channa gachua*
10.	*N. Pharyngocephalus*	*Catla catla*

Table 2 : Host-parasite record of multiple infections

No.	Fish Host	Parasite Species
1.	*Channa gachua*	*Phyllodistomum mansoni, Gyrodactylus spp*
2.	*Labeo rohita*	*Neodactylogyrus dorsalis, N. Mirgali, Dactylogyrus spp*
3.	*Heteropneust fossilis*	*Gyrodactylus gussevi*
4.	*Wallago attu*	*Isoparorchis hypselobagri Isoparorchis sinensis.*
5.	*Mystus vittatus*	*Gyrodactylus mizellei, Gyrodactylus gussevi*
6.	*Calta catla*	*Neodactylogyrus callaius Neodactylogyrus pharyngocephalus*

Identification and Biology of Trematode Parasites

The recorded Trematodes examined to record their possible biological featrues on the basis of the observation and recorded data. Trematode parasites were identified by referring 'Systema Helminthum' Yama Gutti.

Table 3 : Seasonal occurrence of species of Trematode recovered from fishes at indore

No.	Species	Host	Sex	Fish Exam.	Inf. Seasonality	Remarks
1.	*Isoparorchis hypselobagri*	*Wallago attu*	M F	50 50	12 09	Year around
2.	*Gyrodactylus gussevi*	*H.fossilis*	M F	40 60	07 13	Mar. to July
3.	*Dactylogyrus brevitignus*	*Puntius sophore*	M F	66 44	06 21	Year around
4.	*Neodactylogyrus callaius*	*Catla catla*	M F	24 32	07 18	Year around
5.	*Phyllodistomum mansoni*	*Channa gachua*	M F	40 50	01 03	Exceptional

Discussion

Rawson and Rogers (1972b:) observed that water temperature is critical to seasonal occurrence of these species of worms. Gonzelez-Lenza and Alvarez-Pellitero (1982) in their study of population dynamics of *Dactylogyrus legionensis* from *Barbus barbus* bocagei observed that prevalence and intensity of these worms increased with age and length of the barbel. They observed maximum infection in autumn and winter, Zharikova *et al.* (1981) related abundance of *Dactylogyrus spp.* in *Carassius auratus* with immunophysiological status of the host, using antimicrobe effect (Am/eff) and total lipid of blood serum. They observed decrease in antimicrobe/serum effect and total lipods in blood serum during peak period of infection of species. Decrease in abundance followed increase in water temperature.

Table-3 reveal that seasonal occurence of Trematode species at Indore have high incidence and are found almost throughout the year, June to Sept. are warm months at Indore and, in several species of fishes, breeding period corresponds with this period. During breeding season fishes suffer from several stresses like starvation, warmer temp. of water and hormonal stress (several corticosteroids being released into the blood during breeding season.

Conclusion

The author is of the view that water temperature as the controlling factor in seasonal dynamics of Trematode species would be misleading. Trematode infections therefore, will have to be viewed in the total context of infection and immunophysiological status of the affected host.

References

Bychowsky, BE. 1957 (509), Monogenetic flukes, their classification and phylogeny, Moscow Leningrad Acad Sci. USSR press.

Gupta, S.P., 1959, On a trematode phyllodistomum singhas n.sp. of the family Gorgoderidal Loss 1899, from the intestine of freshwater fish Mastgacembelus armatus. *Ind. J. Helm.* 3 (1) pp. 21-28.

Gupta, A.K. Dubey A. and Agrawal SM. 1992, Studies on Monogenean parasites of freshwater fishes at Raipur. Indian Jour. Helminth 44-109-115.

Rizvi, SSH 1964, the parasite fauna of the fish of Rosth erne More, Cheshire Ph.D. Thesis University of Liverpool, England.

14

Hydrobiological Studies of a Tropical Standing Waterbody With Special Reference to Nutrient Status and Primary Production

Ravindra Paul and Satyanarayan Varma

Introduction

Water is one of the most vital resource for all kind of life on earth. Man had been closely associated with water in his surroundings from time immemorial and we are aware that many great civilizations of this world arose and flourished in the vicinity of the sources of freshwater. Although the primary concern of man was thought to be for meeting his basic requirements, it cannot be forgotten that water is the basic for all primary life activities for both terrestrial as well as aquatic organisms.

The total quantity of freshwater of our biosphere is a small fraction of the total hydrosphere (Pennak, 1894; Halbfass, 1933) but is of immense importance in governing the various life activities of all organisms. Inland water covers less than 2 per cent of the earth surface, *i.e.*, approximately 2.5 × 10^6 km (Wetzel, 1983) of which 1.6 million hectares of inland water exist in India alone. In Karnataka, seasonal tanks comprise 2, 13, 971 hectors and perennial tanks about 20,588 hectares of waterspread area (David *et al.*, 1974). The importance if such vast water resources in the dry regions of Deccan Plateau, both for human use and also for fish culture need not be emphasized. It is imperative that we acquire a complete knowledge of the various parameters of these waterbodies, especially in the context of present day stress on these natural environments. Naturally, under such circumstances, the role of limnology and limnologists assumes great significance in understanding the ecological status of any freshwaterbody. However, it is not rather very encouraging to note that, historically, limnological research in India had a late start. The early investigations mostly derived from mere observations. Subsequently, limnology was studied with reference to the organisms, especially plankton (Henson, 1887) which led to the so-called branch - "*Hydrobiology*". Even though this kind of work is still going on, it is felt that there is a need to go a long way in understanding thoroughly the ecological interrelationships of any aquatic habitat. The long held idea that a suitable environment is necessary for any organism since life depends upon the continuous exchange of nutrient cycling and energy flow between the organism and its surrounding environment which will in effect control photosynthesis, food chain, foodwebs and biodegradation, led to extensive works of several investigators during the beginning of the century *viz.*, Fritsch and Rich (1909), West and West (1912), Delf, (1915), Round (1953), Kemmere and Narhold (1969), Muzino and Mori (1971), Lin (1972), Miller *et al.* (1973), Tilman (1977), Russo (1978), Sounder Gaurd and Jensen (1979), Bhatnagar (1982) and Unni (1984) and several others.

India with the strategic position presents unique and varied physical, geographical and meteorological features and is rich in inland water sources which exist as both lotic and lentic systems. In so far standing waterbodies are concerned, those in India and especially in peninsular South India, are unique in their own category, since most of them are man-made unlike the natural lakes in temperate and sub temperate parts of the world. We have considerable information on these waterbodies from the higher latitudes. In contrast, many of these standing waterbodies of south India are quite smaller in dimensions. The geographic situation of these waterbodies and the influence of climatic conditions play a major role in deciding the ecological status of the waterbody. It is worthwhile mentioning here that peninsular India experiences the south-west monsoon for a period of three to four months, from June to September and this rainfall in varying quantum at different regions is the major source of freshwater to all the seasonal and perennial freshwater ecosystems. The topographical situation allows formation of impoundments of water and thus resulting in seasonal standing waterbodies.

From time immemorial, various waterbodies, streams, rivers, lakes and oceans have been the prime victims of the pollution due to disposal of sewage. Potentiality of domestic sewage in enhancing the yield of agricultural crop is well established. Waste water, although acting as source of nutrients to some extent, present problems such as toxicity to fish and their natural food. When the environment of fish is altered due to the introduction of these wastes, the impact on the fish fauna becomes pronounced either in beneficial or adverse ways depending upon the degree and quality of pollution. Incidence of mass mortality of fish due to discharge of domestic wastes have been reported by Joshi (1988). Sublethal concentration of poisonous materials such as heavy metals, oils and detergents, while not affecting the actual survival of the fish, may effect their growth either directly or indirectly (Hepher, 1974). When the discharge of raw or treated sewage exceeds the self purifying capacity of an aquatic body, major alterations in the concentrations of dissolved oxygen, hydrogen-ion-concentration, free carbondioxide, ammonical nitrogen, hydrogen sulphide, chloride, nitrate and nitrite are bound to occur. The changes in these physico-chemical conditions of the waterbody would affect its ecology and further deprive us not only of the future use of such a waterbody but also the available aquatic food such as fish. At the same time, the sewage disposal also leads to the fertilization of the receiving waterbody due to the addition of the phosphate and nitrogen, enhancing the phyto and zooplankton production and ultimately the aquacultural yields. The organisms of these variety of lentic habitat have aroused immense interest of many limnologists and hydrobiologists.

In the recent years a remarkable contribution is made in this field and different trends of study are observed. Some important contributions like those of Ruttner (1953); Singh (1960); Sreenivasan (1964 a,b) and Valecha and Bhatnagar (1989) have initiated a systematic approach to the understanding of hydrobiology. Many others have contributed towards specific aspects of nutrient dynamic studies, *viz..*, Martin and Stroud (1973); Jones and Backman (1976); Cooke *et al.* (1977); Gloss *et al.* (1980) Canfield and Bachmann (1981) to mention few. It is evident from the above mentioned literature that dynamics of lentic waterbodies - large and small are governed by the seasonal fluctuations typical to tropical areas such as that of our country, especially in South India.

Literature pertaining to the nutrient enrichment loading and its dynamics in freshwater ponds is rather scanty in India. Available information pertaining to the limnological aspect of the few standing waterbodies of Gulbarga is considered to be a preliminary use. Gaddad *et al.* (1983) emphasized on the aspects pertaining to sewage pollution of Jagat Tank. Vijaykumar (1992) presented the elaborate picture of physico-chemical parameters of three standing waterbodies of Gulbarga area. Zohra (1994) reported the influence of pesticide pollution on primary productivity. Mallanagouda (1996) studied zooplankton in relation to hydrography. In view of such paucity of information, the present investigation was undertaken to study the nutrient distribution which occur as a result of man-made activities such as by the cattle excrements that get into the pond through cattle washing and the other sources of nutrients are those brought in form the embankments through leaching and the significant quantities of substances carried into the pond with incoming sewage and their dynamic status in a standing waterbody of Gulbarga city *i.e.*, Jagat Tank.

The present study involves investigations on the dissolved nutrient constituents in Jagat Tank with special reference to their spatial distribution and variation in time which constitutes a profile of nutrient budget and further provides an information of energy turnover in the aquatic ecosystem. The study extended over a period of about one year from June, 1996 to March, 1997, though inadequate attempts to understand the role of nutrients in primary production.

Gulbarga city (16^0-12' to 17^0 - 46' latitude and 76^0 - 04' to 77^0 - 42' longitude) located at an elevation 458m above m.s.l., presents the typical climate of Deccan Plateau with semi arid conditions, scanty to moderate rainfall and temperature ranging between 15^0 to 45^0.

The Jagat Tank situated in the middle of the township is subjected to the stress of human activities especially by receiving untreated domestic sewage. This also forms a source of nutrient loading to the waterbody beside the embankments which also release through leaching and other human activities. The tank has total catchment area of about 0.275 sq. km. The mean depth is 2.1 meters with a maximum depth of 3.5 meters during the monsoon season and the minimum being 1.2 meters during the dry period.

The tank is utilized for fish culture activities by co-operative society run by the local fisherman since the last 35 years. The various features namely drainage of sewage into the tank which forms an additional source of nutrient loading attracted our attention to understand the details of hydrobiological status with special emphasis on primary production of this tank and hence the study. This tank assumes importance from economic point of view since it is used for fishery activity and is known for a substantial yield annually.

Presently, an attempt is made herewith to sketch out a picture of the contributions of some important workers from different parts of world and also from various geographical regions of our country. It is worthwhile mentioning that this kind of review is an essentiality in obtaining a clear limnological picture of the standing waterbodies and this would become a continued process especially in view of the ever increasing human activity on the ecological processes.

Considerable information is available on the role of climatic factors in the production of plankton which has been clearly emphasized in the works of Chandler (1944), on lake Erie. In the recent years the studies on standing waterbodies have been carried out with different orientation for specific problems as evidenced by some of the available literature *viz..*, Siegfried *et al.* (1982) who presented information on the Big Bear Lake (USA) encompassing a two years study and explained the interactions of various physico-chemical parameters and their impact on phytoplankton production. Robertson (1984) gave a detailed account of physico-chemical parameters of eigth North American Great Lakes. Fukuhara *et al.* (1985) discussed the interaction of physico-chemical parameters of a Lake in Japan. Vijaykumar and Paul (1991) presented information on the diurnal fluctuations in some physico-chemical parameters of the tank and moat in Gulbaraga, Karnataka. Chakrabarty and Saha (1993) gave an account of physico-chemical and hydrobiological features of three stagnant waters sources in Darjeeling Hills. Dhamija and Yatish (1994) made a study on variations in the physico-chemical characteristics of a lentic waterbody of Jabalpur (M.P.).

Of all the chemical parameters, nutrient elements play a significant role in determining the trophic status of the waterbodies.

It is known that these nutrients help in the synthesis of chlorophyll and act as carriers of essential substances. It is necessary in limnological studies to estimate the amount of major nutrients like phosphorus and nitrogen to assess their role in determining the aquatic primary productivity. A deficiency in the concentration of these major nutrients especially is known to result in a reduced plankton production in natural waters and excess of these nutrients may lead to the eutrophication of the waterbody.

Mortimer (1939), attempted, perhaps for the first time, the estimation of nitrogen balance for large bodies of water in Windermere region. Fogg (1947) reported on the role of some blue green algae as nitrogen fixers. There have been quite a few contributions in the recent times with regard to the estimations of nutrient budget for lakes and pond waters and mention may be made of some important ones, *viz.*, Schelske *et al.* (1971) in their studies in Lake Michigan opined that enrichment by nutrients especially phosphorus show significant effects on chlorophyll levels. Goldman (1972) discuss the correlation between heterocysts and nitrogen fixation in Clear Lake and the same author reported on the role of copper inhibiting the suppression of nitrogen fixation in eutrophic lakes. David and Robert (1979) reported on the functionally distinct classes of complex phosphorus compounds in two eutrophic lakes (East and West Twin Lakes) and humic bog (Crazy Eddie Bog) in northeastern Ohio. Siegfried (1982) reported on the phosphorus loading of Big Bear Lake, California. Singer *et al.* (1984) reported on the role of phosphorus content and zooplankton grazing on phytoplankton bloom in small enclosures. Ferris and Tyler (1985) documented the relationship between total phosphorus and chlorophyll content. Comita (1985) reported seasonal cycles of primary production in Brewer Lake and two blooms per year during the study. Havens and Decosta (1986) investigated the alteration in algal community structure with reference to nutrient addition in two different kinds of waters. Istranovics *et al.* (1986) while studying the mesotrphic Lake in Hungary, concluded that phosphorus enrichment increases the biomass of phytoplankton in enclosures. Schelske *et al.* (1986) reported that nutrient enrichment, particularly phosphorus enrichment, affects silica uptake by diatoms resulting in biogeochemical silica depletion, in their study of Great Lakes.

Strener *et al.* (1995) determined the relative importance of nitrogen and phosphorus to the growth of bacteria and algae in an oligotrophic lake. Rapp and Bishop (1995) have reported the chemistry of 134 lakes in Sweden and changes occurred between 1988 and 1993 in response to a decline in sulphate deposition.

Important contributions have bee made by various investigators from different parts of the country since the last four to five decades. Inspite of such observations it is rather not encouraging to note that research work on nutrient budget, production and energy flow on freshwater ecosystems in India has so far, not received the much needed importance (Gulati *et al.* (1980).

The information contributed by Iyenger (1939), Ganapati (1943), Phillipose (1959) on the periodicity of algae of some muddy rain water pools gained much attention. A few contributions in the recent times which are noteworthy and important are those of the following *viz.*, Patil *et al.* (1975) have investigated phytoplankton and zooplankton succession during prestabilization and stabilization of domestic waste water in facultative pond in Dharwad. Kannan and Job (1980) while observing several changes of physio-chemical parameters in a reservoir reported the status of eutrophication. Singh (1990) while studying correlations with physico-chemical parameters as well as nutrients of the waterbody. Mary (1991) reported on the total nitrogen and phosphate content of two freshwater lakes situated in the vicinity of Hyderabad. Subbamma and Rama (1992) investigated on the water quality characteristics of a temple pond near Machilipatnam and reported low levels of nitrates and phosphates compared to high level of silicate. Chakroborty and Saha (1993) investigated on hydrobiological and physico-chemical features with special reference to their nutrient contents of three stagnant water sources in Darjeeling Hills. Mukherjee *et al.* (1994) reported on the phosphate load from detergent inputs and its effect in freshwater resulting in algal blooms. Vijaykumar and Paul (1994) reported the effect of nutrient enrichment on phytoplankton in the Attikolla pond of Dharwad. Anjana and Kanhere (1995) studied on seasonal variations of abiotic factors of a freshwater pond at Barwani (M.P.). Narender and Mahmood (1995) made a study on nutrient status and biological examination of the water of Hubsiguda pond near Hyderabad, with reference to phytoplankton population. Kaur *et al.* (1996) studied the abiotic and biotic components which together account for the trophic status of the waterbodies of a freshwater pond of Patiala, Punjab. Sabu and Abdul (1996) reported on spatial and temporal distribution of nutrients in the Peppara reservoir on the Western Ghats, Thiruvanthapuram, South India.

The standing waterbodies are known to receive different kinds of pollutants through land drainage during the rainy season and over a period of time accumulate in the aquatic ecosystems. Several other pollutants also find entry into the lentic systems through man made activities such as detergents through washing, organic waste form animal like cattles, etc. Such pollutants interfere with the already existing balance between the physico-chemical parameters, specially the nutrient elements and there by influence the productivity of ecosystems. Studies in this direction have been initiated by some workers in our country and mention may be made of Patil *et al.* (1975); Kannan and Job (1980); Singh (1986); Mary (1991); Vijaykumar and Paul (1991); Mathew (1993) and Kaur *et al.* (1996).

However, it is the feeling of the author that the increasing human activity is the primary cause in creating imbalance of the well established ecological process to the extent of causing severe damage to the lentic systems themselves. Considering this increased devastation of the waterbodies, the available information on this important aspects of limnology, that is productivity governed by nutrient elements is comparatively inadequate and there is need for a continuous monitoring of the ecological status and thus implement corrective measures to preserve the productivity of the standing waterbodies. It is with this view, that the present study was undertaken, with emphasis on nutrient distribution and their correlation of primary productivity of tropical pond ecosystem, namely the Jagat Tank of Gulbarga.

Materials and Methods

The information on total rain fall pertaining to the study area, was obtained from the Gauging Subdivision No. 4, Gulbarga, for the period from June 1996 to March 1997.

Water samples were collected at fortnightly intervals. Collections were made on specific dates of ever first and second fortnight and average recordings were taken as the value of the month. Surface water samples were collected using a clean PVC containers for the study of various physico-chemical parameters.

Water samples were collected from three stations in the Jagat Tank. All collections and observations were made between 08.00 A.M. to 10.00 A.M. hours throughout the period of the study.

Primary Productivity

The primary production was estimated by the much preferred Gaarder and Gran's "Light and dark bottle" method. These studies were carried out at all three stations (I, II and III) 10-15 cms below the water surface. The exposure time in the present studies was 3 hours and usually after 10.00 a.m.

Water samples already collected in designed PVC bottles for the purpose of estimation of various parameters were brought to the laboratory and subjected to analysis immediately as far as possible by using Standard Methods for the Examination of Water and Wastewaters; 16th Ed., 1985 (APHA, AWWA, WPCF, USA).

Observations and Results

Though Gulbarga region experiences limited rainfall in general, well defined weather periods can be seen, *viz.*, southwest monsoon season extending from June to September, northeast monsoon season from October to January and summer from February to May. As per the present observations most of the rainfall was recorded during the southwest monsoon months. The fluctuations of total rainfall during the different months are presented in the Table 1.

A rather uniform amount of rainfall was observed during southwest monsoon season spread over June, August and October, however, a maximum of 216.4 mm was observed during September. A considerably low rainfall was observed during the month of July and November with 59.3 and 11.1 mm respectively. The summer months were totally devoid of rainfall.

The productivity of an aquatic ecosystem depends mainly on the interrelationships of numerous parameters with one another. The various parameters of physical and chemical nature in twin are the result of geoclimatic conditions of the region. The overall physico-chemical nature of an aquatic body indicates its total ecological status and further enables an assessment of its utility for purpose like fish culture etc. The present observations are oriented in this direction.

The atmospheric temperature was recorded at all the three stations of Jagat Tank during of study the data is presented in Table 1.

The atmospheric temperature from the three stations I, II, and III varied from 22 to 28.5^{0}C, 22 to 31.5^{0}C and 23.5 to 33^{0}C respectively. Maximum values were recorded during summer in the month of March at all three stations with 28.5, 31.5 and 33^{0}C, respectively. Minimum values were observed during winter season with 22^{0}C at station I in the month of December, at station II with 22^{0}C in the month of January and at station III with 23.5^{0}C in the month of December. The distribution of temperature followed a seasonal trend, showing an ascending pattern during Summer and decreasing pattern during winter with moderate temperature during rainy season.

The water temperature was recorded at the three stations I, II and III of Jagat Tank during the period of study and the observations are presented in Table 1.

The water temperature from the three stations I, II and III were in the range of 21 to 27.5^{0}C, 21 to 27.5^{0}C and 21.5 to 28^{0}C respectively. Maximum values of 27.5^{0}C and 28.5^{0}C were recorded during Summer in the month of June at all the three stations respectively. The minimum values of 21^{0}C, 21^{0}C and 21.5^{0}C were recorded during winter season, at the three stations I, II, and III in the month of December respectively. Aquatic temperature also followed a seasonal trend with higher temperature in monsoon and there after, was observed to fall during the winter and than again there was a slow increase from the latter period of winter towards summer months.

The values of Hydrogen-ion-concentration of Jagat Tank were observed at three stations I, II, III for different seasons are presented in Table 2.

The values observed at the three stations do not show any significant changes. The Hydrogen-ion-concentration values for the study period at the stations I, II, and III fluctuated from 7.9 to

8.9; 8.15 to 8.90 and 8.10 to 8.85 respectively. Higher values were observed at all the three station, with exception of slight changes during October to March and lowest during August to September, with a slight increase in July. During summer, the value was high with slight fluctuation upto the period of onset on monsoon.

Dissolved oxygen is one of the important physico-chemical parameters. The observations at the three stations of Jagat Tank during period of study are presented in Table 2.

The dissolved oxygen values from the stations I, II, and III , varied from 7.8 to 12.32 mg/*l*, 5.66 to 12.52 mg/*l* and 6.46 to 11.11 mg/*l* on an average respectively. The minimum values of dissolved oxygen were recorded during June at all the three stations at station I (7.68 mg/*l*), at station II (5.66 mg/*l*), and station III (6.46 mg/*l*). The maximum value was recorded during winter with a peak in December at all the three stations, at station I with 12.32 mg/*l*, at station II with 12.53 mg/*l* and 11.11 mg/*l* at station III. The winter high may be attributed to the lowering of water temperature which increases the capacity of water to hold more dissolved oxygen. During the summer due to very high temperatures the dissolved oxygen concentration falls down and average values were noticed in the monsoon, because of moderate monsoonal temperatures. This indicates a positive relationship between water temperature and dissolved oxygen.

Free carbondioxide was recorded at the three stations I, II and III of Jagat Tank during the period of the study. Nil values of CO_2 was noticeable during July and September. The data is presented in Table 2.

There are some obvious variations in the values observed at the three stations I, II and III. The free Carbondioxide concentration at station I varied between 6.6 mg/l in the month of October 24.2 mg/l in the month of March at Station II from 6.6 mg/l in the month of August and 17.05 mg/l in the month of March, and at station III varied between 7.04 mg/l in the month of January and 28.05 mg/l in the month of March. Carbondioxide values did not follow the seasonal trend. Wide variation in values in between stations and also season-wise fluctuations were noticeable conspicuously which is presented in the latter part of the paper.

The ecological status of waterbody and productivity are controlled to a large extent by the quantity and quality of external nutrient loadings. The nature of these nutrient inputs reflect, in

Table 1 : Total Rainfall and Average values of Atmospheric Temperatue and Water Temperature at three selected stations of Jagat Tank

Months	*Rainfall (mm)*	*Atm. Temp. °C*			*Water Temp. °C*		
		Stations					
		I	*II*	*III*	*I*	*II*	*III*
June	142.1	28.0	29.0	31.5	27.5	27.5	28.0
July	59.3	27.5	29.5	30.5	26.0	26.0	26.0
August	144.4	27.0	28.5	30.0	26.5	26.0	26.0
September	216.4	27.0	30.5	30.5	25.5	26.0	26.5
October	157.4	24.5	25.5	27.0	26.5	26.5	27.0
November	11.1	22.5	23.5	24.0	22.0	21.0	22.5
December	0.0	22.0	22.5	23.5	21.0	21.0	21.5
January	0.0	22.5	23.0	24.5	22.0	22.0	22.0
February	0.0	26.5	28.0	29.0	22.5	22.5	23.5
March	0.0	28.5	31.5	33.0	24.0	24.5	25.0

Table 2 : Average values of pH, Dissolved Oxygen and Free Carbondioxide at three selected stations of Jagat Tank

Months	*pH*			*Dissolved Oxygen (mg/l)*			*Free CO_2 (mg/l)*		
	Stations								
	I	*II*	*III*	*I*	*II*	*III*	*I*	*II*	*III*
June	8.20	8.40	8.20	7.68	5.66	6.46	14.30	11.00	15.40
July	8.55	8.75	8.75	7.88	7.07	8.49	0.00	0.00	0.00
August	7.90	8.30	8.15	8.28	8.08	7.88	14.30	6.06	8.25
September	7.85	8.15	8.10	10.51	8.69	10.51	0.00	0.00	0.00
October	8.55	8.50	8.35	9.60	9.09	8.28	6.60	8.80	7.15
November	8.70	8.75	8.75	10.91	9.90	9.70	9.35	9.35	7.15
December	8.50	8.75	8.65	12.32	12.53	11.11	14.30	7.70	15.40
January	8.85	8.90	8.75	11.11	11.32	9.19	10.67	8.25	7.04
February	8.90	8.60	8.80	09.90	11.32	09.30	9.35	11.55	14.85
March	8.90	8.85	8.85	8.89	9.39	8.89	24.20	17.05	28.05

turn, climatic regime and various watershed characteristics. Therefore, it is necessary in limnological studies, to estimate the amount of nutrients. In the present study the four important *viz..*, phosphate, nitrate, sulphate and silicate are studied over a span of about one annual cycle.

The values of phosphate-phosphorus concentration of Jagat Tank recorded at the three stations I, II and III, during period of study are presented in Table 3.

The phosphate-phosphorus concentrations at the stations I, II, and III varied from 0.225 mg/l to 0.645 mg/l, 0.235 to 0.485 mg/l and 0.205 to 0.450 mg/l respectively. Maximum values were recorded during southwest monsoon season at station I (0.645 mg/l) in the month of August and at station II and III (0.485 mg/l and 0.450 mg/l respectively) in the month of October. On the other hand minimum values were recorded during summer period at station I and II (0.225 mg/l and 0.235 mg/l respectively) in the month of January.

The values of nitrate-nitrogen concentration recorded during the study period at three stations I, II and III are presented in Table 3.

The nitrate concentration observed at the three stations I, II and III ranged between 0.055 to 0.465 mg/*l*, 0.050 to 0.285 mg/*l* and 0.060 to 0.260 mg/*l* respectively. Maximum values were recorded during southwest monsoon, at station I with 0.465 mg/*l* in the month of August, at station II with 0.285 mg/*l* in the month of October and at station III with 0.260 mg/*l* in the month of August. The minimum values were recorded during summer period in February and at station II and III with 0.050 mg/*l* and 0.060 mg/*l* respectively in the month of March.

The values of ammonical nitrogen concentration of Jagat Tank recorded at the three stations I, II and III during the period of study are presented in Table 3.

The ammonical nitrogen values were found to vary from 0.340 to 0.850 mg/l, 0.385 to 1.150 mg/l and 0.280 to 0.845 mg/l at station I, II and III respectively. Maximum values were recorded during summer at station I (0.850 mg/l) in the month of June and at station II and III (1.150 and 0.845 mg/l) respectively in the month of March. The minimum values were observed during winter season in the month of October at station I (0.340 mg/l) at station II (0.385 mg/l) and at station III (0.280 mg/l).

The values of sulphate concentration of Jagat Tank recorded at three stations I, II and III during the period of study are presented in Table 3.

The sulphate concentration was observed to be in the range from 4.60 to 8.95 mg/l, 4.70 to 9.75 mg/l and 4.55 to 8.95 mg/l at stations I, II and III respectively. Maximum value of 8.95 mg/l was recorded during summer at station I in the month of February at station II with 9.75 mg/l in the month of March and at station III with 8.9 mg/l in the month of February and March. The minimum values were recorded during early southwest monsoon period at station I with 4.60 mg/l in the month of July at station with 4.70 mg/l in the month of June and July and at station III with 4.55 mg/ l in the month of July.

The values of silicate concentration of waterbody recorded at three stations I, II and III during the period of study are presented in Table 3.

The silicate concentration exhibited fluctuations between 7.20 to 11.90 mg/l, 7.35 to 11.90 mg/l and 6.95 to 11.80 mg/l at stations I, II and III respectively. Maximum value was recorded during summer in the month of February at all three stations while minimum value was recorded during southwest monsoon season in the month of September at all the three stations.

The processes of building up primary organic compounds from inorganic substances at the expense of the external radiant energy is an in-built capacity of any ecosystem including the aquatic ecosystem. However, it is the phytoplankton which are the major contributors to the total photosynthetic production of organic matter in any aquatic ecosystem and hence, the need for evaluation of primary production. Gross primary productivity, Net primary productivity and Community respiration are studied for a period of about one annual cycle and correlation between nutrient status and productivity is proposed to be discussed in the latter part of this work.

The estimated gross primary productivity observed at the three stations I, II and III of Jagat Tank during the period of study are presented in Table 4.

The gross primary productivity observed from the three stations I, II and III varied in the range from 0.697 to 1.848 gC/ m^3/hr, 0.485 to 1.849 gC/m^3/hr and 0.606 to 2.909 gC/m^3/hr

Table 3 : Average values of Phosphate, Nitrate, Ammonical Nitrogen and Silicate at three Silected Stations of JagaTank

Months	Station I					Station II					Station III				
	PO_4	NO_3	NH_4-N	SO_4	SiO_3	PO_4	NO_3	$NH_4$4N	SO_4	SiO_3	PO_4	NO_3	NH_4-N	SO_4	SiO_3
June	0.34	0.11	0.85	5.20	9.20	0.24	0.15	0.75	4.70	9.60	0.28	0.09	0.60	5.50	9.80
July	0.36	0.16	0.76	4.60	9.60	0.32	0.24	0.54	4.70	9.85	0.31	0.20	0.32	4.55	10.15
Aug.	0.65	0.45	0.72	8.10	9.15	0.41	0.16	0.15	8.50	9.60	0.43	0.26	0.41	8.65	10.85
Sep.	0.32	0.20	0.45	5.15	7.20	0.37	0.15	0.47	5.65	7.35	0.33	0.22	0.44	5.35	6.95
Oct.	0.42	0.18	0.34	5.75	9.00	0.48	0.28	0.38	5.50	8.95	0.45	0.20	0.28	5.90	8.80
Nov.	0.39	0.14	0.36	6.05	9.50	0.35	0.12	0.41	6.20	9.45	0.37	0.15	0.33	6.30	9.35
Dec.	0.29	0.09	0.41	7.90	10.30	0.28	0.22	0.41	8.50	10.25	0.31	0.06	0.45	8.10	10.45
Jan.	0.23	0.21	0.35	8.70	11.00	0.26	0.16	0.40	9.10	11.00	0.20	0.20	0.31	8.10	10.90
Feb.	0.32	0.05	0.49	8.95	11.90	0.29	0.11	0.46	8.50	11.90	0.36	0.08	0.42	8.95	11.80
March	0.22	0.27	0.71	8.45	10.80	0.23	0.05	1.15	9.75	11.50	0.23	0.06	0.84	8.95	11.65

Table 4 : Average values of Gross Primary Production, Net Primary Production and Community Respiration at three stations of Jagat Tank

Months	*Station I*			*Station II*			*Station III*		
	GPP	*NPP*	*CR*	*GPP*	*NPP*	*CR*	*GPP*	*NPP*	*CR*
June	1.515	1.212	0.303	1.454	1.273	0.182	1.636	1.394	0.242
July	0.697	0.515	0.182	0.848	0.697	0.152	0.970	0.788	0.182
August	0.131	0.909	0.121	0.485	0.423	0.061	0.606	0.515	0.091
September	1.515	1.424	0.091	1.303	1.212	0.091	1.333	1.182	0.152
October	1.030	0.909	0.121	1.000	0.818	0.182	0.879	0.788	0.091
November	1.242	0.970	0.273	1.152	0.939	0.212	1.212	0.879	0.333
December	1.152	0.879	0.273	1.121	0.940	0.182	1.273	1.031	0.242
January	1.350	1.091	0.243	1.394	1.212	0.243	1.091	0.940	0.152
February	1.394	1.212	0.182	1.515	1.273	0.242	1.485	1.243	0.243
March	1.848	1.667	0.182	1.849	1.636	0.212	2.909	2.697	0.212

respectively. The high values of gross primary productivity were recorded during summer at the three stations I, II and III in the month of March. The minimum values were recorded during the monsoon period at station I (0.697 $gC/m^3/hr$) in the month of July and in the month of August at station II and III (0.485 and 0.606 $gC/m^3/hr$) respectively.

The estimated net primary productivity observed at the three stations I, II and III of Jagat Tank during the period of study are presented in Table 4.

The net primary productivity was observed to vary from 0.515 to 1.667 $gC/m^3/hr$, 0.423 to 1.636 $gC/m^3/hr$ and 0.515 to 2.697 $gC/m^3/hr$ as recorded at the three stations I, II and III respectively. The high values of net primary productivity was recorded during summer at all the three stations I, II and III in the month of March. The minimum values were recorded during the monsoon period at station I with 0.515 $gC/m^3/hr$ in the month of July and in the month of August at the stations II and III (0.423 and 0.515 $gC/m^3/hr$) respectively. This range of net production followed a similar trend as observed for gross production at all the three stations.

The estimated community respiration indicated the utilization of organic carbon by phytoplankton during respiration, observed at all the three stations I, II and III of Jagat Tank during the period of study are presented in Table 4.

The values of community respiration recorded from the three stations I, II and III varied in the range from 0.091 to 0.303 $gC/m^3/hr$ to 0.061 to 0.243 $gC/m^3/hr$ and 0.091 to 0.333 $gC/m^3/hr$ respectively. A high value was observed at station I (0.303 $gC/m^3/hr$) in the month of June at station II (0.243 $gC/m^3/hr$) in the month of January and at the station III (0.333 $gC/m^3/hr$) in the month of November. The low values were observed at station I (0.091 $gC/m^3/hr$) in the month of September at station II (0.061 $gC/m^3/hr$) in the month of August and in the month of August and October at station III (0.091 $gC/m^3/hr$).

Discussion

The evaluation of primary productivity of any standing waterbody is imperative in understanding its ecological status. It is a well known and established fact that various abiotic and biotic components establish vary intricate interrelationships between one another which is again subject to diverse fluctuations due to natural as well as manmade interference. The intricate interplay of physico-chemical parameters further influence the primary producers and subsequently secondary and tertiary producers. This kind of interplay is known to follow a seasonal and cyclic pattern in larger standing waterbodies such as lakes, reservoirs etc., (Hutchinson, 1967) the dynamics of specially the nutrient parameters of smaller dimension standing waterbodies exhibit very interesting inter-relationship which is a subject of the present study.

The area under investigation experienced both southwest monsoon and northeast monsoon rains and the data gathered from June 1996 to March 1997 is presented in Table 1. The total rainfall was in the range of 730.70 mm for the study period of one annual cycle out of which 562.20 mm rainfall was recorded during southwest monsoon months and 168.50 mm rainfall was recorded during northeast monsoon period. The rest of the year remained totally dry. The data indicates scanty rainfall during the study period and is found to be average. Vijaykumar (1992) in an earlier instance during 1986 to 1988 also reported low rainfall in this area and the present observations coincides with his report. Scanty rainfall is typical of this semi-arid region as recurrent data indicate. Undoubtedly, such meagre rainfall influences the climatic conditions and there by the limnological conditions of the aquatic body and ultimately the productivity of the waterbody.

It is well known that the water temperature exerts direct as well as indirect influence on many abiotic and biotic components of any aquatic ecosystems. In the present investigations the water temperature of Jagat Tank followed a trend similar to the atmospheric temperature of the area of all the three stations. Though rainfall is known to influence the temperature distribution, in the present investigation, it is felt that there is clear relationship between rainfall and atmospheric as well as water temperature. The water temperature ranging between a low of 21^{0}C and a high of 28^{0}C coincide with observations of workers investigating similar lentic ecosystems (Patil and Gouder, 1982; Vijaykumar, 1992; Zohra, 1994, Anjana and Kanhere, 1995; Mallanagouda, 1996; Kaur *et al.*, 1996).

Important means of understanding the chemical conditions which prevail in an aquatic ecosystem is through the measurement of hydrogen-ion-concentration (Welch, 1952), pH values recorded during the study period ranged between 7.9 to 8.9, increasing trend was noticed from all the three stations from September to March, with very little fluctuations in monthly values. Vijaykumar (1992) has observed pH upto 10.50 in Jagat Tank of Gulbarga during 1986 to 1988; Narendar and Mahmood (1995) has observed pH upto 9.1 in Hubsiguda pond, Hyderabad; Kaur *et al.* (1996) observed pH upto 8.35 in pond of Patiala; David *et al.* (1974) reported Gulbarga tanks having pH around 9.0 and the present data coincides with his observations. High pH observed during 1986-1988 for the same waterbody indicate the smaller dimensions lentic systems are subject to immediately local conditions and thus exhibits high degree of fluctuations necessitating continuous monitoring.

Dissolved oxygen is a parameter of primary importance in the aquatic ecosystems by virtue of its role in both chemical as well as biological reactions. Dissolved oxygen as observed in the present study fluctuate between the stations which could be attributed to the disturbances of artificial nature as also the influences of related physico-chemical parameters, especially temperature. In the present investigation the dissolved oxygen fluctuate between 5 mg/*l* to 14 mg/*1*. The fluctuations indicate an increasing trend during winter months and decrease during summer months. This observation of seasonal trend corresponds the observations of the workers reporting from similar standing waterbodies *viz..*, Subbamma and Rama (1992); Vijaykumar (1992); Shubha and Shastree (1993); Anjana and Kanhere (1995); Arvind (1995); Narendar and Mahmood (1995); Kaur *et al.* (1996); etc. The moderately high oxygen values noticed during Summer period despite the prevailing high temperatures (*i.e*, during March) may be attributed to addition by photosynthetic activity of aquatic vegetation as opined by Ganapati (1943).

Free carbondioxide is one of the most important parameters of the limnetic ecosystems governing the photosynthetic activity a described by Welch, (1952). The estimated values of free carbondioxide was recorded during the period of study ranged between 6.6 to 28.05 mg/*1*. The high values were observed in the month of March at all three stations and this may be because of higher rate of decomposition in the warmer months. The low values and complete absence during the month of July and September may be due to the increase photosynthetic activity or could be correlated with alkalinity and pH. This observation of seasonal trend corresponds with observations of other workers such as Vijaykumar (1992); Subbamma and Rama (1992); Zohra (1994); Arvind (1995); Mallanagouda (1996) etc.

Of all the chemical parameters, nutrient elements play a very significant role in determining the tropic status of the aquatic body already pointed out. Among the various micronutrients, phosphorus is known to play an important role in governing the primary productivity of any ecosystem (Hutchinson, 1957). In the present investigation the concentration of phosphates recorded during the period of study from all the three stations fluctuated between 0.205 to 0.485 mg/1. The high values during the monsoon may be because of inflow of rainwater and low levels during summer may be due to the fact that many algae including diatoms utilize this vital element. This type of observation of seasonal trend corresponds with the observations of other waterbodies as detailed herewith : Schelske *et al.* (1971) proposed phosphate as the main nutrient parameter to accelerate eutrophication in lake Michigan; David and Robert (1979); Mary (1991) observed similar trend of fluctuations of phosphate; Vijaykumar (1992) reported seasonal trend in phosphate concentrations; Vijaykumar and Paul (1994); Ajnana and Kanhere (1995); Arvind (1995); Narender and Mahmood (1995) and Sabu and Abdul (1996) also reported fluctuations of phosphate concentration according to the seasons. In the present investigations the annual average value was found to be higher than the optimum (0.1 to 0.2), thus indicating eutrophication of the tank and this may be due to large influx of sewage into the tank. The correlation of this parameter with primary production is dealt with later in this work.

Nitrogen is considered as one of the major constituents of cellular protoplasm and forms a major nutrient that affects physiologically and ecologically the growth and development of phytoplankton. It exists both inorganic and organic forms. Major inorganic components are nitrate-nitrogen, nitrite-nitrogen and ammonical-nitrogen which are normally estimated separately and are summed up to give the inorganic-nitrogen content. In the present study, only nitrate-nitrogen and ammonical-nitrogen have been estimated. The concentration of nitrate-nitrogen was recorded during the period of study from all the three stations fluctuated between 0.050 to 0.465 mg/l. The fluctuations exhibit a trend of increase during monsoon period with a maximum of 0.465 mg/l recorded in the month of August and decreasing during the summer. The higher values during monsoon period may be obviously due to the input from the incoming rainfall water, but the low values in the summer may be due to utilization by aquatic biota or due to action of denitryfying bacteria which are quite active at higher temperature.

The concentration of ammonical nitrogen recorded during the period of study from all the three stations ranged between 0.280 to 1.150 mg/l. The fluctuations exhibit an increasing trend during the summer months and a declining trend during winter months. The maximum values were recorded in the month of March, with 1.150 mg/l. The increase may be attributed to increase in temperature and low values were recorded in the month of October and the decline may be due to direct utilization of ammonical nitrogen by the phytoplankton as pointed out by Rao and Govind (1964), observed an increase in the production of phytoplankton during August which was preceded by the flood water rich in nitrate and phosphate draining into the reservoir, as also observed by Ayyappan and Gupta (1981), in Ramasamudra Tank, Dakshina Kannada, Karnataka, who advocated that the months of premonsoon and monsoon give high quantum of nitrate which coincided with increased rate of gross primary production. Similar trend was observed by Mary (1991) investigating the productivity of two small lakes - Banjara and Nadimi - lying in the vicinity of Hyderabad. Kanabur (1990), while describing nitrogen metabolism in the sewage of the moat adjacent to the Jagat Tank during (1971-1979), observed seasonal variations of ammonical-nitrogen with higher values during summer. Other workers who observed similar trend of ammonical - nitrogen were Vijaykumar (1992), Subbamma and Rama (1992), Anjana and Kanhere (1995), Narender and Mahmood (1995), Kaur *et al.* (1996) and Sabu and Abdul (1996).

Sulphate is the second or third most abundant nutrient present in most freshwater bodies (Hutchinson, 1957). Entry of sulphate into freshwater systems is through a variety of sources such as rainwater, sewage and also derived from sedimentary sources. It is also a well known fact that autotrophic organisms and may heterotrophic bacteria require considerable quantity of sulphur from sulphate in the media. This is a matter of considerable importance as regards the productivity of the concerned waterbody. The concentration of sulphate was recorded during the period of study from all the three stations, ranging between 4.55 to 9.75 mg/l. The fluctuations exhibit a trend to gradual increase from the onset of monsoon to the summer period. The high values was recorded in the month of March, with 9.75 mg/l and low values was recorded in the month of July. The high values during summer may be attributed to increased evaporation. Hutchinson (1957) also points out an increase in sulphate content in the enclosed waterbodies of semi-arid regions during summer which is attributed to increased evaporation and hence increase in the concentration of sulphate. Same trend of observations were also reported from Yousuf *et al.* (1986) of a forest lake of Kashmir. Anjana and Kanhere (1995) reported lower values of sulpher during monsoon period and higher values during summer period. It is also presently opined that a high rate of evaporation is an important factor contributing to higher concentrations of sulphate in this waterbody.

Silicate is an essential nutrient for any aquatic body especially in phytoplankton production. It is also well known a fact that autotrophic organisms and may heterotrophic organisms require silicate for their normal growth. Silica is an essential nutrient in the waterbody because diatoms dominate phytoplankton assemblages and have siliceous frustules. Like nitrate, silica is therefore, utilized by diatoms for growth and the amount of silica utilized in the photic zone is a reflection of the trophic states of the waterbody. The concentrations of the silicate recorded during the period of the study from all the three stations was found to be in the range between 6.95 to 11.90 mg/l. The fluctuations exhibit a trend of increase from monsoon period towards summer and decline during monsoon period. The increase or decrease of diatom population appears to be closely related to variations in silicate content indicating dynamics of silicate usage in surface waters. It is also well known that nutrient enrichment, particularly phosphorus enrichment, affects silica uptake by diatoms. Seasonal silica uptake by diatoms is well known from the classical studies of the English

Lake District (Reynolds, 1984). Considerable information is available on the silicate content and its utilization by diatoms and other algae in the surface water from various workers *viz.*, Hutchinson (1967); Schelske and Stoermar (1972); Schelske (1974); Schelske *et al.* (1986); Yousuf *et al.* (1986); Arvind (1995); Narender and Mahmood (1995); Sabu and Abdul (1996) etc.

Gross primary productivity (GPP) of Jagat Tank was observed that GPP varied marginally between the stations I, II and III indicating a rather uniform production for the whole ecosystem during the study period. However, the general trend observed was that of increase in GPP towards early summer months with gradual decrease towards later parts of the summer at all the three stations.

Net primary productivity (NPP) of Jagat Tank exhibited fluctuation between all the three stations and this could be due to distribution of autotrophs. However, and increasing trend could be observed towards early summer at all the three stations.

Community respiration of Jagat Tank exhibited fluctuations between all the three stations, with peak respiration being observed during June at station I and during January at station II and at station III during the month of November, and lowest value of community respiration was observed during the month of September and October.

Gross production was found to increase under the influence of high pH and lesser carbondioxide in the eutrophic lakes of south India as observed by Sreenivasan (1963). Further, the same author points out comparatively lesser production in oligotrophic lakes with lesser pH and low concentrations of phosphate and silicate. The combined effect of several parameters *viz..*, carbondioxide, dissolved oxygen, turbidity or there individual effects seems to be responsible for fluctuations in gross production as put forth by Sreenivasan (1964 a, 1964 b and 1976).

In the present studies,. though there is direct relation of gross and net production with parameters like pH and carbondioxide, no clear cut relation is discernible with the nutrient elements – at least phosphate and nitrate, which is rather unusual. Further, during the middle of monsoon season especially in September, when both phosphate and nitrate were found in higher concentrations, the gross production was observed at its lowest for the whole annual cycle. This could be interpreted as no obvious influence of these two nutrient element on gross production. However, it is worth while mentioning that increased turbidity and less availability of radiant energy during the monsoon as a more influential cause in bringing down productivity. This seems more possible in ponds being influenced by land drainage and sewage influx. However, it is interesting to observe a direct correlation between increasing trend of gross production with higher concentration of other two nutrients, namely sulphate and silicate almost throughout the period of study. Schelske *et al.* (1971) corroborated high concentration of nutrients especially phosphate with higher plankton productivity. While, Ayyappan and Gupta (1985) observed that higher production rates coincides with low pH, higher carbondioxide, low total alkalinity with no mention of nutrients. Khan and Siddiqui (1971) reported bimodel fluctuations in primary productivity showing higher values during post-winter months indicating a close relationships of production with temperature profile and phosphate concentrations. The present observation indicate similarities though with reference to atleast some parameters such as temperature, carbondioxide etc. However, it is also interesting to note that there appears to be a close relationship of productivity with ammonical-nitrogen, sulphate and silicate amongst the nutrient elements. Further, dense algal blooms were observed twice during the study period *i.e.*, during September and there after during the March. In the present studies the combined influence of parameters was not very evident and showed higher production rates with high pH and higher values of free carbondioxide. Valecha and Bhatnagar (1989) reported high primary procution during March to June which is attributed to increase in temperature and solar radiation. The present study on the Jagat Tank shows similar relationship with the observations in high primary production form December to March which is attributed to increase in the temperature. Prasad (1990) observed increased productivity during winter and post monsoon seasons and indicates the role of nutrients as being significant. Such observation are found to be identical to Jagat Tank production in the present investigations. Arvind (1995) reported higher concentration of nutrients phosphate and sulphate during rainy months and lower concentrations during summer. The present investigations correlates with his observations. However, a seasonal variations in the phosphate and sulphate was denied by Ganapati (1943), Anjana and Kanhere (1995) reported lower concentration of phosphate, nitrate and silicate during winter season and higher concentrations during summer. The present investigations coincides with his observations regarding silicate but phosphate and nitrate were higher during winter season and lower during summer. Narender and Mahmood (1995) reported that higher concentration of nitrogen and phosphorus accelerates the algal blooms. The present investigations coincide with their observations of algal blooms during the month of September. Birsal (1996) reported higher primary productivity of phytoplankton during summer period and lower during winter, showing clear cut correlation with the temperature and the present investigation correlates with his observation at least during post monsoon period.

In general the gross production and net production follow a similar trend of fluctuations between the months and seasons. This shows that the specific conditions of micro habitats influence the productivity in the standing waterbody under consideration and it can be safely put forth that the lentic system governed by stressful conditions like sewage influx, human activities, etc. do not exhibit a holistic profile or productivity.

During the course of present study the estimations of biochemical oxygen demand revealed certain interesting observations and it is presently the opinion of the author, that there seems to be at least some relation of biochemical oxygen demand with productivity, though this cannot be stated confirmatively. Kanbur (1990) suggests higher levels of ammonical nitrogen in relation to large influx of sewage for the same waterbody. In the present studies a close relationships is observed between ammonical nitrogen and gross productions which may suggest the influence of biochemical oxygen demand and its role in degradation of organic material and release of nitrogen compounds probably indirectly. This needs to be substantiated with further investigations.

The observations on fluctuations in nutrient concentrations and productivity levels shows that the gross production is influence by several parameters, the important ones being pH, free carbondioxide, dissolved oxygen and the major nutrients.

However, it is interesting to note that from amongst the nutrients phosphate-phosphorus and nitrate-nitrogen do not exhibit any direct relation as observed by several investigators, specially with reference to phosphorus. Further a direct relationship is observed between the increasing trends in concentrations of ammonical-nitrogen, sulphate and silicate along with the interplay of parameters like pH, carbondioxide and dissolved oxygen. There by it is the opinion of the author that during the advance state of eutrophication when concentration of phosphate increase phenomenally, probably the productivity gets affected. And also, it is supposed that the autotrophic components may selectively depend upon nutrients. When the general trends in the temporal variations in the nutrients in different stations of tank are observed, it becomes clear that the rate of cycling of nutrients and there availability is more important rather than the magnitude of the quantity of the nutrient present. This fact determines the process of the eutrophication and the nature of biota in the ecosystem and also primary productivity.

References

APHA, AWWA, WPCF. 1985, Standard methods for the examination of water and wastewater. *American Public Health Assoc. Inc.* New York, 16th Edition, p. 1268.

Anjana, M. and Kanhere, R.R. 1995, Seasonal variations of abiotic factors of a freshwater pond at Barwani (M.P.). *Poll. Res.*, 14(3) : 347-350.

Arvindkumar, 1995, Some limnological aspects of the freshwater tropical wetland of Santhal pargana (Bihar), India. *J.Envi.Poll..*, 2(3) : 137-141.

Ayyappan, S. and Gupta T.R.C. 1981, Limnology of Ramasmudra Tank - Hydrography. *Mysore J. Agri. Sci.*, 15 : 305-312.

Ayyappan, S. and Gupta T.R.C. 1985, Limnology of Ramasmudra Tank - Primary production. *Bull. Bot. Soc. Sagar*, 32 : 82-85.

Bhatnagar, G.P. 1982, Limnology of lower lake Bhopal with reference to sewage pollution and eutrophication. *MAM. UNESW. Tech. Rep. Bhopal.* 1-5 and 7-55.

Birsal, N.R. 1996, The Primary productivity of the Supa Reservoir during its filling phase. *J. Environ. Biol.* 17(2) : 127-132.

Canfield, D.E. and Bachman, R.W. 1981, Prediction of total phosphorus concentration, chlorophyll-a and Secchi disc in natural and artificial lakes. *Can. J. Fish. & Ag.Sci.* 39 : 414-423.

Chakrobarty, D. and Saha, G.K. 1993, The physico-chemical and Hydro-biological features of three stagnant water: Sources in Darjeeling Hills. *Geobios.* 20 ; 61-65.

Chandler, D.C. 1944, Limnology studies of western lake Erie. V. Ralation of limnological and climatic factors to the phytoplankton of 1941. *Trans Am Microsc.Sco.* 63(3) : 203-236.

Comita, Gabrial, W. 1985, The seasonal primary productivity cycles, their normal distribution and the associated chemical and physical cycles in a freshwater impoundment. *Arch. Hedrobiol. Suppl. Biol.,* 72(1) : 1-48.

Cooke, G.D., McComas, M.R., Waller, D.W. and Kennedy, R.H.. 1977, The occurrance of internal phosphorus loading in two small, eutrophic, glacial lakes in northeastern Ohio. *Hydrobiologia.* 56(2) : 129-135.

David, A. Rao, N.G.S. and Ray, P. (1974), Tank fishery resources of Karnataka. *Bull. con. Inlan. Fish. Res. Inst.* Barrackpore. 20 : 87.

David, A.F. Robert Heath, T. 1979, Functionality distinct classes of complex phosphorus compounds in lake water. *Am. Soc. Limnol. Oceangr.* 24(3) : 463-473.

Delf, E.M. 1915, The algal vegetation of some ponds on Hampstead Health. *New Phytol.* 14 : 63-80.

Dhamija, S.K. and Yatish Jain. 1994, Variations in the physical co-chemical characteristics of a Lentic waterbody of Jabalpur (M.P.) *J.Envi. and Poll.* 1 (3 & 4) : 125-128.

Ferris, M.J. and Tyler, P.A. 1985, Chlorophyll-total phosphorus relationships in lake Burrangorang, New South Wales and some othe souther Hemisphere lakes. *Aust. J. Mar. Fresh. Res.,* 36 : 157-168.

Fogg., G.E. 1947, Nitrogen fixation by blue-green algae. *En-deavour.* 6 : 172.175.

Fritsch, F.E. and Rich, F. 1909, A five year observations of the fish pond Abbots Leigh near Bristol *Proc. Bristol nat. Soc.* 2 : 27-54.

Fukuhara, H. Hoshino, S and Ogawa, Y. 1985, Limnological studies of Lakes in Niigata prefecture. II Linmological studies of lakeAmagoiike, Yoshigahira Niigata, Japan. *Biol. Lab. Fac. Niigata Univ.* 9-22.

Gadded, S.M., Prabhakar, M.N. and Rodgi, S.S. 1983, Ecological studies of two polluted bodies of water at Gulbarga. *Poll.Res.* 2(2) : 49-52.

Ganapati, S.V. 1943, An ecological study of a garden pond containing abundant zooplankton. *Proc. Indian Acad. Sci.,* (Bombay), 17 : 41-58.

Gloss, S.P., Kidd, D.E. and Mayer, L.M. 1980, Advective control of nutrient dynamics in the epilimnion of a large reservoir, *Limnol. Oceano.* 24 : 219-228.

Goldmann, C.R. 1972, The role of minor nutrients in Limiting the productivity of aquatic ecosystems. *Am. Soc. Limno. Oceano.* (Sp. Symposium). 1 : 21-38.

Gulati, R.D. and Wurtz, G., Schulz. 1980, Remarks on the present status of limnology in Indian based on the Indian Publications in Hydrobiologia and suggestions for future approach. *Hydrobiologia.* 72 : 211-222.

Halbfass, W. 1933, Seen in Handuch der Geophysik. Berlin, Borntraeger Bd. 7. If. T. 223 : 122-182.

Havens, K.E. and Decosta, J. 1986, A comparison of Phytoplankton responses to nutrient additions in acidic and circumnetural pH Lake water. *Hydrobiologia.* 137 : 211-222.

*Hensen, V. 1887, "Uber die Bestimmung des planktos oder des in meere treibenden materials on Pflanzen. Und. Thieren".. *Ber. Kommn. Wiss.Unters. Dt. Meera.* 5 : 1-109.

Hepher, B. and Schroeder, G. 1974, Wastewater utilization in integated aquacultue and agricultue system. U.S.E.P.A.*Environ.Protec. Tech. Ser.* 660/2 : 9-15.

Hutchinson, G.E. 1957, A treatise on limnology. Vol. 1. John Wiely and Sons, Inc. New York. NY : 1015.

Hutchinson, G.E. 1967, A treatise on limnology. Vol. 2. John Wiely and Sons, Inc. New York. NY : 1115.

Istvanovics, V., Voros, L. Herodek, S., Toth. G.L. and Tatrai, I. 1986, Changes of phosphorus and nitrogen concentration and phytoplankton in enriched lake enclosures. *Limno. Ocean.* 31(4) : 798-811.

Iyengar, M.O.P., Algal problem peculiar to the tropics with special reference to India. *Proc. 25th Indian Sci. Cong.* 25(4) : 141-149.

Jones, J.R. and Backmann, R.W., Prediction of phosphorus and chlorophyll levels in Lakes *J.Wat.Poll. Contr. Fed.*48 : 2176-2182.

Joshi, H.C. 1988, Impact of man-made environmental modifications on the riverine ecology and productivity. Conservation and management of inland capture fisheries resources of India. *Inland Fish. Soc. India* Barrackpore, India : 85-90.

Kanabur, V.V. 1987, Catalase activity in sewage and in sewage treated effluents near Sharanabasaveshwar tank at Gulbarga. *Envi. and Ecotoxicology.* 147-152.

Kanabur, V.V. 1990, Ecological influence on nitrogen metabolism in sewage near Sharanabasaveshwar tank at Gulbarga. *Trend in Ecotoxicology* : 117-121.

Kannan, V and Job, S.V. 1980 : Diurnal depth-wise and seasonal changes of physico-chemical factors in Sathiar reservoir. *Hydrobiologia.* 70 : 103-117.

Kaur, H. Dhillon, S.S. Bath, K.S. and Mander, G. 1996, Abiotic and biotic components of freshwater pond of Patiala-Punjab: *Poll. Res.15*(3) : 253-256.

Kemmerere, A.J. and Narhold, M.N. 1969, A method for gross primary productivity measurement. *Limnol. Oceangr.*14(4) : 607-610.

Khan, A.A. and Siddiqui, Q.A. 1971, Primary production in tropical fish pond at Aligarh, India. *Hydrobiologia.* 37(3-4) : 447-456.

Lin, C.K. 1973, Phytoplankton succession in a eutrophic lake with special reference to Blue-green algal bloom. *Hydrobiologia.* 39(3) : 321-334.

Mallanagouda. S.B. 1996, Zóoplankton studies in relation to Hydrography of a tropical pond ecosystem. *M.Phil Dissertation, Gulbarga University, Gulbarga.* pp 39.

Martin, R.G. and Stroud, R.H. 1973, Influence of reservoir discharge location on water quality, biology and sport fisheries of reservoir and tailwaters. 1968-71. U.S. Army Engineer Waterways Experiment Station, Vicksburg. MS.

Mary Esther Cynthia Johnson. 1991, Nitrogen and phosphate content of two freshwater lakes. *J.Aqua Biol.* 6(1/2) : 19-26.

Mathew Varghese. 1993, Productivity studies of Macrophytes in a tropical pond polluted by domestic effluents. *Envi. Eco.* 11(2) : 317-320.

Miller, W.E., Malony, T.E. and Greene, J.E. 1973 : Algal productivity in 19 Lake waters as determined by algal assays. *Water.Res* 8 : 667-679.

Mortimer, C.H. 1939, The work of the freshwater biological association of Great Britain in regard to water supplies. The nitrogen balance of large bodies of water. *Off. Circ. Br. Wat. Wes. Ass.* 21 : 1-10.

Mukherjee, B., Pankajakshi, G.V.N., Bose, P. Kumar, R. Kumar, P, Kumar, D., Kumar, A., Kumari, R., Singh, S and Konogia,S. 1994, The phosphate load from detergent inputs and its effect in freshwaters. *J.Environ.Biol.* 15(1) : 27:39.

Muzino, T. and Mori, S. 1970, Preliminary hydrobiological survey of some southeast Asain Inland Waters. *Biol.J.Linn. Sco.* 2 : 77-117.

Narender Rao, V. and Mahmood, S.K. 1995, Nutrient status and biological characteristics of Hubshiguda Pond. *J.Envi.Poll.* 2(1) : 31-34.

Patil, H.S., Dodakundi, G.B. and Rodgi, S.S. 1975, Succession in zoo and phytoplankton in sewage stabilization pond. *Hydrobiologia* . 47(2) 253-264.

Patil, C.S. and Gouder, B.Y.M. 1982, Freshwater fauna of Dharwad (Karnakata State, India) : Cladocera. *Karnatak uni. Sci. 27* : 115-126.

Pennak, A. 1894, Morphologie der Erdoberflache, Stuttgart Engelhorn, Vol. I XIV : 471.

Phillipose, M.T. 1959, Freshwater phytoplankton of inland fisheries. *Proc. symp. algology ICAR.* 272-291.

Prasad, D.Y. 1990, Primary productivity and energy flow in upper lake., Bhopal (India) *Indian.J.Environ.Health.* 32(2) : 132-139.

Rao, D.S. and Govind, B.V. 1964, Hydrology of Tungabhadra reservoir. *Indian.J.Fish.* 11(1) : 321-344.

Rapp, L and Bishop, K. 1955, Sulpher deposition and changes in Swedish Lake Chemistry 1988 - 1993. *Water Air and Soil Poll.* 85(4) : 2449-2454.

Reynolds, C.S. 1984, The ecology of freshwater phytoplankton. Cambridge University Press, New York. NY : 384.

Robertson, A. 1984, The present status of research on the Zooplankton and zoobenthos of the Great Lakes. *J.Great Lakes Res.* 10(2) : 156-163.

Round, F.E. 1953, An investigation of tow benthic algal communities in Mahan Tran, Yarkshiva *J.Ecol.* 41 : 174-197.

Russo, A.R. 1978, Some ecological observations on a permanent pond in Southern England, Primary production of planktonic seasonal succession. *Hydrobiologia.* I(60) : 33-48.

Ruttner, F. 1963, Fundamentals of Limnology, 3rd Ed. University of Toronto. Press Toronto, Canada. 295.

Sabu Thomas and Abdul Aziz, P.K. 1996, Spatial and temporal distribution of nutrients in the Peppara reservoir. A man made ecosystem on the Western Ghats, South India. *Poll.Res.* 15(1) : 5-10.

Schelske, C.L. Stoermer, E.F. and Feldt, E.L. 1971, Nutrients : Phytoplankton productivity and species composition as influenced by upwelling in Lake Michigan. *Proc. 14th. Conf. Great Lakes Res.* 102-113.

Schelske, C.L. and Stoermer, E.F. 1972, Phosphorus, silica, and Eutrophicaiton of Lake Michigan *Reprinted from Nutrients and Eutrophication special symposia.* Vol. I : 157-171.

Schelske, C.L. 1974, Nutrients inputs and their relationship to accelerated eutrophication Lake Michigan. *E.J. Monke* (ed) : 59-81.

Schelske, C.L. 1985, Silica depletion in lake Michigan. Verification using Lake Superior as an environmental reference standard. *J.Great Lakes Res.* 11(4) : 492-500.

Schelske, C.L. and Stoermer, E.F., Fahnenstiel, G.L. and Haibach, M. 1986, Phosphorus enrichment, silica utilization and biogeochemical silica depletion in the Great Lakes. *Can.J.Fish.Aquat.Sci.* 43(2) : 407-415.

Scott, F.M. 1927, Introduction to the limnology of Searsvilla Lake. Standford Univ. *Publ.Biol.Sci.* 5 : 3-83.

Shuba Pathak and Shastree, N.K. 1993, Physico-chemical status of the lentic hydrosphere of Rukmini and Visar Sarovar of Gaya, Bihar. *Envi. Ecol.* 11(1) : 57-66.

Seigfried, C.A. 1982, Phosphorus loading in a Mountain reservoir in Southern California. *Wat. Res. Bull.* 18(4) : 613-620.

Seigfried, C.A., Herrgesell, P.L. and Kopache, M.E. 1982, Limnology of eutrophic reservoir Big Bear Lake, California. *Calf. Fish Game* 68 : 90-108.

Singer, R. Evans, G.L. and Pratt, N.C. 1984, Phytoplankton limitation by phosphorus and zooplannkton grazing in an Acidic Adirondack Lake. *J.Fresh Ecol.* 2(5) 423-434.

Singh, V.P. 1960, Phytoplankton ecology of the inland waters of Uttar Pradesh. *Proc. Symp. Algal* ICAR, New Delhi. 243-271.

Singh, V.P., Davis, F. 1986, Relationship between primary productivity and environmental parameters of tropical lakes : A statistical analysis. *Poll. Res.* 5(3 & 4) : 103-109.

Singh, Ravindra. 1990, Correlation between certain physico-chemical parameters and primary productivity at Jamalpur Munger. *Geobios.* 17(5/6) : 229-234.

Sounderguard, M. and Sand-Jensen, K. 1979, Physico-chemical environment, phytoplankton biomass and production in oligotrophic soft water lakes, Kalgoard, Denmark. *Hydrobiologia.* 63(3) : 241-253.

Sreenivasan, A. 1963, Primary production in three upland lakes of Madras State, India. *Curr. Sci.* 3 : 130-132.

Sreenivasan, A. 1964 , : Limnological features and primary production in polluted moat at Vellore, Madras State. *Envi.Health.* 6 : 237-245.

Sreenivasan, A. 1964 b, The limnology, primary production and fish production in a tropical pond. *Limno and Ocean.* 9(3) : 391-396.

Sterner,R.W., Chrzanowski, T.H., Elser, J.J. and George, N.B., 1995, Sources of nitrogen and phosphorus supporting the growth of bacteria and phytoplankton in an oligotrophic Canacian Shield Lake. *Limno. Ocean.* 40(2) : 242-249.

Subbamma, D.V. and Rama, S. 1992, Studies on the water qualtiy characteristics of Temple Pond, near Machilipatnam, (A.P.) *J.Aqua.Biol.* 7(1/2) : 22-27.

Tilman, D. 1977, Resources competition between planktonic algae. an experimental and theoritical approach. *Ecology* 58(2) : 338-348.

Unni, K.S. 1984, Limnology of a sewage polluted tank in Central India. *Int. Revue. Hydrobiol.* 69(4) : 553-565.

Valecha, V and Bhatnagar, G.P. 1989, Primary productivity of Phytoplankton in eutrophic lower lake, Bhopal, India. *Envi & Ecol.* 7(1) : 202-205.

Vijaykumar, K. and Ravindra Paul. 1991, Diurnal fluctuations in some physico-chemical parameters of tank and moat in Gulbarga, Karnataka. *J. Nat. Con.* 3(2) : 109-113.

Vijaykumar, K. 1992, Limnological studies of perennial and seasonal standing waterbodies of Gulbarga area. *Ph.D thesis.* Gulbarga University, Gulbarga pp.160.

Vijaykumar, K. and Ravindra Paul. 1994, Effect of nutrient enrichment on phytoplankton production. *Proc.Acad.Environ Biol.* 3(2) : 171-175.

Welch, P.S. 1952, Limnology McGraw Hill Book Company, New York, Toronto and London (IInd Ed) pp.538.

West, W. West, G.S. 1912, On the periodicity of phytoplankton of some Irish Lakes *Trans.Roy.Irish.Acad.* 33b : 77-116.

Wetzel, R.G. 1983, Limnology. Sounders College Publishing Philadelphia, PA : pp. 767.

Yousuf, A.R., Masood, H., Balkhi, A and Qadri, M.Y. 1986, Limnological features of forest lake of Kashmir. *J.Zoll.Sco.India.* 38(1 & 2) : 29-42.

Zohra Jabeen. 1994, Studies on primary productivity of two tropical pond ecosystems. *M.Phil Dissertation, Gulbarga University, Gulbarga.* pp. 44.

15

A Comprehensive Study of Water Quality Parameters in the River Ganga Between Narora and Kannauj: Primary Production and Plankton

Asif A. Khan, Aftab Alam and Rajeev K. Gaur

Introduction

River Ganga after originating from snowfed mountains, flows through a long course draining an area of 8,61,404 km2 and sustain 37 per cent of the Indian population. Usually rivers are used as repositories for the disposal of sewage, industrial effluents and other wastes and river Ganga is not an exception to it (Kudesia, 1990; Krishnamurti *et al.*, 1991; De, 1993; Saha *et al.*, 1996). Human interference has reached to such an extent that the malais has put the holy river in an ecological assault (Datta and Sinha, 1993). An understanding of the hydrobiological aspects of the river is, therefore, a very important tool for monitoring the extent of pollution, in which primary production studies provide the opportunity to evaluate the capability of the system in order to compute efficiency of energy within the system and better understanding of first level of aquatic food chain.

Present communication contains an account of seasonal variations in gross—and net primary productivity alongwith total phyto—and zooplankton in the selected stretch of river Ganga from Narora to Kannauj for a period of three years.

Materials and Methods

The study was conducted in 236 km stretch of the river Ganga between Narora and Kannauj. Four sites were selected as sampling stations with their distances in kilometres down stream, namely Narora (0.0), Kachla (65.0), Fatehgarh (181.0) and Kannauj (236.0) Fig.1. All four stations are important sites for ceremonial baths during different occasions. At Kannauj, Ramganga, a tributary of Ganga river system, and a highly polluted river, Kali Nadi join the main Ganga river. In addition a sewage drain also enters into the river at Kannauj.

Samples were collected from each station during four seasons *viz.*. post-monsoon (September - October), winter (December - January), spring (March - April) and summer (May - June).

Primary production was measured by the light and dark bottle method at an incubation period of 4 hrs. (APHA, 1985). Winkler's method was followed for the estimation of dissolved oxygen. The values of dissolved oxygen were converted into carbon values by multiplying with a conversion factor of 375 and results are expressed as mgC/m^3/hr. Phytoplankton were collected in polyethylene containers from each station and 500 ml of samples were fixed Lugol's solution (1ml/100 ml). Zooplankton samples were collected by using conical plankton net made up of No. 20 botling silk and preserved in 5 per cent formaldehyde solution. Plankton counts were made using Sedgwick-Rafter cell.

Results and Discussion

Records of seasonal variations in gross— and net primary productivity of the river Ganga at all the four stations are computed in Table 1, whereas seasonal fluctuations in primary production and its relationship with phyto— and zooplankton at different stations are depicted in Figs. 2-5.

The values of GPP and NPP are generally higher at Kannauj followed by Fatehgarh, Narora and Kachla respectively in decreasing order. The overall maximum value of GPP (437.5 mgC/m^3/hour) as well as NPP (375.0 mgC/m^3/hour) were found at Kannauj during post-monsoon months of 1989, while the minimum value of GPP (75.0 mgC/m^3/hour) was recorded at Fatehgarh during winter of 1987 and of NPP (25.0 mgC/m^3/hour) at the same station in the winter of 1989. All the stations showed minimum values of GPP during winter of 1987.

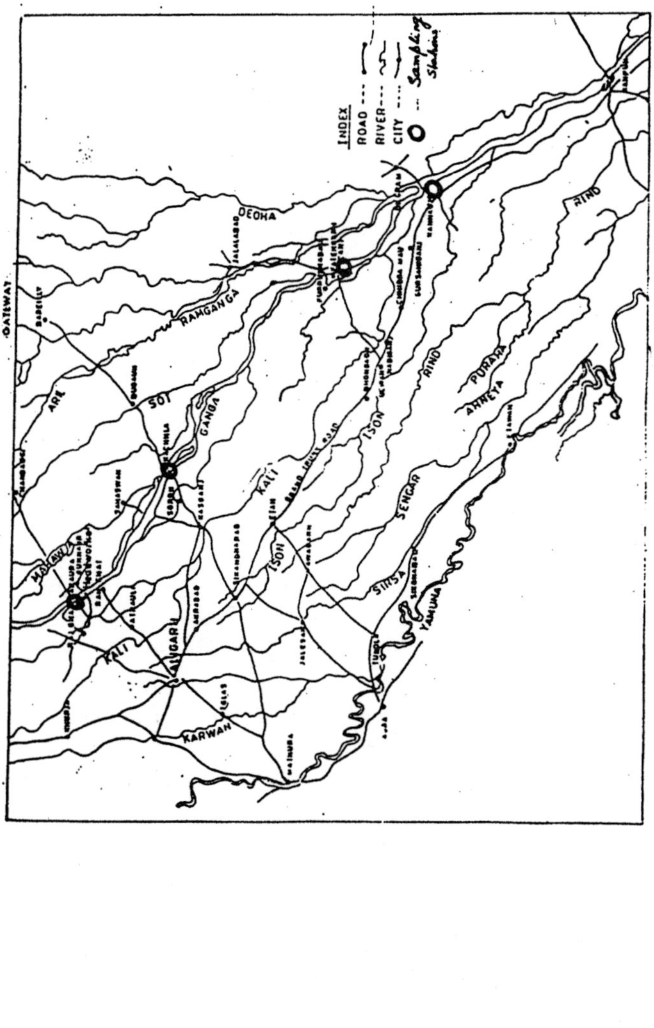

Fig. 1 : The river stretch from Narora to Kannauj showing sampling stations

Phytoplankton in the studied stretch of Ganga belonged to five groups namely, blue-green algae, diatoms, desmids, green algae and flagellates. Their number varied from 111 to 928/ml. Highest phytoplankton number was recorded during post-monsoon months with a decreasing trend during winter, spring and summer. In terms of grand averages for different stations, the abundance was highest at Kannauj followed by Fatehgarh, Narora and Kachla.

Zooplankton comprised of three groups, namely rotifers, cladocerans and copepods. Their number ranged from 29 to 624/L, being highest during post-monsoon months at Kachla and lowest at Fatehgarh during spring months.

Temperature, light and nutrients are important limiting factors for primary productivity, contributing to the great variability in its values in a given water-body. Ali and Khan (1978) have reported higher values of production in warm season and lower values in cool months while Arvola (1983) has noticed spring peak and then a decline in primary productivity in summer. Generally riverine systems are considered to be much productive due to the limiting effects of nutrients which influence primary productivity after controlling phytoplankton production (Jhingran, 1983). However, in polluted rivers an ample supply of nutrients through sewage does not permit nutrients to be a limiting factor for primary productivity which, therefore, becomes more dependent on other variables.

Attempts to measure primary productivity in terms of phytoplankton standing crops have been made by Wright (1959) who observed significant correlation between primary production and phytoplankton population. Present study also show that the total phytoplankton correspond with the GPP and NPP at all the stations (Figs. 2-5) which indicate that phytoplankton are the main organisms affecting energy assimilation process of the ecosystem and are relatively good index of productivity. Phytoplankton cells lost an appreciable amount of assimilated carbon during metabolic activities particularly through respiration and excretion (Fogg *et al.*, 1965) and due to this fact the values of GPP were always found to be higher than the values of NPP.

Zooplankton in turn are dependent upon phytoplankton, as the later forms basic food for the former and, therefore, zooplankton followed phytoplankton in abundance at all the stations during different seasons.

Table 1 : Seasonal Variations in Gross and Net Primary Production at Sampling Stations (1987-90)

Seasons		*S T A T I O N S*							
		Narora		*Kachla*		*Fatehgarh*		*Kannauj*	
		GPP	*NPP*	*GPP*	*NPP*	*GPP*	*NPP*	*GPP*	*NPP*
PMON	'87	168.75	168.75	140.63	112.50	93.75	75.0	187.5	131.25
WINT	'87	112.5	56.25	93.75	37.50	75.0	56.25	131.25	93.75
SPRG.	'88	187.5	112.5	131.25	93.75	187.5	112.5	375.0	325.0
SUMM	'88	181.25	146.75	150.0	56.25	243.75	150.0	150.0	56.25
PMON	'88	131.25	93.75	112.5	93.75	214.29	171.43	187.5	131.25
WINT	'88	150.0	112.5	103.13	93.75	200.0	162.5	262.5	243.75
SPRG.	'89	150.0	56.25	150.0	93.75	150.0	93.75	187.0	150.0
SUMM	'89	243.75	187.0	150.0	56.25	262.5	150.0	262.5	187.5
PMON	'89	200.0	125.0	150.0	56.25	281.25	187.5	437.5	375.0
WINT	'89	175.0	50.0	150.0	75.0	100.0	25.0	150.0	131.25
SPRG	'90	131.25	56.25	112.5	56.25	ND	ND	175.0	125.0

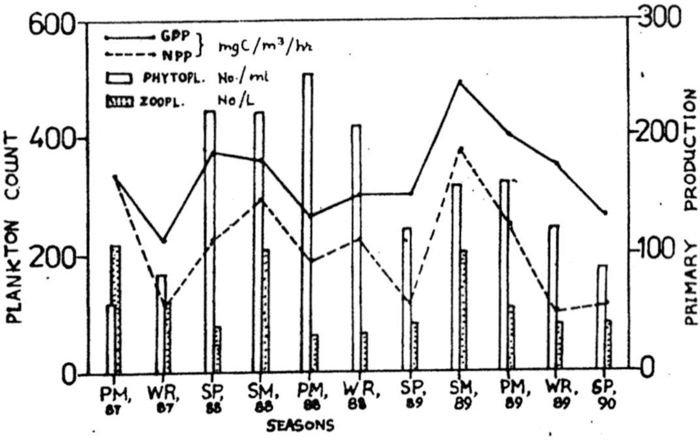

Fig. 2 : Seasonal variation in plankton and primary production at Narora.

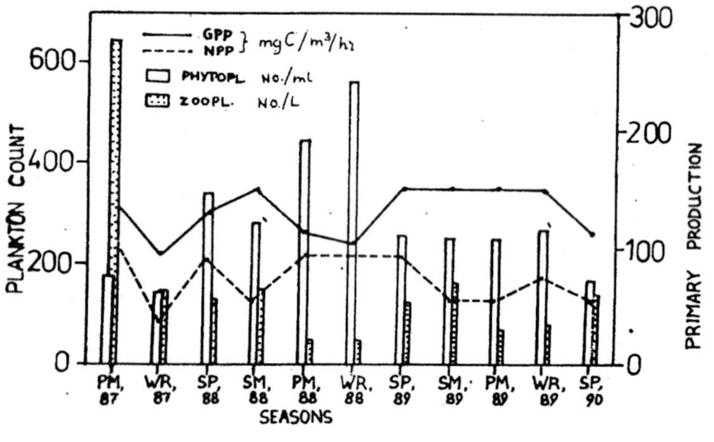

Fig. 3 : Seasonal variation in plankton and primary production at Kachla.

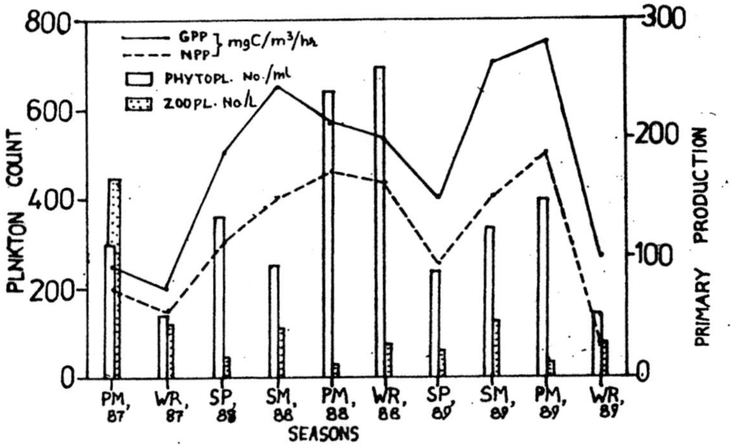

Fig. 4 : Seasonal variation in plankton and primary production at Fatehgarh.

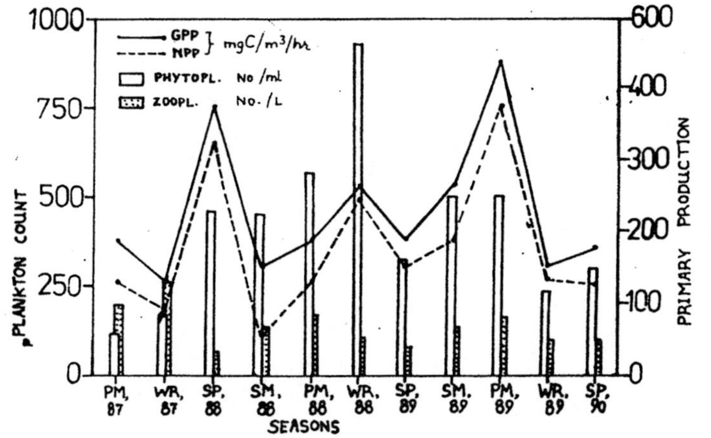

Fig. 5 : Seasonal variation in plankton and primary production at Kannauj.

Acknowledgement

Authors are grateful to Chairman, Department of Zoology, for providing facilities during this work.

Reference

Ali, M. and Khan, A.A. 1978. Studies on the primary production in two ponds at Aligarh. *J. Asiatic Soc. Bangladesh (Sc)*. 3 (2) : 93-98.

APHA, 1985. Standard methods for examination of water and wastewater 16th ed. Washington D.C.

Arvola, L. 1983. Primary production and plankton in two small polylimnic forest lakes in Southern Finland. *Hydrobiologia*. 101 : 105-110.

De, A.K. 1993. Environmental chemistry. Wiley Eastern Limited, New Delhi.

Dutta, D.K. and Sinha, G.M. 1993. Harnessing the Ganga river resources with reference to commercially viable fish species. In Advances in Limnology. H.R. Singh Ed. 347-354. Narendra Publishing House, Delhi.

Fogg, G.E., Nalewajo, C. and Watt, W.D. 1965. Extracellular products of phytoplankton photosynthesis. *Proc. Royal Soc. B*. 162 : 517-534.

Jhingran, V.G. 1983. Fish and fisheries of India. Hindustan Publishing Corporation Press, New Delhi.

Krishnamurti, C.R., Bilgrami, K.S.; Das, T.M. and Mathur, R.P. 1991. The Ganga (A scientific study). The Ganga Project Directorate, Northern Book Centre, New Delhi.

Kudesia, V.P. 1990. Water pollution. 3rd revised edition, Pragati Prakashan, Meerut.

Saha, B.K. ; Rao, R.J.; Behera, S.K. and Pandit, R.K. 1996. Diel fluctuations of some water quality parameters of the river Ganga (Rishikesh-Kanpur) during June, 1994. *Poll. Res.*, 15 (1) : 61-65.

Wright, J.C. 1959. Limnology of Canyan Ferry Reservoir. II. Phytoplankton standing crop and primary production. *Limnol. Oceanogr*. 4 (3) : 235-245.

16

Water Quality Management in Aquaculture

Amita Saxena

Air and water exists as fundamental media which divides the world into 2 major environments, terrestrial and aquatic. The media are not quite by isolated from each other however, some moisture is present almost everywhere in the atmospheres. Those characteristics of intermixture play a role in subdividing the terrestrial environment into arid and humid climate and the aquatic environment into stagnant and running aerated waters.

Air is composed of 79.1 per cent nitrogen, 21 per cent oxygen, 0.03 per cent carbondioxide and several other gases in much smaller quantities. These gases are not chemically combined but exists as a simple physical mixture. Water by contrast, consists primarily of single compound H_2O, it influences almost all the biophysiological process of flora and fauna present in it. It also influences almost all the biophysiological processes of plants. It also influences the availability and uptake of nutrient from the soil and helps to maintain the temperature and resistance to temperature change in the acquatic life because it has higher specific heat, latent heat of fusion and latent heat of evaporation than any common substance. Scientific water management aims at utilization of available water resources in aquaculture without affecting the soil health adversely. The lakes and oceans freeze only at the surface even pond rarely freeze to bottom.

Temperature may drop upto 0^0C but the metabolic reactions go on in normal fashion. In the culture of fishes because soil and water interaction taken place so importance of soil is important, soil should be free from vertical and lateral seepage such as impervious soils are ordinary heavy clay, clay loametic sites to be avoided are rocky out crops share ledge, sand, gravel and limestone areas.

A rough field method to test the soil as to its water relativity is by squeezing a moist chunk into a light ball and if it does not crumble after some handling or squeezing, it may regarded as satisfactory. In case of porous soils the pond bottom can be treated with bentosite, clay or other soil sealants. Water retention in ponds could also be satisfactory in somewhat percolative soils if the water table of areas is high. If the soil is porous and water tableton then sites are unsuitable pond. The degree of porosity determines the requirements of additional water to maintain the required levels. Except in ponds based on water soluble or jointed or faulted rocks, the seepage rate generally gets reduced with passage of time since in compacted and moistened soils the clay particles smell and fill up interestices of soil particles.

The site must have adequate and dependable source of water. The ususal sources of water supply are reservoirs, streams, springs, canal surface run off, wells tubeswells etc., stream of canals are the good source of water supply in plains, if they fill the requirement like :

(a) a flow is enough to fill the pond.
(b) it maintains a fairly constant water level.
(c) the streams are not subjected to excessive flooding.
(d) the water shed is well vegetated.
(e) stream carries only a slight salt load such that even during rainy season the water remains clear, small streams are subjected the flood during heavy rains should be avoided as they usually carry a considerable amount of silt reducing the fertility of ponds. If stream is used as a water source the ponds are built some distance away from the stream valley so that the water from the fish ponds does not seep through into the stream and even during floods there is neither any danger of pond embankments being washed off nor the farm itself being flooded. In case of excessive flow of water through the stream a high embankment between the stream and the farm may be built to control the volume of water for 8 hectare pond 5 hectare of water area is required. (The maximum utilization of the land should be made by creating 70 per cent water area).

Artisian well are used for water supply in areas where they conform to certain geological contents. Underground seepage is often used as the watersupply for ponds formed in abandoned mining pools, rock quarries; burrow pits etc. In these deep excavations, the bottom is often below the existing water table. In addition to seepage water, there is usually a considerable amount of underground drainage that enters the pond during wet period and helps maintain the water level.

Plan well and construct pond, the slope of the embankment is meant the distance in horizontal axis for each foot of height. For good quality, soil slope should be 2:1. There should be 60 cm of free board in ponds of water areas 0.4 to 1.2 ha and 90 cm of free board for larger waterbodies. In areas of heavy rain fall a 60 cm freeboard is necessary for nursery and rearing ponds also, but in the regions with a low average rainfall a 30 cm free board would serve the purpose. For every 2 cm of average rainfall, a free board of 1 cm is workable. The embankment should be 10 per cent higher than required, excluding the free board to provide for settlement of earth.

The drain pipe is highly desirable for efficient pond management such as pipes should be installed at the lowest point of every pond. The drainage pipe should be large enough for quick drainage, outlet and inlet should be made, drain pipe are asbestos, cement, cast iron, corrugated metal, fibre glass, galvanized iron, terracotta concrete tiles etc. Ponds upto 1.2 ha in area should have 153-203 mm diameter drain pipe while for larger water areas 203-305 mm pipe is necessary. Inlet pipe should be 152 types of inlets (i) simple furrow type; (ii) made of concrete with slots for sliding for a control board and screen. The control board should be strong to withstand the pressure of incoming and outgoing water compact the embankment by ramming and use caustic soda, sodium carbonate in combination raw cowdung benetonica apply to control. A seepage mechanical sprill way is necessary for all block ponds with water sheds of over 4 ha.

The newly constructed ponds generally have muddy water for considerable length of time. Some ponds may be muddy indefinitely due to (i) type of soil or (ii) the activities of burrowing animals or bottom feeding fish, clay particles from certain soil types tend to remains in suspension indefinately to clear such waters of these clay particles agriculture gypsum which is hydrated calcium suplhate and has been reported to have no effects either on the water or fish life is recommended to be applied at the rate of 192 kilogram per 1000 cubic meter of water volume by scattering the chemical evenly over the entire pond surface.

Plantation along embankment has manifold benefits such as :

(i) it decreases soil erosion which increase age of pond;
(ii) provide shade for fish;
(iii) reduce variation in water temperature;
(iv) alling leaves act as measring;
(v) Giving crop and beautification.

Depth

Depth influences the physical and chemical properties of water. A shallow pond allow more light penetration and increases productivity. 1 to 1.5 meter depth is essential for survival of fish in summers, excessive heat causes water evaporation.

Shore Conditions

The length and the nature of shore water plays a significant role as it enhances the productivity. But shady shore trees, surface and submerged plants and turbidity due to still lower the productivity by cutting the light.

The density representative of pure water is 1.000 g/cc at 4^0 C while that of pond water is 1.001 g/cc. The density representative of sea water (at 35 ppT salinity) is 1.0289 g/cc and protoplasm is 1.028 g/cc. the density of protoplasm is closely similar to that of sea water and only slightly greater than that of freshwater. The density influence differences in pressure inertia, viscosity and mobility of media. The pressure is same inside the body as outside it support and resistance to motion in air is less than that of water since water has nearly the same density as protoplasm, where as in air it is very much less dense, water furnishes much more buoyancy in air.

The coefficient of viscosity of water is 60 times that of air at the same temperature because of relatively great inertia of water medium, some animals such as jelly fish and scallops can propel themselves in our direction by pumping water in opposite direction. The squid can dart backward with remarkable rapidity by ejection water from its siphon.

Variations in temperature has an important influence on all the organisms including fishes. How temperature reduces metabolic activity. Feeding, breeding, respiration and all other physiological activities are influenced by temperature so cold water fishes cannot survive in plain. Temperature also effect the chemical changes in soil and water and the content and pressure of dissolved gases. Oxygen contents in water decreases with rise in temperature.

As temperature reduces below optimal growth slows down because the rates of metabolic reactions decrease. Quick decreases in temperature are often fatal to aquatic organisms although many can adapt slow reductions from coastal temperature environments where natural seasonal changes in the temperature may be drastic and it is beneficial such as.

Temperature can also affect the chemical composition of an organism. The saturation of fatty acids is known to vary with the temperature and although this may not be concern in most cases it can be critical when raisin feed organisms going to support a primary crop. At the same time use of thermal effluent in commercial lobster culture is suggested.

Movement of water also causes erosion of soil and rain floods increased turbidity affecting fish production. Turbidity is due to suspended matter like salt and clay or due to planktons Indian major crops are able to tolerate relatively high turbidity. Fishes live permanently in turbid water and show various adaptations like reduced eyes and mucus is capable of coagulating suspended matter to protect the gells light varies with geographical location of pond. It affects the turbieity in shallow ponds light reaches upto the bottom and causes heavy growth of vegetation. Penetration of light is responsible for distribution and production of plantation and photosynthetic activity of plants and dissolved oxygen of the pond. It is essential for photosynthesis. In the presence carbonic acid, lime stone dissolves forming calcium bicarbonate Ca (HCO_3) which influence the organic matter of water.

pH

pH of water of pond may bė acidic, basic or neutral, we have to make adjust with time pH is not constant and varies in relation to other chemical in water pH being alkaline in afternoon and most acidic in morning. The concentration of hydrogen ion H^+ refers to the number of moles of hydrogen ions in a latter solution.

$$pH = -\log H \text{ or } H^+ = 10 \quad pH$$

Alkalinity

Alkalinity commonly results from CO_2 and water attaching sedimentary carbonate rocks and dissolving out of some of the carbonate to form bicarbonate solution Alkalinity, measures the buffering capacity of water and content of CO_2 in its various forms are involved in this carbonate-bicarbonate-carbonic acid buffering system. Alkalinity was found to increase in water containing hyacinth plants.

Hardness

Hardness depends on the amount of calcium and magnesium salts dissolved in water. The hardness may range from zero to hundreds of mg/litre in terms of $CaCO_3$ depending on source and treatments to which the water has been subjected hardness increases in water in presence of aqualicplants.

The chemical conditions depends upon the several chemical substances of which dissolved gases like oxygen CO_2, inorganic salts and nutrients. Henry's law stand as :

$$C = k_g$$

where C — Concentration of gas in solution.
— partial pressure
— constant of solubility

The solubility of gas depends upon the following conditions:

(i) Rise in temperature reduces solubility.

(ii) Rate of solubility is greater when the gases are dry than when they contain water vapour.

(iii) Exchange of gases takes place due to water action and agitation.
(iv) Increasing concentration of dissolved salts diminishes solubility.

Carbondioxide

It is present in water through respiration from atmosphere as partial pressure CO_2 in air is low the amount which remains in solution in water at given temperature is low.

Dissolved Oxygen

During day time plants consumes CO_2 and release O_2 while at night, they consume oxygen and release CO_2 through respiration. All animals consume oxygen during respiration to the balance of O_2 is maintained. Excess of oxygen due to over CO_2 assimilation causes mortality due to over metabolism. Defficiency of O_2 causes infection and attack of parasite.

Decomposition of organic matter - Aerobic bacteria use up the oxygen of water while decomposing organic matter. Chemical oxidation of sediments also takes place. In nature gases like CO_2, methane and H_2S often accumulate in large amounts and excess amounts rise in the form of bubbles removing the dissolved oxygen. Significant increases in heat are also often fatal because that are quick decrease in DO while at the same time raising BOD by increasing the metabolism of organic.

In the natural environment there is normally enough oxygen in water to support the animals that are living there. But under aquaculture conditions especially in intensive culture we often find ourselves with very unnatural conditions when the demand for oxygen by the fish or other organisms is extremely high because

1. The cultured organisms are stocked in high densities.
2. Microbial breakdown of excess wastes in the pond phytoplantation macroalger and higher plants also require O_2.

Although aeration is the normal means used to increase the amount of oxygen in pond farmers sometimes add calcium hydroxide or potassium permaganate when the threat of fish kill

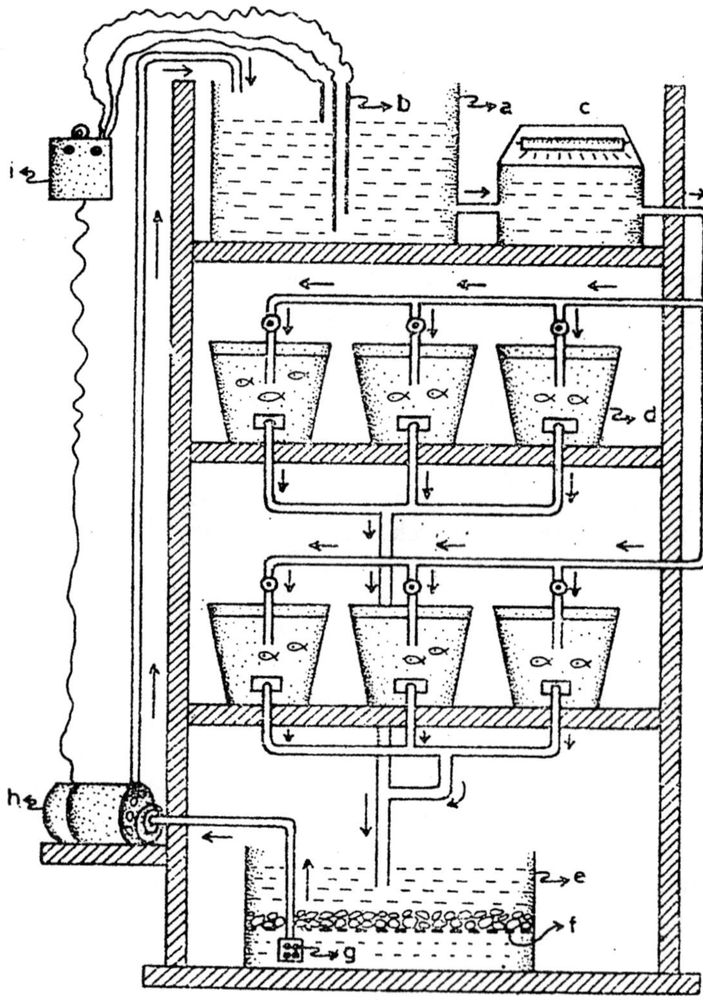

Fig. A : Indoor water recirculation system.

(a) Upper reservoir; (b) Probes; (c) U.V. Chamber; (d) Experimental trough; (e) Lower reservoir; (f) Biological filter; (g) Foot valve; (h) Pump; (i) Water level controller.

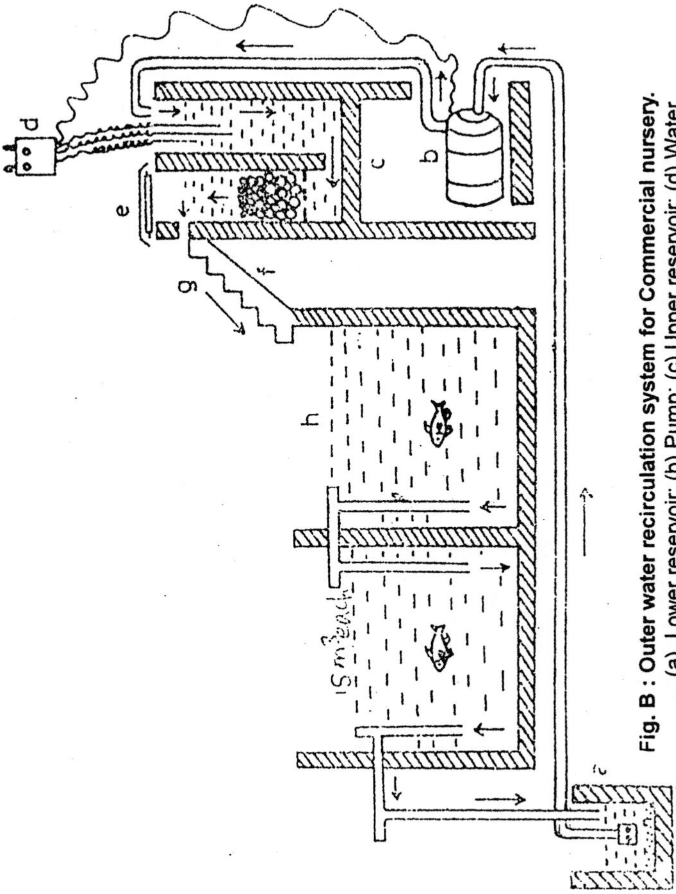

Fig. B : Outer water recirculation system for Commercial nursery.
(a) Lower reservoir; (b) Pump; (c) Upper reservoir; (d) Water level controller; (e) U.V. Light; (f) Biological filter; (g) Cascader; (h) Fish tank.

os bear because these will oxidise the organic matter and lower the demand for dissolved oxygen.

One can have the oxygen supply system by compressor and pump and the oxygen from outside/atmosphere and get the pond water agitated by which it can maintain O_2 level. Fig. A and B at 3-60 0' clock A.M. is the mortality phase. Abnormally 14-16 ppm oxygen is present in the fish ponds and 22-24 ppm is the super saturated condition but best level O_2 is 7-8 to 14-16 ppm. Above the below that is change of stress. Animal understress needs more oxygen. Approximately 10±2 ppm is maintained by pumping water.

Perforated pipes are fitted 1 feet below the surface of water. If it is filtered at the bottom the metabolities get agitated and more oxygen is required.

The presentation should be in a spiral shape wheel or paddle may also be used at the pipe this give churning .

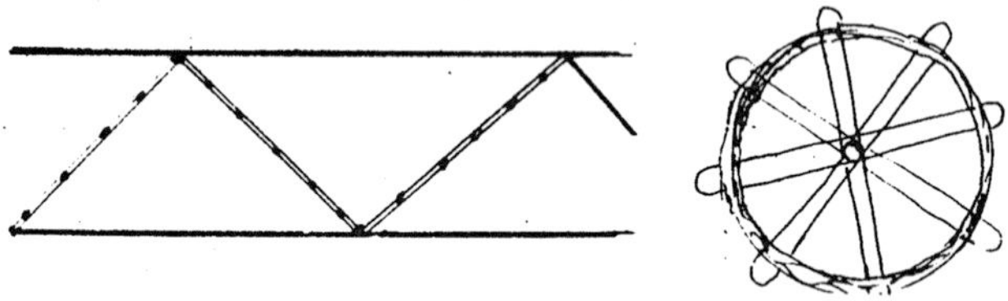

Fig. C : Spiral shape wheel or paddle

The water containing metabolities are pumped into biological filter fitted on the embankment. Then this water is put on stars so that atmospheric oxygen get mixed and the water get oxygenated, then use the water for pond. The water put into pond with shower system and water should through 3-4 feet height. Andhra Pradesh people are putting a straight pipe 1-2 meter above the water surface in the middle of tank alongwith length. The water is pumped from same tank or from supply water based on the size of tank.

Accumulation of metabolities like NH_4. Ammonium sulphate, H_2 result in increase in BOD which hampers fish pond. The methane bacteria are obligate anaerobes. They decompose organic compounds with production of CH_4 in marshy ponds and eutrophic lakes during summer.

$$CH_4 + 2O = CO_2 \pm H_2O$$

Hydrogen sulphide is toxic aerobic organism because it inactivates the enzyme cytochrome oxidase.

Accumulation of metabolities like NH_4, Ammonium sulphite, H_2S result the increase in BOD which hampers the fish production.

If the bottom is agitated and metabolities are allowed to come up. Due to agitation they get oxidised. At the time of agitation, BOD increases supply proper oxygen to enhance the growth but this process is not so easy. In the traditional ponds this method was adopted to bath cattles and other animals if by net, agitation is done the fish disturbs 40-50 per cent are recycled and 50-60 per cent remain there and get settled again.

Nitrogen is used by organisms in a large number of chemical reactions and is an essential components of aminoacids, the building blocks of proteins. Nitrogen can occur in two oxidation states +5 and +3 and found in nutrients forms of

(i) Ammonia or ammonium (NH_3 or NH_4 respectively).
(ii) Nitrite (NO_2).
(iii) Nitrate (NO_3).
(iv) Organic molecules.

It is important for certain nitrogen fixing organisms (Cynobacteria). In most established ponds there is a natural nitrogen cycle that will convert the organic nitrogen to a form that plants can assimilate. Nitrogen compounds can also be toxic to plants and especially to animals when these compounds are present at high concentration. Besides certain organic nitrogen compound is ammonia. Ammonia and less toxic ammonium are in an equilibrium that depends upon pH; the more acidic water is that the more free hydrogen ions that are present, the more the equilibrium shifts from NH_3 to NH_4^+. As temperature rises but as salinity increases, this ratio decreases in general temperature and salinity are less important than pH in determining the form of the molecule. For trout fingerling the tolerance for ammonia in water is 30 per cent less at a DO level of 5 mg O_2 than at 8.5 mg O_2/lit since nitrite and especially nitrite are less toxic than ammonia and because they are found in lower concentrations in culture systems they are generally of less concern to culturiest than is ammonia. They can cause unwanted algal blooms and nitrite is known to oxidise hemoglobin in fish blood to methemoglobin. Thereby reducing the blood's ability to take O_2 as water passes over gills. Phosphorus although is present in natural water in less quality a critical factor in the maintenance of pond fertility. Phosphorus is necessary to all living organisms as it is the constituents of ATP (adenosine triphosphate), the energy carrier and present in nucleic acid as phosphate groups of nucleotide.

Phosphorus is available in water combined with a number of ion the more common forms being phosphates of iron and calcium is varies from less than 1 mg/m^3 to very high. Main supply of phosphorus in natural water comes from the weathering of phosphorus bearing rock and from leaching of soils.

The efficiency of phosphate fertilizers in relation to pond productivity are found triple superphosphate (40 percent P_2O_5) to be more efficient in maintaining a higher level of single phosphorus in water phase which single superphosphate (16 per cent P_2O_5). Maintained a higher level of available phosphorus in soil phase. Triple superphosphate at 25 kg/ha when applied to slightly acidic soil (pH 6.0) gave the best results in terms of survival and growth.

Various nitrogen fertilizers are utilized for pond management.

(i) **Chile saltpetre** (sodium nitrate) : Chile sulphate has about 16 per cent nitrogen and readily soluble in water. The optimum doze for the use of saltpetre is 200 kg/ha.

(ii) **Ammonium sulphate**: It promotes the growth of phyto and zooplankton, bottom biota and fish to some extent at the rate of 400 kg/ha.

(iii) **Ammonium Nitrate** (Norway Saltpetre): It contains 20 per cent pure nitrogen in ammonium form and 15 per cent nitrogen in nitrate form and promotes the growth of bluegreen algae.

(iv) **Ammonia Liquor** : 20 per cent nitrogen is present in it and affects as ammonium sulphate. The ammonia raises the pH of pond immediately applied 450 kg/ha.

(v) **Acidic Ammonium Carbonate** : This salt has 16 per cent nitrogen and 50 per cent CO_2. It is applied for monoculture of green algae 400 kg/ha gives best results.

(vi) **Urea** : It has 45-46 per cent pure nitrogen. Apply 200 kg/ha/year.

(vii) **Calcium Ammonium Nitrate** : The predominant compound of calcium is $CaCO_3NO_2$. Thus there is an inseparable relationship between carbonic acid, CO_2, pH and anion CO_3 in water. Calcium Ammonium Nitrate is a granular fertilizer containing 15.5 per cent nitrogen and 27 per cent CO, Of is application at the rate of 500 kg/ha was the best result.

Potassium Fertilizers, potassium is not limiting factor in fish pond. In fish extracellular Na^+ often surpasses potassium . There is some evidence that highly concentrated water with a pronounce potassium content are lethal to aquatic fauna, the Na/K ratio being less than ten.

Mureate of Potash 38-62 per cent, potassium sulphate 48 per cent, potassium nitrate 44 per cent an kainit 12-16 per cent, potassium oxide are important fertilizers soluble in water. The optimum doze of nutrient of potash in fish pond is 150 kg/ha/year.

Magnesium Fertilizers

Magnesium is beneficial in fish ponds, where the soil or land drainage is deficient in the element or in situations in which the level of other element has raised to point where Mg becomes factor tunnting further increase in fish production Forsterite (Mg_2SiO_4) can be changed by carbonic acid to form silica, carbonate and serpentine ($H_4MgSl_2O_9$) $2Mg_2SIO_4 + 4H_2O + 4CO_2$ $2H_4Mg_3Sl_{12}O_9 + 4\ MgCO_3 + SIO_2$. Dolomite Ca Mg $(CO_3)_2$. Eposomsalt ($MgSO_4.7H_2O$). 150 times more suluble than gypsum ($CaSO_4.2H_2O$) Mg salts are useful in peaty muds. Sitca is utilised most by diatoms and a major decrease in abundance occurs in epitimnetic water following diatom blooms. Abundance of silica is regulated by redox potential and is possibly caused by reduction of ferric silicate formed in superficial layers of bottom mud when the latter comes in contact with oxygenated water.

Trace Element Fertilizers

The effect of trace elements such as Zn, Mn, Fe, Cu, Bo, Mo, I_2 etc. occur in very small amount with a great significant in parting productivity to waters. The use of Zn, Mo, Co, Cr, Ba, Se and Sn at 0.05 ppm and that of Zn, Mo, Co and Bo at the rate 0.5 ppm increases the plankton production.

Organic Manuring

(1) *Liquid manure* : At the rate of optimum doze of 300 kg/ha/ year. It stimulates the growth of phytoplankton and hardwater plant.

(2) *Farmyard manure* : Cow and pig dungs are commonly used. It conditions the soil. The organic matter in the manure helps in formulation of pond mud, promotes nitrification and formation of a good colloidal structure.
11,000-22,000 kg/ha/year. Half of it is added 15 days before stocking is of fish seed.

3. *Green Manuring* : Its importance are :
   - (i) it supplies organic matter, which besides acting as food for soil organism induces microbial activity;
   - (ii) forms humus;
   - (iii) restores soil nitrogen in creates conserving influence on soil nutrients;
   - (iv) increases the availability of phosphoric acid in soil
   - (v) promotes the growth of periphytons which act as food stock in heap with a mus topping to prevent their drifting away and checking deoxygenation in pond water 1650 kg/ha green manure mixed with dry cowdung, stable refuse, poultry manure and oil cake gives good results.

   Cowdung : 20,000 kg/ha/year cowdung and 600 kg/ha NPK in fish pond in installment necessary for biological activity productivity of water.

   Compost : At the rate of 5000 kg/ha/year gives good productivity.

Ideal Quality of Water

pH	-	6.8-8.6 slightly alkaline pH is preferable
Temperature	-	20^0 - 30^0
Dissolved O_2	-	6 - 8 ppm

CO_2	-	15 ppm
Phosphate	-	0.2 - 0.4 ppm
Silicate	-	0.05 - 0.1 ppm
Calcium	-	52 ppm
Bicarbonate Alkalinity	-	About 100 ppm
Ammonia	-	0.2 ppm
Chlorine	-	0.003 ppm
Seepage	-	1 meter/year

Application of lime according to soil pH for pH maintenance.

pH		**Lime (kg)**
4 - 5	-	2000
5 - 6	-	1000
6 - 6.5	-	800
6.5 - 7.5	-	200- 250

Prod-ucti-vity	*pH*	*Organic carbon per cent*	*Available N_2 Mg/100 gm soil*	*Phos. phorus*	*Nutr. Requr.(Kg/ha) C*	*N*	*P*
Low	5.5	0.5	25	3	600-720	200-250	100-150
Med.	5.5-6.5	0.5-1.5	25-50	3-6	480-600	150-200	75-100
High	6.5-7.5	1.5-2.0	50	6	300-480	100-150	50-75

Total Fertilizer Required kg/ha/year.

Produc-tivity	*Cowdung*	*Ammonium or sulphate*	*Urea*	*Singlesuper phosphate*	*Triplesuper phosphate*
Low	22000	1500	670	845	350
Medium	12000	1000	440	585	236
High	10000	750	335	845	175

If phosphate and available nitrogen are 0.25 and 2.0 ppm then there is no need of P and N fertilizers.

17

Biomonitoring of Water Quality of River Ramjan (at Kishanganj) in Relation to its Impact on Biological Components

B.N. Pandey, P.K.L. Das, S. Vatsa Dubey, & S. Hussain

Introduction

Water is the essential ingredient for the survival of organisms. Next to air, water is the most essential requirement for human life. Besides it is also needed for developmental needs, may it be irrigation, hydro-power, industries, municipal supplies, fisheries and recreation etc. Its availability in nature gradually in terms of quantity and quality of usual water for diverse purpose is scarce and nearly constant. Fresh water is essential for sustenance of life, human, animal and plant. About 65 per cent of human body is water, 9 per cent of it is to be replaced everyday. Water serves as medium in which the plants and animals of lakes and characteristics are gradually affected due to discharge of domestic, municipal, industrial and other several factors (Sahu *et al.*, 1994 and 1996) during day and night which may be fatal for aquatic life.

Rivers have been most important source of water. The river ecosystem receives water primarily from their water sheds, marginal run off and domestic sources. These waters contain excess of organic matter, excess of nitrogen and phosphorus, suspended solids, chloride and lot of garbage, effluent and sewage. Urban rivers are polluted with discharge from sewage treatment plants and overflowing sewage caused by rainfall. These waste waters accelerate eutrophication of aquatic environment. The pollution has become a serious problem through out the world. It is unfortunate that the rivers are being increasingly used as natural dustbin for discharge of wastewater. India has 14 major rivers along with their numerous tributaries. Studies show that most of these rivers are polluted to a great extent by domestic sewage and industrial wastes (Ajmal *et al.*, 1984; Kar *et al.*, 1987; Singh *et al.*, 1988 Kulshreshtha *et al.*, 1988; Shukla *et al.*, 1989; Sinha *et al.*, 1989, and Goel and Atude 1995). The importance of limnological studies is well realised in view of its value in estimating the productive nature of waterbodies. The limnological characteristics of water of different regions are of immense practical value in fish culture programmes. References on lotic environments are comparatively less in India. However, there have been some systematic studies on the physico-chemical and biotic factors of running water system of Bihar (David and Roy, 1966; Pahwa and Mehrotra, 1966; Bilgrami and Siddiqui. 1983; Dora and Roy, 1987; Towheed *et al.* 1988; Jhingran *et al.*, 1989; Pandey *et al.*, 1992 & 95). However, our knowledge on the limnological aspect of river Ramjan is still meagre. Keeping this fact in mind, the present work was undertaken which deals with the limnological aspect of river Ramjan.

Study Area

Kishanganj is a district (25°54'0" to 26°35'0" North latitude and 87°38'12" to 88°12'3" East longitude) of the Kosi division of the Bihar. The climate is monsoon having an annual rainfall of 2000 mm. Formerly it was popularly known as 'Kala Pani' and maleria was prevalent. The distance of Bangladesh, Nepal and China from Kishanganj town is 23 km, 30 km and 204 km respectively. Its population is about 9,85,993 having an area of 1938.4 sq. miles. River Ramjan is the main river of the district.

The River Ramjan

Inspite of impressive work on river ecosystem, nothing is known about the limnology of river Ramjan of Kishanganj. It originates from the left side of old channel of Dok river near east of Belwa (15 km) north from Kishanganj and terminating to Sudhani river below Majhia a bridge (7 km from Kishanganj). It receives drainage and garbage of whole municipal and Sarogi Paper Mills are also being discharged into this river. The river is rich repository of flora and fauna and grow fish in abundance.

Material and Methods

The present investigation was carried out in the river Ramjan at five sampling stations (Fig. 1) selected in the stretch of 15 kms. The water samples were collected at the monthly intervals between Jan. 1996 and Dec. 1996 from each sampling stations in a polythene bottle of 3 ltrs. capacity. Water temperature was recorded with an ordinary mercury thermometer graduated from 0°C to 50°C. The analysis of pH, dissolved oxygen, free carbondioxide, carbonate, bicarbonate, chloride and phosphate was done according to the methods of APHA (1975) and Trivedy and Goel (1984).

For the plankton, 50 ltrs. of river water was taken and filtered through the boilting silk plankton net (No. 30) with 77 mesh/sq. cm. The samples were taken from different areas of the stations regularly. The plankton concentrate on the plankton net was preserved with 5 per cent formalin. Separation and counting of the plankton was done by taking 1 ml. of sub-sample into a Sedgwick Rafter plankton counting chamber of 1ml capacity. All organism were counted according to the procedure described by Welch (1952).

RESULT AND DISCUSSION

The physico-chemical analysis of the river is shown in Table 1.

Temperature

It is well known fact that the physico-chemical analysis of the river is shown in table 1, the water temperature exerts direct as well as indirect influence on many abiotic and biotic compositions of aquatic ecosystem. Temperature has variable effects, it may kill some organisms and may stimulate the growth of other (Kotpal and Bali, 1989). In the present investigation, the water temperature followed more or less similar trend of the air temperature. The water temperature varied between 19.5 - 30.7 °C.

pH

Changes in pH concentration of water will bring about suitable changes in the functional and as well as structural variation in the organism of water body. Growth of algae and

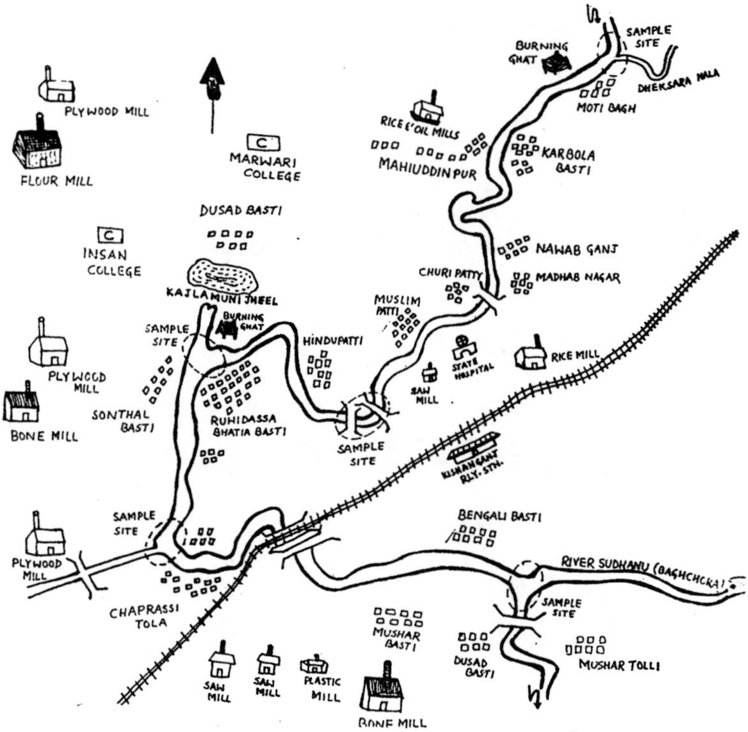

Fig. 1 : An outline map of Kishanganj showing the sampling sites in the river Ramjan

Table 1 : Physico-chemical analysis of river Ramjan studied during January 1996 to December 1996

Month	Temp °C		pH	Trans	Do	Free Co_2	Co_3	Hco_3	No_3-	Po_4	Cl-
	Air	Water									
Jan.	22.5	19.5	8.5	72.4	9.4	0.9	0.0	88.5	0.35	0.09	24.5
Feb.	24.6	19.9	8.4	69.3	9.2	0.8	0.0	88.6	0.40	0.07	22.5
Mar.	28.9	25.6	8.1	55.6	9.0	1.0	0.0	87.2	0.42	0.80	27.8
Apr.	32.8	23.1	7.9	49.9	8.8	1.2	0.0	90.9	0.48	0.11	28.4
May	34.9	28.4	7.5	45.7	8.4	1.5	0.0	99.5	0.52	0.12	30.5
June	35.8	29.6	7.2	42.5	8.2	1.9	0.0	70.2	0.54	0.14	32.4
July	35.9	30.7	7.5	38.2	7.9	1.12	0.0	60.1	0.00	0.16	32.1
Aug.	34.6	28.9	7.3	53.1	7.7	1.15	0.0	52.2	0.57	0.16	28.2
Sept.	33.4	26.6	7.4	29.2	6.9	1.16	0.0	49.5	0.60	0.17	26.1
Oct.	31.8	23.2	7.6	34.5	6.8	1.2	0.0	50.6	0.62	0.15	25.2
Nov.	29.1	21.2	7.8	46.5	7.9	1.1	0.0	55.7	0.45	0.12	26.1
Dec.	24.1	20.1	8.0	62.3	7.7	1.0	0.0	61.8	0.44	0.11	24.1

macrophytes depend upon ionic composition of water body. The variation in pH is often linked with the species composition and life processes of animal and plant communities inhabiting them. pH of the river varied between 7.2 to 8.5. Thus the river was alkaline throughout the year. The highest value of pH during winter appeared to be influenced by low water level, number of phytoplankton and the highest value of oxygen (Shyam Sunder, 1988; Pandey *et al.*, 1993). The low value of pH during monsoon might be due to high turbidity and elevated water temperature. Klein (1973) has pointed out that the pH values between 5.0 and about 8.8 are detrimental. Thus the present value of water may be considered suitable for fish production.

Dissolved Oxygen

Dissolved oxygen is one of the most important parameters in water quality assessment and reflects the physical and biological process prevailing in the water. Any water polluted by organic wastes suffer from decline in dissolved oxygen (Butcher, 1947). In the present investigation the dissolved oxygen varied from 6.8 - 9.4. The high value of oxygen during winter was due to abundance of vegetation. The low value of oxygen during rainy season might be due to higher growth of bacterial which utilized oxygen for metabolic activities. However, Shukla and Bais (1990) and Vijay Kumar (1995) have reported higher concentration of oxygen during summer season and have suggested that higher value of DO_2 is due to increased radiations.

Free Carbondioxide

Free carbondioxide is an extremely necessary constituent in an aquatic ecosystem (Welch, 1952). The respiratory activity by aquatic organisms produces considerable amount of CO_2 in the water, which dissolves partly in water as carbonic acid and part of it diffuses to atmosphere. The free CO_2 value of the river ranged from 0.8 to 1.9. The high value of CO_2 during rainy season might be due to high rate of composition of organic matters by bacteria resulting in rapid production of CO_2. High CO_2 content is indicative of high pollution (Todda, 1970;

Cole 1979). Verma *et al.*, (1988) suggested that high carbondioxide content influences the amount of dissolved oxygen and affects the number of aquatic organisms.

Turbidity

Turbidity restricts the penetration of sunlight and hence reduce photosynthetic activity which is turn is related to the productivity of water mass. The suspended particles causing turbidity may also absorb considerable amount of nutrient element like phosphorus, nitrogen, potassium in their ionic forms making them unavailable for plankton production. Turbidity makes the water unfit for domestic purposes. In the present study the turbidity varied from 29.2 to 72.4. The high turbidity during rainy season may be attributed to the land drainage contributing to increase in particulate suspended matter. Roy (1955) reporting on the hydrology of an irrigation tank from Bangalore, Karnataka opines that high turbidity adversely affects primary production. The highest transparency during winter season was due to close competition between macrophytic vegetation and phytoplankton. Kaul *et al.* (1972) and Wetzel (1975) found that macrophytes were capable of storing nutrients, suspended in water in more quantity that the phytoplankton could do. Zutshi and Vass (1973) have similar observations while studying the limnology of Kashmir lake.

Alkalinity

Alkalinity of water is capacity to neutralise a strong acid and is characterised by the presence of all hydroxyl ions capable of combining with hydrogen ions. Rai, (1980) has reported that waters, rich in free CO_2 are comparatively less alkaline where as water deficient in the gas are more alkaline. In the present investigation carbonate alkalinity was absent through out the year. Bicarbonate alkalinity ranged from 49.5 - 99.5. Bicarbonate alkalinity was highest during summer and lowest during rainy season. On the basis of alkalinity value of the river water can be classified under high category of nutrient type after the classification of Philipose (1959) as the alkalinity value 100 mg/l. On the basis of Moyle (1946) be termed as hard water type.

Chloride

Chloride is generally undertaken as major factor to aquatic cation-anion balance of ecosystem. Chloride content in the river under investigation was found highest during summer and lowest during winter. The highest value of chloride during summer months might be due to gradual decrease in the amount of water and increased amount of excreta laid down by various aquatic fauna (Cole, 1979; Pandey *et al.*, 1995). The low value of chloride during rainy season might be due to increased amount of water (Jana, 1973; Swarananlatha, 1994).

Nitrate

Nitrate parameter is an excellent parameter to judge the organic pollution. Thresh *et al.* (1994) have attributed the nitrogen richness of freshwater body to the pollution of animal origin. The level of the total nitrogen concentration is an index of carbon budget of a freshwater body (Munwar, 1970; Pillai and Serrnivasan, 1975). In the present study the nitrate value ranged from 0.35 to 0.62. The low value of nitrate during winter season might be due to high macrophytic and phytoplankton production. Lee *et al.* (1975) have indicated that in plants, the decrease in nitrogen is probably associated with entire uptake of this element by these plant communities. The high value of nitrate during rainy season might be attributed to rain showers, decompositon of organic matters and influx of flood water (Rao and Govind, 1966).

Phosphate

Among the various micro-nutrients phosphorous is found to play an important role in governing the primary productivity of any ecosystem (Hutchinson, 1957). The primary concern to phosphorus lies in its ability to increase the growth of nuisance algae and eutrophication. In the present investigation the phosphate value was usually low, the minimum being 0.07. The minimum value of phosphate during winter may be due to abundance of phytoplankton which utilize it. Such findings have also been reported by Kataria *et al.* (1996). The maximum value of phosphorous during rainy season might be due to agricultural run off which brings phosphate from the catchment areas.

Phytoplankton

Phytoplankton is an important component of aquatic flora and play key role in maintaining proper equilibrium of aquatic ecosystem between abiotic and biotic component of aquatic ecosystem. They serve as food for fishes directly or indirectly and are the primary producers of phytoplankton aquatic ecosystem.

The phytoplankton communities of the river were represented mainly by four groups of algae.

(a) *Chrolorophyceae* : Chlorophyceae was encountered as the most significant group of phytoplankton with a contribution of 37.60 per cent (Table 2) to the annual total production. It exhibited maximum density in the months of May, April and December and least in September (Fig. 2A &B). The group includes *Spirogyra* sp., *Clostridium* sp., *Chlamydomonas* sp., *Volvox,* Netrium, *Pandorina Zygnema* sp., *Chara* sp. and *Pediastrum* sp.

Table 2A : Monthwise distribution of Phytoplankton groups in Ramjan River (cells/lit.)

	Phytoplankton			
Months	*Chloro-phyceae*	*Myxo-phyceae*	*Bacillario-phyceae*	*Eugleno-phyceae*
January	1670	1370	625	165
February	1120	1475	645	185
March	1295	3170	580	210
April	1585	2055	395	180
May	2475	1795	860	275
June	1375	1585	175	315
July	1290	1075	450	170
August	1085	1245	810	185
Ssptember	695	575	525	325
October	1575	1380	695	475
November	1380	995	595	385
December	1875	1580	775	650
Percentage	37.60	39.43	15.36	7.58

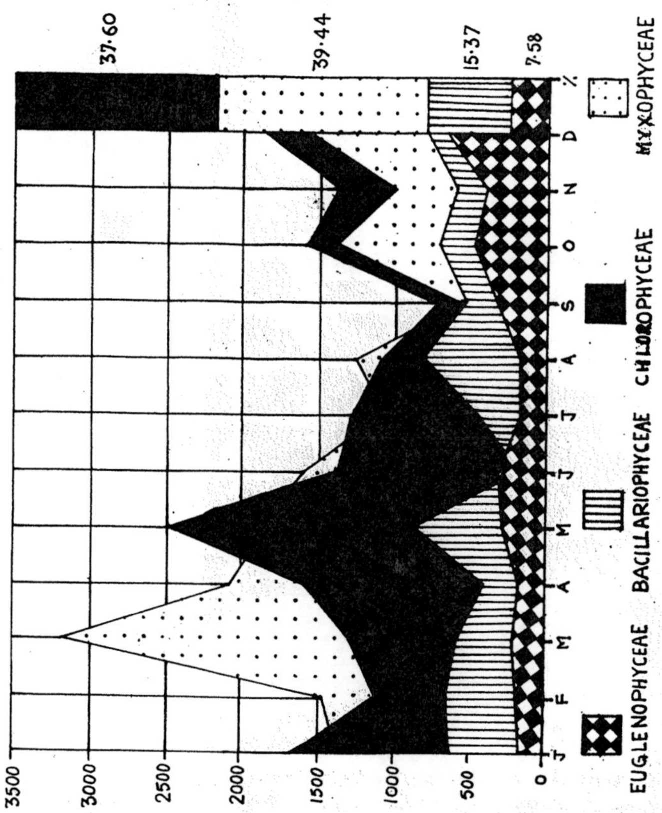

Fig. 2A : Monthwise distribution of Phytoplankton

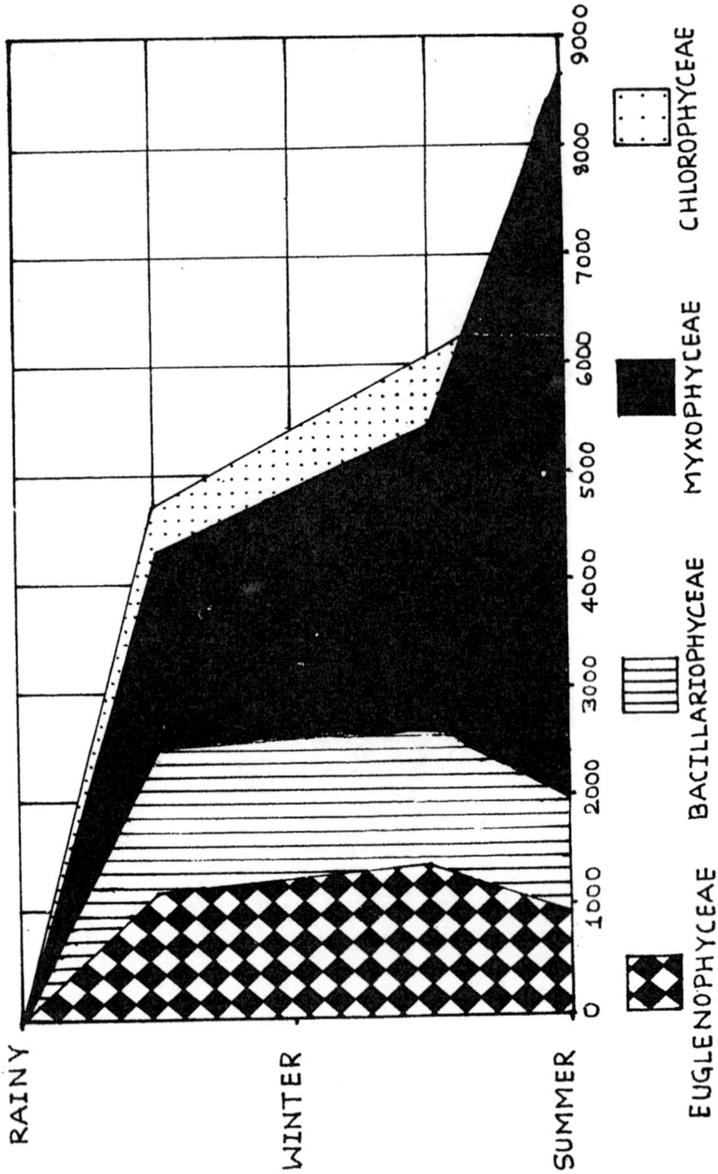

Fig. 2B : Seasonal variation of Phytoplankton

Table 2B : Monthwise distribution of Zooplankton groups in Ramjan River (cells/lit.)

	Zooplanktons		
Months	*Rotifera*	*Cladocera*	*Copepoda*
January	280	150	125
February	210	120	90
Mmarch	460	135	80
Aapril	480	750	150
May	375	550	160
June	305	290	95
July	290	190	98
August	310	185	75
September	335	150	80
October	280	155	85
November	270	165	76
December	375	150	120
Percentage	48.45	36.49	15.06

(b) *Myxophyceae* : It was the most significant group of phytoplankton having a contribution of 39.44 per cent (Table-2A & B) to the total population. It exhibited maximum density in the months of March and April and least in September (Fig. 2A &B). This group includes *Anabaena spherica, Oscillataria limentica, Microcystis robusta, M. aeruginosa, M. protecystes Gloeocapsa, Formidium fragilaria* and *Meriosomopedia.*

(c) *Bacillariophyceae* : It accounted for a contribution of 15.36 per cent (Table 2A) of the total annual phytoplankton production. Its maximum density was noticed during May and August and least in June (Fig. 2A &B). This group includes *Synedraacus, S. splendenis, Navicula phygamea, Melosira Meridion, Cymbella, Amphora* and *Fragilaria.*

(d) *Euglenophyceae* : It contributed 7.58 per cent (Table 2A) of the total annual phytoplankton production and was represented by *Euglena viridis.*

Several authors have emphasized the importance of water temperature in periodicity of blue-green algae. Vyas and Kumar (1968) found temperature to be determining factor in phytoplankton productivity. In the present study a direct correlation of phytoplankton production with temperature was observed. Seasonal variation of phytoplankton along with temperature changes may be due to oxygen and carbondioxide variations along with other chemical characteristics of water. In the present study peak of phytoplankton was observed during summer followed by winter and rainy which might be due to increased water temperature (Yogamoorthy and Devi, 1996) have also reported summer peak of phytoplankton. Several authors have emphasized the importance of water temperature in the eperiodicity of blue-green algae (George, 1976; Singh and Swarup, 1979 and Reynolds *et al.*, 1981). Rainfall and high turbidity produced by high wind velocity during rainy season had a direct hearing on phytoplankton reducing these to minimum numbers.

The river harbours have many pollution indicator diatoms like *Navicula, Fragillaria, Cyclotella*,etc. Algae in general because of their photosynthetic activity not only contribute to oxygenation of water but also constantly form a biological film and act beneficially in maintenance of water quality. They also pose problems of clogging taste and odour and impure water quality. Some of the algae which have been reported in the present study are known to produce problems in drinking water. *Chlamydomonas* produce musty, greasy and fishy odours. *Microcystis* can produce toxins and make the sludge jelly like causing problems in sedimentation tanks. *Anabaena* is filter and screan clogging algae which also produce characteristic odour and interferes with coagulation. *Euglena* is a quite problematic algae producing fishy odour and sweet taste in water. It can also produce a slime and can interfere with coagulation.

According to Palmer (1969) presence of Cyanophyceae, euglenophyceae and certain diatoms are characteristics of eutrophication and orgnaic pollution. Kumar and Singh (1994) have reported Bacillariophyceae as dominant phytoplankton followed by Myxophyceae, Chlorophyceae and Euglenophyceae in their study of a tropical shallow pond at Dumka.

Zooplankton

Zooplankton is an important component of secondary production in an aquatic system. They are an important food source of higher organisms including fishes. The quantitative analysis of zooplankton do not only reflects energy transfer in the system but it also certainly provides the information about the happenings in the foodcycle. Several zooplankton species have been classified as indicator of polluted conditions.

The zooplankton community of the river consisted of three groups (rotifera, copepoda and cladocera) comprising 29 species. The three groups of zooplankton were :

(a) *Rotifera* : *Bracionus angularis, B. forticula, B. rubens, B. fulcatus, B. caudatus, Keratella troipa, K. valga, Filinia longiseta, F. terminalis, Monostyla lunaria, M. bulla, Horaella* sp., *Leopadella* sp. *Polyarthra* sp., *Cephalodella forticula, Rotaria* sp., *Brachionus monostyla*.

(b) *Cladocera* : *Moina micrura, Adona longicanda, Chrodorus gibbus, Alonella* sp., *Macrothrix* sp., *Daphnia* sp., *Ceriodaphnia* sp. *Ophruoxus sp., Mesocyclops* sp.

(c) *Copepoda* : *Heliodiaptomous vidunus, Mesocyclops* sp., *Nauplius* sp., *Diaptoms*.

Rotifers showed superiority over other groups both in terms of species and genera and population density. Dominance of rotifera over otehr groups has also been reported in other waterbodies of the world (Pennak, 1944; Alikunhi, 1957; Michael, 1968; Singh and Sahai, 1978). The sequence of dominance of various groups were Rotifera (48.45 per cent) > Cladocera (36.49 per cent) > Copepoda (15.06 per cent) as shown in Table 2B an Fig. 3A &B.

According to George (1966) the abundance of rotifers followed by cladocera is an indication of the eutrophic nature of the waterbodies. The dominance of rotifera may be attributed to their dependence on phytoplankton detrital nature and bacterial as food (Sarwar and Praveen , 1995)

The zooplankton peak was found during summer followed by winter and rainy. Dominance of rotifera in seasonal data of zooplankton as observed in the present study is in accordance with the findings of Michael, 1968, Saha *et al.*, 1971; Pandey *et al.*, 1992, 1995; Sarwar and Parveen, 1995. The peak value of zooplankton during summer might be due to optimal thermal and nutritional condition and higher concentration of oxygen (Singh, 1991). The lowest zooplankton recorded during winter may be related to low temperature. Marshall and Orr (1972) also observed minimum zooplankton population at lower temperature.

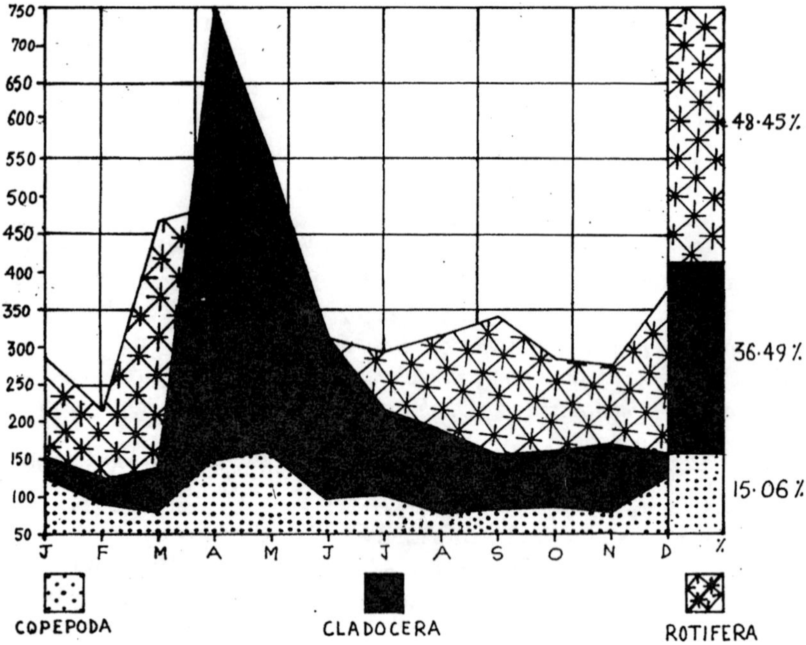

Fig. 3A : **Monthwise distribution of Zooplankton.**

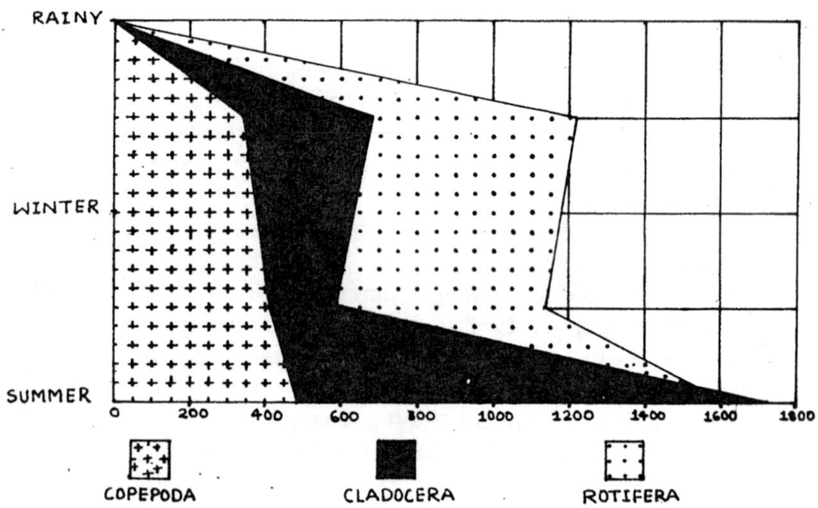

Fig. 3B : **Seasonal Variation of Zooplankton.**

Several zooplankton species have been classified as indicator of polluted conditions. *Brachionus monostyla, Cephalodella, Keratella, Adona, Cyclops, Chydorus, Cerodaphnia,* and several members of the group Protozoa (Rawson, 1956; Davis, 1961; Patalas, 1972; Tundisi and Tundisi, 1976; Widutu, 1967; Pennak, 1978 and Pandit, 1980). Several of the species identified in the present study fall into the category of pollution indicators. *Cerodaphnia* and *Chydorus* grow abundantly in the polluted conditions. According to Aycock (1942), *Mesocyclops* can tolerate low level of oxygen indicating the organic pollution. *Keratella* is also an indicator of polluted conditions.

Macroinvertibrates

The macro-invertebrates are one of the major biological components of the aquatic ecosystem. They usually occupy a higher trophic level in a food chain. The macro-invertebrates are the most valuable indicator of environment condition in an aquatic ecosystem because of their life cycle stages and comparatively stable mode of life.

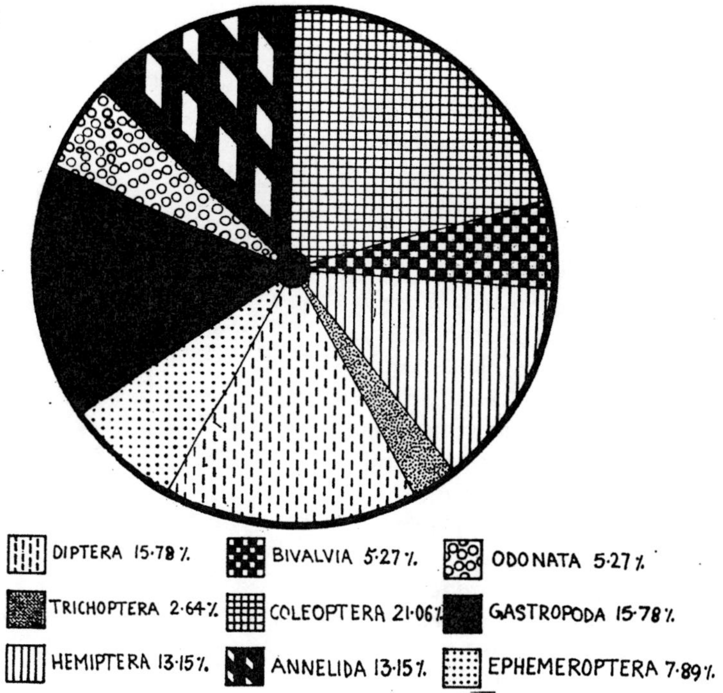

Fig. 4 : Percentage distribution of macroinvertebrates.

The macro-invertebrates fauna belonged to nine major groups (Fig. 4) with a total of 38 species. Following the groupwise break up of different species of Macro-invertebrate along with their percentage encountered in Ramjan River.

Macro Invertebrates	*Species*	*Percentage*
Annelida	5	13.15
Odonata	2	5.27
Gastropoda	6	15.78
Ephemeroptera	3	7.89
Diptera	6	15.78
Hemiptera	5	13.15
Coleoptera	8	21.06
Trichoptera	1	2.64
Bivalvia	2	5.27
Total	38	

Macrozoobenthos

1. Annelida
   - (a) *Hirudinaria granulosa*
   - (b) *Tubifex tubifex.*
   - (c) *Styleria sp.*
   - (d) *Pheritma.*
   - (e) *Nais.*
2. Coleoptera
   - (a) *Hydrophilus sp.*
   - (b) *Laccophilus sp.*
   - (c) *Agabinus sp.*
   - (d) *Haliplus sp.*
   - (e) *Guignotus pradhani.*
   - (f) *Berosus pluchellus.*
   - (g) *B.Indicus.*
   - (h) *Enochrus sp.*
3. Hemiptera
   - (a) *Micronecta proba.*
   - (b) *Micronecta marpoe.*
   - (c) *Corixa promontoria.*

(d) *Lithocerus sp.*
(e) *Notonecta sp.*

4. Diptera
(a) *Chrinomus sp.*
(b) *Antocha sp.*
(c) *Monopelopia sp.*
(d) *Chaoborus sp.*
(e) *Clinotanypus sp.*
(f) *Eristalis sp.*

5. Trichoptera
(a) *Hydropsyche sp.*

6. Ephemeroptera
(a) *Ephemeralla sp.*
(b) *Caenis sp.*
(c) *Heptagenia sp.*

7. Odonata
(a) *Hagenius sp.*
(b) *Enallagma sp.*

8. Gastropoda
(a) *Lymnaea acuminata.*
(b) *L. klutsola*
(c) *Viviparus benglensis*
(d) *Gyraulus convexius*
(e) *Allocinma orcula*
(f) *Planispira proxima*

9. Bivalvia
(a) *Corbicula stiatella*
(b) *Lamellidens marginalis*

For biomonitoring of water quality use of bio-indicators gaining greater applicability and popularity. Community structure of macrobenthic invertebrates population has been used to ecological health of rivers (Kolwitz and Marsan, 1908, 1909; Carpenter, 1926; Butcher, 1946; Pentlow *et al.*, 1938; Beck, 1955, Hawkes, 1979; Hymes, 1960; Chandler, 1970). Presence of bio-indicators, chrinomous larvae and *Tubifex* species indicate the effect of pollution (Sinha, 1991). Tarzwell and Gaufan (1956) and Coker (1985) reported that *Fristalri sp.* and Tubifex occurred in strong sewage pollution and anoxic waters. In the present investigation also these species were abundant near point of discharge of domestic sewage in the river water. *Hydropschyche* were abundant at cremation site, as they feed on dead organic matter. The molluscs usually do not tolerate pollution as high as do *Tubifex* and *Chironomids*, but a few like *Lymnea* and *Physa* can tolerate very low oxygen conditions (Harman, 1974). They remain richer in the water with regard to the number of species as also reported by Anwar and Siddiqui (1988). Hemipterans predominated by adults and nymphs of water bugs were associated with floating macrophytes of the river. These forms are mostly plant juice suckers and take shelter amongst the thick hyacinth root masses and submerged vegetation. Coleopteran insects of aquatic habitat has been regarded as water scavengers. Maximum coleopterans were found during the season of high litter decomposition of hyacinth mat.

The species diversity index was found to be 2.34. Thus on the basis of species diversity inJex it may be concluded that the river water is of moderate quality. Based on the criterion laid down be Whilm and Dorris (1966) the stretch of river water studied is moderately polluted.

Coliforms

The quality of water intended to be used for public water supply can be best judged by microbial assessment. The presence of various coliforms in water demands attention. Microbial parameters are quite significant for health risks and have been introduced in several water quality assessing standards for diverse uses. The presence of coliform in water cause frequent outbrust of waterborne epidemics. The number of coliform bacterial varied from 400-1400/ 100 ml in the river. The drinking water standard recommended by ICMR for coliform group is 1 per 100 ml of water. Hence, the river water found to be contaminated with coliform group. Garud (1983) has reported the presence of coliforms in the river.

Macrophytes

Aquatic macrophytes include a vast majority of aquatic vascular plants. They are found mainly in the shallow regions of ponds, lakes rivers, swamps and stream etc. They are of considerable ecological and economical importance. Macrophytes is an important factor for helping in maintaining ecological balance. Aquatic macrophytes play a pivotal role in maintaining primary productivity of river ecosystems. Aquatic macrophytes are the main primary producers of organic matter on which fishes thrive. They are also source of oxygen. They also respond to the changes in water quality and have been used as indicators of pollution in several cases (Westtake, 1981 and Best, 1982).

Following Macrophytes were listed in the present investigation :

1. *Hydrilla verticellata*
2. *Ceratophyllum demersum*
3. *Utiricularia flexuosa.*
4. *Utricularia stellaris.*
5. *Pistia*
6. *Azolla*
7. *Limnophila*
8. *Cyperus cephalotes.*
9. *Cyperus platystylis.*
10. *Jussiea repens*
11. *Caesulia axillaris*
12. *Eichhornia crassipes*
13. *Neptunia oleracea*
14. *Polygonum sp.*
15. *Lemna*
16. *Trapa*
17. *Ipomea*
18. *Nelumbo sp.*
19. *Potamegaton sp.*

The macrophytes surveyed during present investigation can be categorised into four types :

A.	*Floating*	:	*Eichhornia crassipes*, Pistia, Azolla, Lemna.
B.	*Emergent*	:	Trapa, Ipomea, *Nelumbo sp.*, Potamegaton sp.
C.	*Marginal*	:	*Jussiea repens, Cyperus platystylis, Cyperuscephalotes, Neptunia oleracea.*
D.	*Submerged*	:	*Utricularia flexuosa, Utricularia stellaris, Hydrilla verticellata, Ceratophyllum demersum.*

In the present investigation *Eichharnia Crassipes* was the most abundant in river water. Frank (1976) and Kar (1939) have reported that Eichhornia provides both the habitat and food for several harmful animals and other vector of diseases like malaria, encephalitis etc. They also impart undesirable taste and colour to water (Krishnamoorthi *et al.*, 1968), Krishnamoorthi and Rajagopalan, 1970). Similar observations were also noticed during the present survey of river water. During the period of present investigation it was realized that the peoples of nearby villages of river water suffers from a number of diseases such as malaria, filaria, kala-azar, cholera, amoebic dysentry etc. Further, mass mortality of fishes was also observed in river waters heavily infested with water hyacinth. As water hyacinth provide ideal site for the growth of snail which are intermediate host of river fluke. Infection of liver fluke in this area is very common. Due to this parasites the cattles suffer from liver rot resulting in the less production of milk which effect the economy of the people. In the Koshi zone most of the lands are low where water logging is a common feature. These lands are also infested with water hyacinth which interferes with the germination of seeds leading to less production of the crop, such results have been also reported by the other workers (Basak, 1948, Haque and Sharma, 1980). Floating islands were also observed which were due to decomposition of dead plants which occurs very slowly. The decomposing organic matter remains buoyant and suspended just below the mat (Trivedy *et al.*, 1978).

Considering pollutants, pollution in the river Ramjan may be categorised by three broad groups-physical deterioration, chemical deterioration and micro-biological deterioration. Chemical deterioration may be grouped into two categories - organic and inorganic. In concluding remark it can be said that the stress of deterioration of the river Ramjan is due to the addition of sewage, agricultural run off as well as human interferences. People of this region are using this river as a natural dustbin for dumping various types of wastes without taking care of bad effects of these wastes. Soil erosion as well as burning of corps on both the sides of the river bank are also the major cause of deterioration of water quality of the river. The study indicates that both chemical and biological aspects reveal grossly polluted nature of the river.

Summary

The paper deals with an ecological study with special reference to biological components on river Ramjan at Kishanganj (Bihar). A large proportion of the catchment is comprised of urbanized and human activity influencing the river.

The results revealed well defined seasonal variation. On the basis of physico-chemical parameters the river appears to be moderate trophic level and hard type. The bulk of phytoplankton was shared by Chlorophyceae (9), Myxophyceae (8), Bacillariophyceae (8) and Euglenophyceae (1). Many of the algal species out of the total 26 reported from river like *Euglena, Microcystis, Chlamydomonas, Anabaena* and *Spirulina* were recognised as pollution indicator. The bulk of Zooplankton was shared by rotifera (16), cladocera (9) and copepoda (4). In all 29 species of zooplankton were identified. Zooplankton also comprised some pollution tolerant species like Brachionus, Keratella and Cyclops etc. The macro-invertebrates fauna belong to 9 major groups with a total of 38 species. Presence of *Chironomids* and *Tubifex* indicated the build up of substantial quantity of organic matter. The species diversity index was found to be 2.3 indicating moderate quality of water. 19 species of macrophytes and 17 species of fishes were identified.

The study indicated that both chemical and biological aspects reveal grossly polluted nature of the river.

Acknowledgement

S.Hussain is thankful to Dr. Asad Imam, Director, Millia Educational Trust, Purnea for his help in carrying out this work.

References

Ajmal, M., Nomani, A.A. and Khan, M.A. (1984), Pollution in the Ganges river India Wat. Sci. Tech. 5(7), 347-358.

Alikunhi, H.H. (1957), Fish culture in India. Farm Bulletin, New Delhi, No. 2 ICAR p.6.

Anwar, S. and Siddiqui, M.S. (1988), On the distribution and dynamic of macro-invertebrates fauna of the river Kali in northern India. J. Environ, Hlth. 9(3 suppl.), 333-341.

APHA (American Public Health Association) American Water Works Association and Pollution Control Federation (1975). Standard methods of examination of water and wastewater 14th edition. APHA, Washington, USA.

Aycock, D. (1942), Influence of temperature on size and form of cyclops. Vernalis Fischer. Y. Elisha Mitchell Sc. Soc. 58 : 84-93.

Basak, M.N. (1948), Water hyacinth compost. Dept. Agri. West Bengal Bull. Alipore P. 14.

Berk, W.M. (1955), Suggested method for reporting biotic data. Sewage Indust. Wastes, 27 : 1193-1197.

Best, E.P.H. (1982), Effects of water pollution on freshwater submerged macrophytes. 27-56. In : (C.K. Varshney ed.) Water Pollution and Management Reviews, S. Asian Publ. Pvt. Ltd. New Delhi.

Bilgrami, K.S. and Siddiqui, E.N. (1983), Biomonitoring of the river Ganga in Mokama-Barauni Industrial Complex. Biol. Bull. Ind. 5(3) : 278-281.

Butcher, R.W. (1946), The biological detection of pollution. J. Inst. Sew. Purif. 2 : 92-102.

Butcher, R.W. (1947), Studies on ecology of river VII. The algae of enriched water. J. Ecology, 35 : 186-191.

Carpenter, K.E. (1926), The lead mine as an active agent in river pollution Ann. Appl. Biol. 13 : 395-401.

Coker, R.E. (1985), Streams, Lakes and Ponds, Univ. of N. Carolina Press, 327 pp.

Cole, G.R. (1979), A text book of Limnology 2nd Ed. New York. The e.v. Mosley Co. London.

Chandler, J.R. (1970), A biological approach to water quality management, Wat. Pollution Control. 4 : 415-422.

David, A. and Roy, P. (1966), Studies on the pollution of the river Daha (N. Bihar) by Sugar and distillary wastes. Environ. Hlth. 8(1), 6-35.

Davis, C.C. (1961), The biotic community in the great lakes with respect to pollution. Proc. Conf. on water pollution and the great lakes. Depaul Univ. Chicago, 80-87.

Dora, M.M. and Roy, N.N. (1987), Investigation of water quality of Swaranrekha river for irrigation. Ind J. Environ.Hlth. 29(4), 292-298.

Frank, P.A. (1976), Distribution and Utilization research on tropical and subtropical aquatic weeds in the United States p. 353-359. In : Aquatic Weeds in S.E., W. Junk, The Hague.

Garud, J.M. (1983), Studies on Water Quality problems in Karad and Kolhapur city of Maharashtra. M.Sc. Thesis, Shivaji University, Kohlapur.

George, J.P. (1976), Hydro-biological studies on lower lake of Bhopal with special reference to productivity of economic fishes. Ph.D.Thesis Bhopal University.

George, M.G. (1966), Comparative plankton ecology of five fish tanks in Delhi, India, Hydrobiol. 27 : 81-108.

Goel, P.K. and Autade, V.B. (1995), Ecological studies on the river Panchganga at Kolhapur with Emphasis on biological components. Recent Researches in Aquatic Environment. (Eds. Ashutosh Gautam, and N.K. Agarwal).

Hymes, H.B.N. (1960), The limnology of polluted waters. Liv. Un. press, 202 pp.

Haque, A. and Sharma, S. (1980), Water hyacinth to fight water pollution. Science Reporter, Dec.

Harman, W.N. (1974), Snails (Mollusca : Gasttropoda) Pollution Ecology of Fresh-Water Invertebrates. Eds. C.W. Hart and S.L.H. Fuller, 274-312. Academic Press, New York and London.

Hawkes, H.A. (1979), Invertebrates as indicators of river water quality. In (Biological Indicators of Water Quality). Ed. by A. James and Lillian Evison. John Wiley and Sons, New York.

Hutchinson, G.E. (1957), Treatise of Limnology. John Wiley and Sons, New York.

Jana, B.B. (1973), Seasonal periodicity of plankton in a fresh water pond, West Bengal, India. Int. Rev. Geo. Hydrobiol. 58 : 127-143.

Jhingran, V.G., Ahmad, S.H. and Singh,A.K. (1989), Application of shannon-wienner index as a measure of pollution of river Ganga at Patna, Bihar, India, Curr. Sci. 13 : 99-111.

Kar, G.K. Mishra, P.C. and Das, R.C. (1987), Pollution studies in river Ib III : Plankton population and primary productivity. Ind J. Environ. Hlthy. 29(4) : 322-329.

Kataria, H.C., Quereshi, H.A., Iqbal, S.A. and Shandilya, A.K. (1996). Assessment of water quality of Kolar reservoir in Bhopal, M.P. Poll. Res. 15(2) : 191-193.

Kaul, V., Zutshi and Vass, K.K. (1972), Biomass productivity of some macrophytes in Shrinagar lakes. In P.M. Colley and F.B. Golley (Eds). Tropical Ecology Emphasing Organic production. pp. 295-311.

Kolkwitz, R. and Marsson, M. (1908), Okologie der pflanzlichen saprobein, Ber. Dtsch. Bot. Ges. 26 : 505-519.

Kolkwitz, R. and Marsson, M. (1909), Okologie der tierischen saprobein. Int. Tev. Ges. Hydrobiol. 2 : 126-132.

Kotpal, R.L. and Bali, N.P. (1989), Concept of Ecology 4th Edn. Vishal Pub. Jalandar, India.

Klien, I. (1973), River pollution II causes and effects (5th Imp.) Butter Worth and Co.Ltd.

Krishnamoorthi, K.P., Shende, S.W. and Hbdulappa (1968), On a collection of some molluscs (bivalves and gastropods) from some waterbodies of Nagpur with ecological note. Env. HH. 10 : 43-50.

Krishnamoorthi, K.P. and Rajagopalan (1970), Survey of molluscs nuisance in some water supplies of Calcutta. Proc. Symp. of Mollusca pt. II : 746-754.

Kulshrestha, S.K. Adholia, U.N., Khan Altaf A., Bhatnagar Anita, Baghail Manju (1988), Macrozoobenthose in river Chambal and its tributaries. Ind. Assoc. Water. Polln. Cont. Techl. Ann. 15 : 39-41.

Kumar, A and Singh, A.K. (1994), Studies on composition and seasonal population density of plankton in a tropical shallow pond at Dumka. J. Mendel 11(3-4), 153-154.

Lee, G.F., Bentley, E. and Amundsen. N. (1975), Effects of marshes on water quality. In "Coupling of land and water system" — A.D. Hasler, 105-126. Spinger Verlag Berlin, Heid. New York.

Marshall, S.M. and Orr. A.P. (1972), The charophytes of the Dal lake, Kashmir, Proc. 21st Ind.Sci. Congr. 295 (Abst.)

Michael, R.G. (1968), Studies on zooplankton of a tropical fish pond. Hydrobiol. 32 : 47-68.

Moyle, J.B. (1946), Some indices of lake productivity. Trans. Am. Fish.Soc. 76 : 322-334.

Munwar, M (1970), Limnological studies on fresh water ponds of Hyderabad, India II. The biotope. Hydrobiol. 35(1) : 163-172.

Pehwa, D.V., and Mehrotra, S.N. (1966), Observation on the fluctuation in the abundance of plankton in relation to certain hydrobiological conditions of river Ganga. Proc. Nat. Acad. Sci. India, 36 B(2) : 157-189.

Palmer, L.M. (1969), Comparative rating of algal tolerating organic pollution. J. Phycol. 5 : 78-82.

Pandey, B.N. Lal, R.N., Mishra, P.K. and Jha, A.K. (1992), Seasonal rhythms in the Physico-chemical properties of river Mahananda (Katihar), Bihar. Env. & Eco. 10(2), 354-357.

Pandey, B.N., Mishra, A.K., Das P.K.L. and Jha, A.K. (1995), Studies on hydrobiological condition of river Saura in relation to its imparct on biological health. Recent Researches in Aquatic environment, (Eds. Ashutosh Gautam and N.K. Agarwal) Daya Publishing House, New Delhi.

Pandey, B.N., Jha, A.K., Das P.K.L., and Pandey, A.K.(1993), Physico-chemical characteristics of swamps of Libri river, Purnia (Bihar), Acta. Ecol. 15(2): 98-102.

Pandey, B.N., Mishra, A.K., Jha, A.K. and Lal, R.N. (1992), Studies on qualitative composition and seasonal fluctuation in plankton composition of river Mahananda, Katihar (Bihar). J.Ecotoxicol. Env. Monit. 2(1) : 93-97.

Pandey, B.N., Jha, A.K., Das P.K.L., Pandey, P.K. and Mishra, A.K. (1995), On the Seasonal abundance of phytoplankton in relation to certain ecological conditions in the stretch of Kosi river. Poll. Res. 14(3) : 283-293.

Patalas, K. (1972), Crustacean plankton and eutrophication of St. Lawrence Great Lakes. J. Fish Res. Bd. Can. 20 : 1451-1462.

Pennak, R.W. (1944). Diurnal movement of zooplankton organisms in some Colorado mountain lakes. Ecology. 25: 387-403

Pennak, R.W. (1978), Freshwater invertebrates of the United States. John Wiley and Sons, New York.

Pandit, A.K. (1980), Biotic factor and food chain structure in some typical wetlands of Kashmir. Ph.D. Thesis, University of Kashmir, Srinagar.

Pentelow, F.T.K., Butcher, R.W. and Grindley, U. (1938), An investigation of the effects of milk wastes on the Bristol Avon. Fish Invest. Lond. Series 1, Vol. 4 No.1.

Pillai, K.V. and Sreenivasan, A (1975), A note on the carbon and nitrogen status of some lakes reservoirs and pond of Tamil Nadu. J.Inland Fish. Soc. India Vol. II (195), 95-97.

Philipose, M.T. (1959), Freshwater phytoplankton of inland fisheries. Proc. Symp. Algology. 4 : 275-291.

Reynolds, C.S. Jawarski, G.H.M., Chiech, H.A. and Leedala, G.F. (1981), On the annual cycle of the blue-green algae Macrocystes aeruginosa Kutz. emend Eleukin. Philosoph. Transactions Roy. Soc. London Biol. Sci. 293 : 419-477.

Rai, D. N. (1980), Ecological structures of certain swamps of Darbhanga (Bihar), Ph.D. Thesis, Bhagalpur Univ., Bhagalpur.

Rao, D. Subha and Govind, B.V. (1966), Hydrobiology of Tungbhadra reservoir, Indian J. Fish (1), 321-344.

Rawson, O.S. (1956), Algal indicators of trophic lake types. Limnol. Oceanger. 1 : 18-25.

Roy, H. (1995), Plankton Ecology of river Hooghly (West Bengal). Ecology, 36 : 169-175.

Saha, G.N., Sehgal, K.I., Mitra, E. and Nandy, A.C. (1971), Studies on the seasonal and diurnal variation in physico-chemical and biological conditions of a perennial fresh water pond. J. Inland Soc. India. 3 : 79-102.

Sahu, B.K. Rao, R.J. and Behara, S.K. (1994), "Studies of some physico-chemical characteristics of the Ganga river water (Rishikesh - Kanpur) within twenty four hours during winter, 1994", Ecology, Environment and conservation, Vol. 1 (in Press).

Sahu, B.K. Rao, R.J., Behara, S.K. and Pandit, R.K. (1996), Diet fluctuations of some water quality parameters of the river Ganga (Rishikesh - Kanpur) during June, 1994.

Sarwar, S.G. and Parveen, A. (1995), Community structure and population density of zooplankton in two inter-connected lakes of Srinagar, Kashmir, Poll. Res. 15(1) : 35-38.

Shukla, S.C. Kant, R. and Tripathi, B.D. (1989), Ecological investigation on physico-chemical characteristics and phytoplankton productivity of river Ganga at Varanasi, Geobios. 16(1), 20-27.

Shukla, S.N. and Bias, V.S. (1990), Changes in the physico-chemical profiles of the Bila Reservoir during winter season. Trends in Ecotoxicology. 37-41.

Shyam Sundar (1988), Monitoring the water quality in stretch of river Jhelum, Kashmir, Ecology and Pollution of Indian Rivers. Edi. U.P. 131-162. Ashish Publishing House, New Delhi.

Singh R. (1991), Rotifer fauna of Jamalpur, Munger, Bihar. J Curr. Biol. Sci. 8 : 45-48.

Singh, S.B. and Sahai, R. (1978), Seasonal fluctuation of zooplankton population in relation to certain physico-chemical factors of a pond. Geobios, 5 : 228-230.

Singh, S.R. and Swarup, K. (1979), Limnological studies of Suraha lake Ballia 11. The periodicity of phytoplankton. J. Indian Bot. Soc. 58 : 319-329.

Sinha, A.K. (1991), River Ganga in the stretch between Kalakankar (Pratapgarh) and Phaphamau (Allahabad). A comprehensive study. The Ganga. A scientific study. Northern Book Centre, New Delhi.

Sinha, A.K., Srivastava, S. and Srivastava, K.N. (1989), Physico-chemical studies of river Ganga water at Kalakankiar (Pratapgarh). Indian. J. Environ. Prot. 9(3) : 194-197.

Swarananlatha, N. (1994), Seasonal rhythms of various physico-chemical characteristics of a lake. J. Mendel, 11(1), 65-66.

Tarzwell, L.M. and Gaufan (1956), Sewage and Industrial Waters 28 : 906-924.

Thresh, J.O. Beale, J.P. and Suckling, E.V. (1994), The examination of water supplies. Ed. Taylor, EM. 1949.

Todda, B.K. (1970), Water Encyclopedia. Water Information Centre. Port Washington. New York.

Towheed, M.A. Singh, R.K. and Singh, B.N. (1988), Physico-chemical factors of swamps of Kosi region and main Kosi rivers of North-eastern Bihar in relation to yield of air breathing fishes. Env. & Ecol. 6(2), 386-389.

Trivedy, R.K. and Goel, P.K. (1984), Chemical and Biological methods for water pollution studies. Environmental Publications.

Trivedy, R.K. Sharma, K.P. Goel, P.K. and Gopal, B. (1978), Some ecological observations on floating islands. Hydrobiol. 60 : 187-190.

Tundusi, T.M. and Tundusi, J.C. (1976), Plankton studies in lacustrine environment. Oecologia (Berl), 25 : 265-270.

Verma, A.M., Dudani, V.K., Kumar, B., Kumar, S. and Pandey, S. (1988), Proc. Nat. Acad. Sci. India, 58(B) : 11, 259.

Vijay Kumar, K. (1995), Limnology of Freshwater Pond at Gulbarga. In Recent Researches in Aquatic Environment. 87-88. Daya Publishing House, New Delhi.

Vyas, L.N. and Kumar, H.D. (1968), Studies on phytoplankton and other algal of Indra Sagar Tank, Udaipur, India. Hydrobiol. 31 : 421-434.

Welch, P.S. : (1952), Limnology. McGraw Hill Book Company, New York. Toronto and London (2nd Edi) p. 538.

Westtake, D.F. (1981),Temporal changes in the aquatic macrophytes and their environment, Villeret, pp. 110-138.

Wetzel, R.G. (1975), Limnology. W.B. Saunders Co. Phildelphia, U.S.A. 742.

Whilhm, J.L. and Dorris, T.C. (1966), Species diversity of benthic macro-invertebrates in a stream receiving domestiv and oir refinary effleunts. Am. Midl. Nat. 76 : 427-449.

Widutu, (1976), Zooplankton in the Kortow Skil Lake after 18 years of the experimental restoration of the lake. Pol. Arch. hydrobiol. 42(1) : 91-106.

Yogamoorthi, Y. and Devi, R.R.S.P. (1996), Phytoplankton ecology of Pondicherry marine environment from South East Coast of India. J. Ecobiol. 8(3) : 167-174.

Zutshi, D.P. and Vass, K.K. (1973), Variation in the water quality of some Kashmir Lakes. Trop. Ecol. 14(2) : 182-196.